The Continental Crust
A Geophysical Approach

Two week loan
Benthyciad pythefnos

FEB

This is Volume 34 in

INTERNATIONAL GEOPHYSICS SERIES

A series of monographs and textbooks

Edited by WILLIAM L. DONN

A complete list of the books in this series is available from the publisher.

The Continental Crust
A Geophysical Approach

Rolf Meissner

Institut für Geophysik
Christian-Albrechts-Universität zu Kiel
Kiel, Federal Republic of Germany

 1986

ACADEMIC PRESS, INC.
Harcourt Brace Jovanovich, Publishers

Orlando   San Diego   New York   Austin
London   Montreal   Sydney   Tokyo   Toronto

ACADEMIC PRESS, INC.
Orlando, Florida 32887

United Kingdom Edition published by
ACADEMIC PRESS INC. (LONDON) LTD.
24–28 Oval Road, London NW1 7DX

Library of Congress Cataloging in Publication Data

Meissner, Rolf.
    The continental crust.

    Bibliography: p.
    Includes index.
    1. Earth–Crust. 2. Continents. I. Title.
QE511.M43 1985   551.1'3   85-6115
ISBN 0–12–488950–6 (alk. paper)
ISBN 0–12–488951–4 (paperback)

PRINTED IN THE UNITED STATES OF AMERICA

86 87 88 89   9 8 7 6 5 4 3 2 1

Contents

Preface   *vii*
Acknowledgments   *ix*

**Chapter 1   Introduction: Crust, Mantle, Lithosphere,
                and Asthenosphere**

1.0   Introduction   1
1.1   Definitions of Crust   1
1.2   Oceanic and Continental Crust   4
1.3   Definitions of Upper Mantle   5
1.4   Definition of Lithosphere and Asthenosphere   6

**Chapter 2   The Crust as the Product of a Planetary
                Differentiation Process**

2.0   Introduction   8
2.1   The Lunar Crust   12
2.2   The Crust on Mercury, Mars, and Venus   14
2.3   The Center of Mass–Center of Figure Offset and Its Relation
        to Crustal Thickness   15
2.4   The Bipolarity of Surfaces and the Hypsographic Curves
        of Terrestrial Planets   16
2.5   A General Planetary Outlook   18

**Chapter 3   Collecting Physical Properties
                of the Earth's Crust**

3.0   Introduction   20
3.1   Isostatic Compensation and Density Calculations
        from Gravity Measurements   20

3.2   Scismological Mcthods   35
3.3   Seismicity   41
3.4   Controlled-Source Seismology (Explosion Seismology)   58
3.5   The Electric Conductivity of the Crust   84
3.6   Stress Measurements and the State of Stress in the
      Continental Crust   91
3.7   Crustal Geomagnetism   112
3.8   Crustal Geothermics   124

Chapter 4   Contributions from Laboratory Experiments

4.0   Introduction   141
4.1   Seismic P- and S-Wave Experiments   142
4.2   Attenuation Experiments   151
4.3   Fracture and Friction Experiments   162
4.4   Creep Experiments   169
4.5   Special Implications of Creep Data for the Rheology
      of the Continental Crust   181
4.6   Absolute Age Dating and Its Implications   195

Chapter 5   The Composition of the Continental Crust

5.0   Introduction   213
5.1   Rock-Forming Minerals and Their Elements   215
5.2   Igneous Rocks   220
5.3   Metamorphism and Metamorphic Rocks   228
5.4   Sedimentation and Sedimentary Rocks   233

Chapter 6   Crustal Structure in Various Geologic Provinces

6.0   Introduction   242
6.1   Precambrian Shield Areas and Platforms   243
6.2   Paleozoic and Mesozoic Areas   272
6.3   Cenozoic Mountain Belts   301
6.4   Continental Grabens, Rifts, and Margins   316
6.5   Active Margins   351
6.6   Specific Crustal Structures   355

Chapter 7   The Evolution of the Continental Crust

7.0   Introduction   368
7.1   Physical and Some Petrological Boundary Conditions   369
7.2   Geological Observations   372
7.3   The Mechanism of Crustal Growth   379

References

                                                                 394

Index

                                                                 417

Preface

The past 20 years have yielded a wealth of new data on the continental crust from all branches of geoscience and from many diverse regions of our planet Earth. Thus, the time appeared right for an attempt at a comprehensive treatment of the new data and insights gained so far. Compared with the oceanic crust, the continental crust shows such a wide variety in age provinces and tectonic patterns that a synoptic view of its structure and development is very much needed. The framework of plate tectonics with its concept of rigid (oceanic) plates definitely must be modified and supplemented by observations made in the continental crust. The restriction of earthquakes to the upper part of the continental crust and the similarity of viscosity values in the continental crust and in the oceanic mantle are only two examples of the significant differences between oceanic and continental environments.

This book is primarily written for graduate students and researchers in the Earth's sciences who have a special interest in the structures and peculiarities of the continental Earth's crust. It should also attract the attention of geoscientists working in exploration and exploitation who might want to find relationships between shallow and deeper structures and between present and past tectonic styles and patterns of evolution.

The aim of this book is to develop an integrated and balanced picture of today's knowledge of the continental crust. In this context, geophysical arguments dominate throughout the book; whenever possible, the geophysical parameters are related to geological knowledge and reasoning. Relatively little space is given to detailed mathematical expositions and then only to those topics for which it seemed to be unavoidable. The general earth scientist should understand (and it is hoped appreciate) nearly all sections of this book.

Some words should be spent on the organizational plan. Chapter 1, the first of the two introductory chapters, deals with definitions of crust and lithosphere, and Chapter 2 describes the formation of crusts as a general planetary phenomenon. Chapter 3 is devoted to the background and methods of geophysical studies of the Earth's crust and to the collection of related geophysical parameters. Although some contributions from laboratory measurements are described in this chapter, it was felt that important laboratory experiments should be treated separately. Hence, in Chapter 4 emphasis is given to creep and friction experiments and the various methods of radiometric age dating. Chapter 5 differs somewhat from the other chapters in that it presents an elementary geological treatment and summary of elements from a number of textbooks. It might nevertheless be useful for the geophysical reader as an introduction to the long Chapter 6, which deals with geophysical (mainly seismic) and geological investigations of the crustal structure in various age provinces of the continents. Apart from the ordering according to ages, specific tectonic structures such as rifts, continental margins, and geothermal areas are also discussed. Finally, Chapter 7 tries to give a comprehensive view on the evolution of the continental crust and to collect and develop arguments for crustal accretion and recycling.

Acknowledgments

This book could not have been written without the help of many supporters. In particular, I would like to thank my friend and colleague Professor Lajos Stegena for critically reading the manuscript and making abundant suggestions for changes, omissions, and additions. Many scientists and graduate students from the Geophysical Institute in Kiel volunteered to help with collecting data, plotting diagrams, and organizing material. Among them were Dr. Joachim Voss, Siegfried Moritz, Thomas Wever, Andreas Kathage, Rolf Schade, and Petra Sadowiak. My thanks also go to Lisa Bittner, Annemarie Gabeler, and Hauke Matthiesen who were responsible for the photographic work and drafting. Much valuable editorial and typing support were provided by Volker Lopau, Kathleen Helbig, and Renate Ross.

I would like to thank most sincerely all those who helped me for their great effort.

Last but not least, I want to express my gratitude for the generous financial support I received from the Christian-Albrechts-University of Kiel.

Chapter 1 | Introduction: Crust, Mantle, Lithosphere, and Asthenosphere

1.0 INTRODUCTION

The Earth with its continents and ocean basins, as well as all other solid-state planets in our solar system, has a crust. The Earth and the terrestrial planets have silicate crusts, which means that light silicon dioxide, SiO_2, is dominant in the outer shells. The formation of a crust, hence, is a global and even planetary phenomenon. It may be understood as a very general differentiation process by which—as in a hot furnace—the lighter material separates from a parent material and finally settles on top as a residue. Densities, as well as melting points and the chemistry of partial melts, determine the final composition of the crust after differentiation. Convection may cause a certain remixing or recycling, a further complication in a warm and still restless planet like the Earth. Hence, terrestrial crusts—and especially the continental crust of the Earth—will not be uniform but will differ from place to place.

1.1 DEFINITIONS OF CRUST

The term "crust" may be defined in several ways. Ideas developed in the last century considered the Earth's crust as the outer layer of the Earth, which was supposed to overlie a molten interior. Today the seismological definition

1

seems to be the clearest one in spite of some overlapping in the seismic velocities of cold, deep crustal and young, warm mantle material. It goes back to the Croatian–Hungarian, later Yugoslavian, seismologist Mohorovičić. While studying a Balkan earthquake in 1909, Mohorovičić noticed that the velocity of compressional seismic waves increased considerably at some tens of kilometers depth (Mohorovičić, 1909). Since then a tremendous amount of data about the transition from crust to mantle has been gathered by seismic and seismological methods. The boundary between crust and mantle has been named the Mohorovičić discontinuity or simply Moho. Based on very many data from earthquake observations and observations of artifical sources, the seismological definition of crust is given in terms of the seismic velocities as follows:

(1) crust = outer shell of a terrestrial planet in which the velocity of compressional waves (P-wave velocity = V_p) is smaller than about 7.6 km/s or the velocity of shear waves (S-wave velocity = V_s) is smaller than about 4.4 km/s.

Much earlier than Mohorovičić, Bouguer in the late 18th century found significant differences in another physical parameter, the density, by applying pendulum measurements. These differences were primarily related to those of mountain ranges and the whole Earth. Pratt and Airy in the middle of the 19th century put forward ideas of density compensation, Airy favoring the idea of crustal roots and density differences between crust and mantle. Refined gravity measurements as well as careful sampling and density determination of crustal rocks led to a *density definition* of the crust:

(2) crust = outer shell in which the density of rocks is below 3.1 g/cm^3 = 3.1 to/m^3.

Densities and velocities may be obtained from Bullen (1975), Dzievonski *et al.* (1975), Wohlenberg (1982), and Gebrande (1982). Many new data have helped to establish a correlation between seismic velocities and densities of crustal and mantle rocks, so that the two "geophysical definitions" are compatible with each other and even interrelated. Nafe and Drake (1959) and Woollard (1969) provided important data for such a correlation.

It was known for a long time that most of the continents and ocean floors are covered with sediments. Shale is the most abundant sedimentary rock, followed by limestone and sandstone. Below the sediments the crystalline basement begins. It is granitic or gneissic in the continents and basaltic below the oceans. Today, so much is known about the rocks of the crust that a third definition can be based on the type of rocks:

(3) crust = outer shell which consists predominantly of sediments, gneisses, granite, granodiorite, gabbro, amphibolite, and granulite for the continental crust and of sediments, basalts, gabbros, and some serpentinites for the oceanic crust.

In addition, the continental crust may consist partly of volcanic material such as basaltic or acidic magmatites, either as a layering or in a veinlike pattern.

Because rocks are made of minerals (and/or mineraloids) and because the naturally occurring substances are made of molecules or atoms in a three-dimensonal periodic array, it is evident that mineralogical and chemical descriptions can also be used to define crustal rocks. From the mineralogical point of view there are the feldspars, which are especially abundant in crustal rocks:

(4) crust = outer shell where the proportion of feldspars to other minerals is more than 50%. It is 50.2% (Ahrens, 1965) and consists predominantly of orthoclase $KAlSi_3O_8$ and albite $NaAlSi_3O_8$ (together 31%) and plagioclase ($NaAlSi_3O_8$ and $CaAl_2Si_2O_8$)(19.2%).

The other minerals are mentioned in Chapter 5, where differences between oceanic and continental crusts are described. Such differences are evident when looking at the average chemical composition of crusts (Wyllie, 1971; Ronov and Yaroshevsky, 1969):

(5) crust = outer shell where SiO_2 reaches about 60 wt. % in continental (49% in oceanic) rocks, Al_2O_3 about 15% (16% in oceanic rocks), CaO 6% (more than 12% in oceanic rocks), and Fe_2O_3 below 4% (more than 6% in oceanic rocks).

Clearly, the oxides of the three elements silicon, aluminum, and calcium dominate in the crust. A more thorough inspection must certainly explain why not all of the lighter elements such as Mg, C, and S with a similar "cosmic" abundance appear in the crust, a subject of Chapters 2 and 5.

In spite of the great abundances of silica and alumina, the older definition of the continental crust as a "sial" (silica–alumina) layer, originally considered to overlie a molten "sima" (silica–magnesia) layer, certainly is too simple in view of the abundance of modern data. The definition of the crust as the outer shell of the Earth on top of the Moho is certainly correct but is identical to our first definition.

Some more definitions or descriptions based on magnetic, electrical, rheological, and seismicity data could be given, but all these would not be as unique as the previous ones. Magnetic, especially ferromagnetic, phenomena depend strongly on the Curie temperature, so that varying temperatures will cause a rise or fall of the Curie depth. It may lie in the middle or deeper continental crust or even in the mantle for oceanic environments. The electrical properties and their relation to temperature, crustal structure and composition, pore fluids, and partial melts are still rather poorly understood, and the rheological properties strongly depend on temperature, water content, and composition of crustal material. Their relation to the occurrence of earthquakes, the seismicity, is rather strong and will shed some light on changes of crustal properties inside the crust. Such properties are the subjects of later chapters and cannot serve as a firsthand definition of the Earth's crust.

1.2   OCEANIC AND CONTINENTAL CRUST

As mentioned already in definitions (3) and (5), the oceanic crust contains much more mafic material than the continental crust. Moreover, the oceanic crust is much thinner, in general only 5 km thick below the ocean bottom; toward the continent its thickness grows only by an increasing sediment load, the result of continental erosion. It is important to note that the thickness of

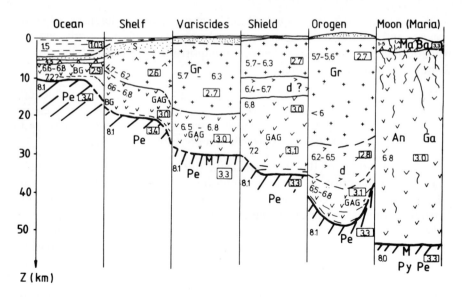

Fig. 1.1   General crustal structure of six different geologic units. Numbers in boxes represent densities in grams per cubic centimeter (to/m³ = 10³ kg³/m³); other numbers represent velocity of P waves in kilometers per second; BG, basaltic–gabbroic in oceanic crust; GAG, amphibolitic and granulitic in continental crust; Pe, Peridotitic–ultramafic; d, dioritic (?), possibly amphibolitic; An Ga, anorthositic gabbro; Ma Ba, mare basalts; Py Pe, pyroxenitic peridotitic; S, sediments; Gr, granitic–gneissic upper crust; and M, Mohorovičić discontinuity (Moho).

the crystalline part of the oceanic crust remains about constant from the time of its creation in the oceanic ridges toward its greatest age of about 180 million years in the vicinity of the continents, not considering hot-spot features, aseismic ridges, island arcs, and similar oceanic anomalies (McLain, 1981; East Pacific Rise Study Group, 1981). The oceanic lithosphere, on the other hand, as defined in the next section, increases considerably in thickness from the oceanic ridges toward the continents.

The continental crust not only contains more light granitic and gneissic rocks and more silicic material but, in general, is also thicker than its oceanic counterpart. Its thickness varies from less than 20 km in shelf areas to more than 70 km in the collisional belts of the Himalayas, under the plains of Tibet, or the Altiplano in Peru (Fig. 1.1). It is more variable and inhomogeneous than the oceanic crust; shows complex interaction with the mantle, magma generation and differentiation of its own, great fault patterns, folding, rifts, grabens, and mountain belts and is—last but not least—the home of our hydrocarbon and mineral resources. The six different geologic units depicted in Fig. 1.1 can only provide a rough picture of the heterogeneity of crusts ranging from an oceanic to a lunar environment. We may note already here that with the exception of orogenes—a short-lived and transient structure of our Earth—crustal depth in the continents increases with age.

1.3 DEFINITIONS OF UPPER MANTLE

After the crust has been characterized it is appropriate to delineate the underlying mantle and distinguish both units from the expressions "lithosphere" and "asthenosphere." Based on the definition of crust and the sequence used in the introductory section we may formulate:

(1) mantle = zone below the crust with P-wave velocities larger than 7.6 km/s (usually larger than 7.8 km/s) and S-wave velocities larger than 4.4 km/s (usually larger than 4.5 km/s)

(2) mantle = zone with density larger than 3.1 g/cm^3 (= 3.1 to/m^3)

(3) upper mantle = zone which consists predominantly of ultramafic rocks like peridotite (even dunite) and garnet-bearing eclogite

(4) upper mantle = zone where the minerals olivine, forsterite (Mg_2SiO_4), and fayalite (Fe_2SiO_4) dominate with major contributions of pyroxene and/or garnet

(5) upper mantle = zone where SiO_2 decreases below 44% and Al_2O_3 below 2% while MgO reaches more than 37% and FeO more than 7%.

It is evident from these items that the boundary between the crust and the mantle, the Moho, really is one of the most prominent boundaries of our planet Earth (another one is the core–mantle boundary). It is definitely a boundary in the elastic, density, petrological, mineralogical, and chemical senses.

1.4   DEFINITION OF LITHOSPHERE AND ASTHENOSPHERE

As will be discussed in Section 3.3, the Moho may also be a minor rheological boundary. But certainly the rheological properties are the foundations on which the expressions lithosphere and asthenosphere are based. On Earth, the lithosphere consists of the crust and a certain part of the upper mantle, sometimes called the "lid". It is generally accepted that its boundary with the asthenosphere may not be a sharp one in the sense of the Moho but a very gradual and wide transition zone where an approach of the temperature to the melting point causes a weak rheology with a possibility of creep movements (Watts *et al.*, 1983)

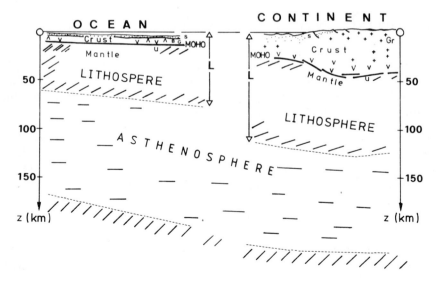

Fig. 1.2   Schematic cross-section of crust, mantle, lithosphere, and asthenosphere for oceanic (left) and continental (right) structures; Moho, Mohorovičić discontinuity (crust–mantle boundary); Gr, granitic–gneissic material of upper crust; BG, basaltic–gabbroic lower oceanic crust; S, sediments; U, uppermost part of mantle (ultramafic material); and L, lithosphere.

Defining the lithosphere we may formulate:

(1) lithosphere = plate = lithospheric plate = upper, rigid, highly viscous rock unit of the Earth with a thickness strongly dependent on temperature, generally 50–100 km thick below the oceans and 100–200 km below the continents.

At greater depths the lithosphere transforms to the:

(2) asthenosphere = zone where the solidus of mantle rocks is reached (or nearly reached) and the material is in a more ductile state with an enhanced possibility of reacting by creep upon any kind of stress; apparently identical with Gutenberg's "low-velocity channel" (Gutenberg, 1959) and with a "low-viscosity zone" (Stacey, 1977a) and a zone of high electrical conductivity.

Figure 1.2 gives a survey of the crust, upper mantle, lithosphere, and asthenosphere for oceanic and continental environments.

Chapter 2 | The Crust as the Product of a Planetary Differentiation Process

2.0 INTRODUCTION

As mentioned before, low-density crusts are present in all solid-state planetary bodies of our solar system. It is widely accepted that the terrestrial planets such as Mercury, Venus, Moon, Earth, and Mars are similar in their composition, although certain differences have been inherited from the accretion in a rotating dust–gas cloud at about 4.6×10^9 years. Since the different accretion temperatures were a function of the distance from the center of the cloud, certain chemical reactions between condensed and volatile particles took place, provided the time for such reactions was comparable to or longer than the time of accretion. According to Levis (1974), there is ample evidence for such an "equilibrium condensation" process in the solar system. The process of condensation was accompanied and followed by the accretion process, in which matter accumulated in a large center of gravity (the sun) and in some secondary centers (the planets). During this process the kinetic energy of the impacting material increased with the increasing radius and gravity of the growing body. The process culminated in the "accretional peak", resulting in extremely high surface temperatures and huge magma oceans. Finally, the terrestrial planets differentiated into heavy metallic cores, mantles, and light Al–Ca-rich crusts, a process that took place in the first 500–600 million years (0.5–0.6 Gyr) (Toksöz et al., 1975). The mantles gradually solidified from below because of the steeper gradients of the adiabats compared to those of the melting-point curves, the solidus, and the liquidus. The crusts, however, the light product of mantle differentiation, lost much of their heat by outward radiation into space or into the atmosphere and, therefore, cooled and solidified from the top to the bottom.

The Earth, as the largest body of the terrestrial planets, was so hot and convection was still so rapid in the beginning that the first crust was unstable and must have been destroyed and remixed into the mantle as soon as it formed (De Paolo, 1981a,b). The first crustal rocks that have survived until today are dated back to 3.8 Gyr. The situation may have been similar on Venus. On the Moon, however, *some* crustal rocks ("breccias") formed 4.6 Gyr ago can be found today in the old highlands, the "terrae". The "accretional tail", a heavy meteorite bombardment which hit all the planets, also caused a certain remixing and great destruction of the early planetary crusts. Very large impacts broke through the whole crust and left huge circular basins, later to be filled with basaltic lavas, a magmatic process that created a new crustal unit, the "mare crust", on the Moon. Similar, but stronger, magmatic activity formed the lowland areas on Mars and Mercury. As the planets cooled—the small bodies much faster than the larger ones —lithospheric thickness grew. As a rheological boundary the transition between the lithosphere and asthenosphere certainly was still inside the crust in the early Earth and possibly still is today on Venus.

On the Moon, Mercury, and even Mars, it soon moved to great depth. At the end of the strong meteoritic bombardment at 3.9 Gyr, when the first permanent rocks on Earth started to settle at the surface, the Moon, Mars,

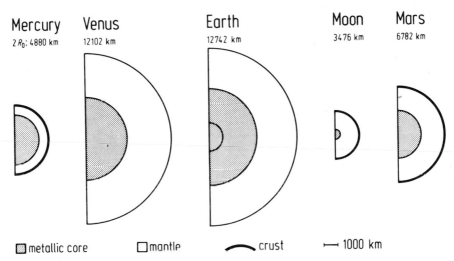

Fig. 2.1   Cross-section of terrestrial planets. [From Meissner and Janle (1984).]

and Mercury had already developed an extensive and permanent highland crust.

The formation of crusts, like that of metallic cores, is certainly one of the most energetic processes in planetary evolution. Figure 2.1 shows a cross section of the terrestrial planets, the different crustal thicknesses being only slightly exaggerated. After the general and dramatic differentiation of the planets into crusts, mantles, and cores, the development of crusts was not yet finished. The larger planets especially had so much internal energy that fast convection hampered the differentiation process considerably. On the other hand, convection also brought less-differentiated material from the deep mantle close to the surface and supported a continuous crustal growth and

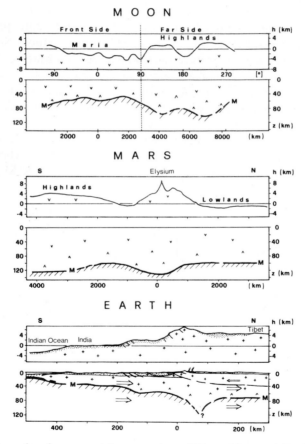

Fig. 2.2 Examples of some crustal cross-sections of Moon, Mars, and Earth after seismic (Earth) and gravity (Moon and Mars) assessments.

differentiation. The result of these two opposing effects of convection is difficult to predict and is mainly dependent on the internal energy, the cooling history with different growth rates of rigid lithospheric plates, and the individual inhomogeneities of terrestrial bodies. Hence, it is no surprise that there is no simple correlation between a planet's size and its crustal thickness. The generally thicker (and older) crusts of the Moon and Mars compared to that of the Earth, as seen in Fig. 2.2 and in Table 2.1, may have been caused by more thorough extraction of material from the mantles of these smaller bodies, a process which was not as strongly hampered by vigorous convection as on Earth, where it has led to a recycling of crustal material or to a

Table 2.1

Selected Physical and Geological Data
for Terrestrial Planets[a]

	Moon	Mercury	Mars	Earth	Venus	References
Crustal thickness (km)	50–100	—	80–130	10–70	—	Voss (1978) Janle and Ropers (1982) Talwani and Langseth (1981)
Lithospheric thickness (km)	1100	—	250–300	20–200	10–50	Toksöz et al. (1974b) Carr et al. (1977) Meissner and Lange (1977) Head and Solomon (1981)
Center of mass–center of figure offset (km)	2	—	2.5	1.1	0.44	Ferrari and Bills (1979)
Ratio of volcanic plains to highland areas	0.2	0.3 (of $\frac{1}{3}$)	0.7	3	8	Talwani and Langseth (1981) Head and Solomon (1981)
Number of peaks in hypsographic curves	1	1 (of $\frac{1}{3}$)	3	2	1	Masurski et al. (1980)
Shape of peaks	broad	broad	broad	1 broad 1 sharp	sharp	Meissner (1983)

[a] From Meissner and Janle (1984).

disappearance of oceanic crust in modern plate tectonic processes. Consider-able thickening of the continental crust on Earth today takes place in continent–continent collisions by obduction and interstacking of crustal material, a process which may also have taken place in the early days of the smaller planets. Also, a thorough and stronger differentiation of the crust itself has taken place on Earth, leading to the very light granitic and gneissic upper part of the continental crust and to recycling of heavy cumulates to the mantle. Such a strong and repeated differentiation, mostly supported by water in various forms, has not happened to the smaller planets. Their crusts are denser and heavier, more basic and primitive than the Earth's continental crust.

2.1 THE LUNAR CRUST

The *Moon* had already acquired its present crustal thickness below the highlands at about 4 Gyr, and basaltic lava from depths of 100–400 km started to fill the depressions that tectonics and the great impacts had left. The transition from a compressional to a tensional regime and the many deep cracks in the lunar lithosphere contributed to easy access of partial melt to the surface. Magmas were rich in Fe and Ti and were of low viscosity.

The process of mare volcanism lasted up to about 3.1 Gyr. The maximum depth of these dense magmatites in the lowlands is not much more than 5 km (Janle, 1981). This is enough to create large gravity maxima in the circular basins, the so-called mascons, but it did not change the crust as a whole to a large degree. It is important to note that the lunar crust had assumed its present thickness at a time when our own crust was still recycling. The lunar lithosphere, on the other hand, as a rheological feature, continued to grow and has acquired a thickness of about 1100 km. Figure 2.3 shows a lowland and a highland area, i.e., a mare- and a terra-type surface of the Moon, considered to be a typical example of the early stages of small bodies, which were not influenced much by later tectonic processes. The difference between the mare plains, with their small impact crater density, and the heavily cratered landscape of the terrae is remarkable.

Thicknesses of the crust and lithosphere on the Moon were calculated by the same methods as those used on Earth and are based mainly on seismological, seismic, and gravitational calculations, which is a subject of Chapter 3. Their application to the Moon and the results concerning its crustal and lithospheric structure were among the major achievements of the U.S. Apollo Program (Nakamura *et al.*, 1976; Toksöz *et al.*, 1974a). While the terra crust is more than 80 km thick, the mare crust has a thickness of "only"

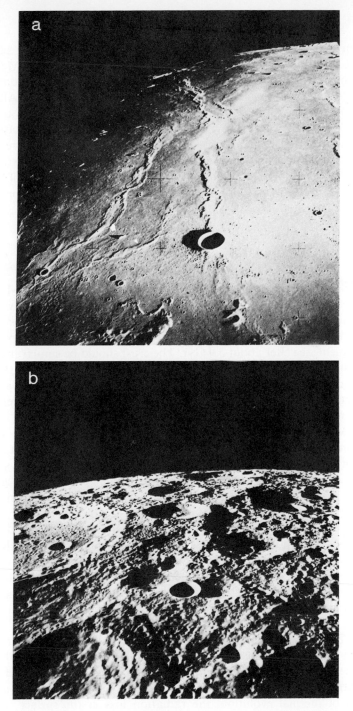

Fig. 2.3 Example of (a) a lowland and (b) a highland area on the Moon's surface.

50–60 km (Voss, 1978). By some process, possibly an early kind of plate tectonics, crustal material was transported to the highland area (Meissner, 1983). The highlands consist of anorthositic rocks, which are composed almost totally of plagioclase, a light mineral series ($NaAlSi_3O_8$, $CaAl_2SiO_2O_8$) which is also in the Earth's crust, but here makes up only about 30% (see also Chapter 3). The extensive plagioclase fractionation on the Moon is most probably a consequence of (1) a favorable pressure–depth environment, (2) the existence of a huge magma "ocean" with the extraction of a very high degree of (partial) melt from the mantle at the time of crustal fractionation, and (3) the lack of repeated (or multiple) melting processes (Anderson, 1981).

2.2   THE CRUST ON MERCURY, MARS, AND VENUS

Processes similar to those on the Moon may have taken place on Mercury, although tectonic activity has lasted slightly longer. There are also volcanic plains and highlands, but compressional features—lobate scarps—cross many of the old craters and must have been active *after* the period of heavy bombardment, i.e., after about 3.9 Gyr. The crust should be similar in size to that of the Moon; the lithosphere should be smaller. The presence of a magnetic dipole field and Mercury's large metallic core could be indications of convective motions at depths of 800–1200 km, which are inside Mercury's metallic core.

Mars, which is about half the size of Venus and Earth, suffered a similar bombardment and underwent a crustal and lithospheric development similar to that of the Moon and Mercury. Its crustal thickness might be slightly larger than that of the Moon, but this is estimated only from gravity modeling. Although most of the tectonic activity, e.g., the formation of the lowlands and their filling with partial melts from depths, was similar to that of the Moon and may even have developed in the same time period, much tectonic activity and modification of the crust took place much later (Neukum and Hiller, 1981; Head and Solomon, 1981). The formation of large rift structures such as Valles Marineris in a tensional regime and the "buckling" of the Tharsis plateau in a compressional environment require a lithospheric thickness of "only" 100–250 km and some recent activity of solid-state convection (SSC) in the asthenosphere (Carr *et al.*, 1977).

Venus, Earth's sister planet, and (like the Earth) having a diameter about twice that of Mars, is expected to show an internal structure and thermal development much like those of the Earth. However, its crustal and lithospheric structure especially deviates considerably because one boundary condition of thermal evolution, the radiation loss at the surface, is very different from that on Earth. A 450° higher temperature keeps radiation losses small. Therefore, the temperature near the surface will be higher, the temperature gradient smaller, and the Venusian lithosphere thinner than that of Earth (Head and Solomon, 1981; Solomon *et al.*, 1982).

Not much is known about the crustal thickness of Venus. Many surface features, as observed in the radar pictures from Earth and from the Pioneer Venus mission (Phillips and Lambeck, 1980), show Earth-like tectonic structures such as large rift-graben zones with elevated shoulders and even mountain belts, apparently folded. In the continents a limited amount of plate tectonics has occurred, but it is doubtful whether or when a planetwide remixing of crust with the mantle has taken place. The present situation on Venus might be comparable to that of the early Earth, where in Archean times the lithosphere was partly still inside the crust and the crust itself was limited to a few "continental", basaltic greenstone belts. Some thick crustal sections, being unstable and only dynamically supported, were created by lateral movements of material.

2.3   THE CENTER OF MASS–CENTER OF FIGURE OFFSET AND ITS RELATION TO CRUSTAL THICKNESS

As seen in Table 2.1, crustal and lithospheric thicknesses of terrestrial planets correlate with the center of mass–center of figure (CM–CF) offset. Although the crustal and lithospheric data for Mars and Venus are based only on models of gravity, flexure theory, and rheology and may have some premature character, it is evident that the larger the crustal thickness, the larger is the CM–CF offset. This correlation can be easily understood if we take into account the bipolarity of surfaces observed on all terrestrial planets (Head and Solomon, 1981). There is no uniform distribution of crustal material all over a planet, but rather a concentration and enlargement in the highlands or continents (continental crust) and a small crustal thickness in the volcanic plains of the lowlands (maria, ocean basins). This bipolarity of the surfaces and crustal thicknesses makes the lighter highlands or continents stand up considerably over the mean radius of a planet in spite of isostatic adjustments. If continents or highlands cover a large area of the surface of a planet, they constitute a "bulge" on the mean radius, or the mean mathematical surface, and cause the CM–CF offset. The CM–CF offset and "highland bulge" may be inherited from a very early period of crustal differentiation and lateral compaction which was rapidly frozen in, as in the case of the Moon; they may grow by some convective activity below a rather thick crust and lithosphere, as on Mars; and they may be continuously maintained by the creation of new oceanic lithosphere and collisional structures, as in case of the Earth. We may also formulate that the stronger the differentiation of a planet into thick and light crusts and heavy mantles, the greater is the possibility of regional topographic features and hence the higher is the amount of the CM–CF offset. Of course, *some* lateral movements of crustal material must be involved in these processes.

2.4 THE BIPOLARITY OF SURFACES AND THE HYPSOGRAPHIC CURVES OF TERRESTRIAL PLANETS

In the last section, the bipolarity of surfaces was introduced as describing the difference between the basaltic and plain lowlands and the light and thick highland (or continental) areas (see Fig. 2.3). In this section, more weight will be attributed to the different material of the two units, to isostatic processes, and to the different peaks of hypsographic curves. The volcanic plains on the Moon, Mercury, and Mars as well as the oceanic sea floor of the Earth have several things in common. They are flat, simple, and rather homogeneous plains; they consist of basaltic material; and they are younger than the surrounding highlands or continents, as seen by radioactive dating (Moon and Earth) and/or by impact crater population (Moon, Mercury, and Mars).

When the plains were formed, relatively low-viscosity basaltic magma erupted predominantly from elongated rift structures and filled the lowlands or depressions. These magmas were partial melts from the underlying mantle, mobilized by heat sources. In the molten stage the density of these magmas is lower than that of the surrounding highland rocks, but after crystallization the magmatites are generally denser. The lithosphere became rigid enough to sustain the additional load to a large degree, giving rise to the lunar mascons (see also Section 7.3.3). For the Moon, Mercury, and Mars heat sources were provided by the downward penetration of the accretional heat peak and the concentration of long-lived radioactive elements on top of the solidifying mantle and below an already formed lithosphere. For the Earth linearly arranged plumes and convection processes delivered the thermal energy to activate the volcanism along rifts. Such processes may also have played a major role in the later stages of Mars in forming tremendous rift structures like the Valles Marineris and bulges like the huge Tharsis plateau, which stands out 4 km above the mean Martian surface. Such a "buckling" seems to be caused by convective activity below a thick lithosphere.

Flat plains, old highlands, and great bulges are reflected in a planetary hypsographic curve. Such a curve shows the percentage of surface that has a

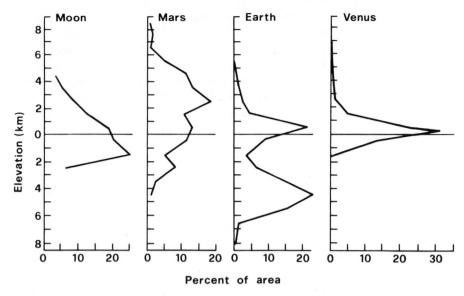

Fig. 2.4 Hypsographic curves for four terrestrial planets. [From Meissner and Janle (1984).]

certain elevation with regard to the average surface, i.e., an ellipsoid or a sphere. We might think that the bipolarity of the planet, i.e., the division into lowlands and old highlands, would provide one maximum each, giving a hypsographic curve with two peaks. Looking at the hypsographic curves of the terrestrial planets in Fig. 2.4 and at Table 2.1, we can see that only the Earth has a curve with two peaks. While the rather sharp upper peak of the Earth's curve is due to sedimentation and the presence of a hydrosphere (i.e., it is caused by the fast continental erosion of elevated areas and the sedimentation in the lowlands and the shelves), the lower, broader peak really marks the ocean bottom, i.e., the top of the oceanic crust (the new "basaltic plains"). The physical reason for the position of this peak at about −5 km below mean sea level (NN) is related to an isostatic adjustment of the newly formed lithosphere with respect to the surrounding continents and will be discussed again in Section 3.1.1. The broad shape of this peak is due to the general increase in depth of the ocean bottom from the ridges toward the continents.

The Moon and probably Mercury have only one maximum, although it is a broad one. The bipolarity of surfaces has provided a lunar CM–CF offset but not two peaks on the hypsographic curve. The reason for the broad distribution of elevations is found in the very early formation of a rather thick crust and lithosphere, still in the period of heavy bombardment between 4.6 and 3.9 Gyr. The random size distribution of impacting bodies resulted in a great scatter of impact-induced elevations like those of crater floors, multiple ringed basins, ejecta blankets, hummocky rims, and secondary craters. In the fast-cooling lithosphere with its increasing rigidity and viscosity these old impact structures survived up to now. As seen in Table 2.1, the ratio of volcanic plains (formed *after* the period of heavy bombardment) to the highland area is rather small for the Moon and Mercury. Hence, mare surfaces do not make a big contribution to the hypsographic curve. Moreover, the great impact basins filled with basaltic material really have different elevations.

The three peaks of the hypsographic curve of Mars can be easily explained. The upper peak corresponds to old heavily cratered highlands, the middle peak to the lowlands created later, i.e., the volcanic plains, which cover much more of the planet's surface than those of the Moon and Mercury (Table 2.1), and the lower peak to some of the great impact plains and some other low-lying plains in the northern hemisphere. The uppermost area in the hypso-graphic curve corresponds to the youngest structure, the Tharsis plateau, which represents a bulge of thick lithosphere, most probably caused by internal convection patterns, possibly in connection with lateral lithospheric movements.

The sharp peak of Venus's hypsographic curve in connection with the small CM–CF offset reflects the general smoothness of the Venusian surface. Not many strong deviations from the mean radius are present; no large areas of highlands were frozen in the warm surface. These observations together with gravity modeling indicate a small lithosphere, which may be still inside the crust as it was in the early Earth. However, no reliable information on the generally basaltic plains and their crustal thickness is available so far. Maxwell Montes in the north, with a peak elevation of 11 km above the mean radius, seems to resemble an orogenic feature like the Himalayas, marked by lateral compressional movements, folded mountain ranges, and elevated rims in the case of the Lakshmi Planum plateau.

2.5   A GENERAL PLANETARY OUTLOOK

Many data on crustal and lithospheric thicknesses and composition of other terrestrial planets are still vague. Gravity modeling, for instance, and a thorough inspection of topographic and gravity anomalies of all wavelengths cannot supply a firm answer to the question of whether a topographic feature with some corresponding pronounced gravity anomaly is supported by (1) a strong crustal or lithospheric flexural rigidity or (2) a dynamic process from below, e.g., an upwelling or downward-directed part of a convection cell, or whether the depth of isostatic compensation is extremely (often unreasonably) deep. A missing regional gravity anomaly between large topographic units like basaltic plains and adjacent highlands or continents may indicate an Airy-type compensation with a greater crustal thickness below the elevated areas, as is mostly the case on Earth. Although, in theory, compensations of the Airy and Pratt type can be distinguished by observing the fine structure and the spectrum of the gravity curve in the area of transition from lowlands to highlands, often the resolution is insufficient, and the errors are too large.

There is no doubt, however, that the formation of light crusts in a giant differentiation process is a most general planetological phenomenon, and many features of crustal structure and composition of the terrestrial planets resemble those on Earth. While the crusts—chemically different from the mantles—preserve their thickness once the period of magmatic process has died out, the lithospheres—as thermally/rheologically defined units—grow thicker and thicker as the planets cool.

Chapter 3 | Collecting Physical Properties of the Earth's Crust

3.0 INTRODUCTION

Although most details of crustal structure have been found by seismic methods, the assessment of isostatic processes and of densities plays a key role in the historical as well as in the tectonic field. Also, stress measurements and magnetic, electromagnetic, and geothermal methods have made significant contributions to our present knowledge of crustal features. The concept of isostasy, however, as developed in the last century, was a continuous challenge for investigating compensation mechanisms. The existence of density differences, in the vertical and in the horizontal direction, provides a powerful tectonic engine in the form of rising diapirs and plumes as well as sinking slabs. Modern additions to the concept of isostasy are provided by the growing knowledge of viscosity and rigidity as limiting factors for isostatic mechanisms.

3.1 ISOSTATIC COMPENSATION AND DENSITY CALCULATIONS FROM GRAVITY MEASUREMENTS

Bouguer's pendulum measurements in 1737–1740 in Peru/Ecuador were the first experimental efforts to measure the ratio of the average density of the Earth $\bar{\rho}_E$ to that of a mountain range $\bar{\rho}_M$. The first $\bar{\rho}_E/\bar{\rho}_M$ value of 4.5 and similar ones were obtained by measuring the gravity g at the foot and on top of high mountains. The first pendulum measurements gave values of the density ratio that were much too large, because the mountain range with a density of about 2.5 g/cm³ was assumed to be situated uncompensated on top of the surrounding lowlands. However, in 1772 a value for $\bar{\rho}_E/\bar{\rho}_M$ of 1.8 was calculated at a rather peaked and isolated mountain in Scotland, resulting in reasonable values for $\bar{\rho}_E$ (Bullen, 1975). It was nearly 100 years later, around 1855, that Pratt and Airy put forward ideas on density compensations below the Earth's surface, and about 200 years later the first quantitative calculations on the isostatic behavior of topographic anomalies of different size were formulated.

In the early 19th century comparisons between distance measurements by geodetic triangulation and by astronomical methods were carried out, for instance, during a great land survey of India, south of the Himalayas. The astronomical determinations were based (as usual) on the angles of stars with respect to the vertical, which is defined by a plumb line. Certain discrepancies between the geodetic and the astronomical methods were observed. Apparently the plumb line was affected by the gravitational attraction of the mountain range. However, when calculating this additional gravity effect of the visible shape of the mountain range and assuming reasonable density values, it was found that the deviation should have been much larger and the discrepancy between astronomical and geodetic measurements much greater. Apparently the mass of the mountain range was compensated somehow by a less dense material below the surface.

3.1.1 ISOSTATIC MODELS

Sir George Airy proposed in 1855 that the visible mountain ranges could not be supported by the rigidity of the subsurface but have light roots and "float" in a denser substratum much as an iceberg floats in (denser) seawater. Pratt suggested at the same time that the high elevations were sustained by virtue of their lower density. Both ideas are illustrated in Fig. 3.1. They provide guidelines for gravity as well as for seismic investigations of crustal structure. While Heiskanen and Vening–Meinesz (1958) still estimated that on the average 63% of continental elevations are compensated by a *crustal root* and 37% by density differences à la Pratt, modern views of isostasy distinguish between the size and width of a mountain range, its age, its thermal regime, and the mechanism by which possible roots can develop. Small isolated mountains like the Harz in Germany have not (yet) reached an isostatic balance and have no roots. Young mountain ranges, still warm, often show a lateral displacement of their roots (Artemyew and Artyushkov,

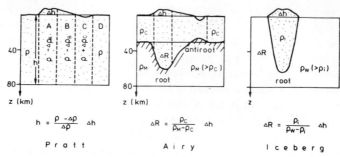

Fig. 3.1 Models of isostatic compensation. Here $\Delta\rho$ is the density difference with regard to average density ρ_1, ρ_M the density of the mantle, ρ_C the density of the crust, ρ_i the density of ice, ρ_W the density of water, ΔR the thickness of the crustal root and Δh the excess elevation.

1967), which seems to be related to subduction or other kinds of decoupling processes in the lower crust/upper mantle. In general, only young, Cenozoic mountain ranges have roots at all. The Caledonides, the Appalachians, and the European Variscides do not have roots like the Alps (Theilen and Meissner, 1979). In some orogenic features the isostatic balance is achieved by an anomalously hot mantle or is violated by dynamic forces from below. For reviews of the combined use of seismic and gravity data see Woollard (1969), Demenitskaya and Belyaevsky (1969), Bullen (1975), and Bott (1982).

Calculations of the gravity effect of the topography and that of an assumed root zone result in a combined theoretical gravity effect which must be compared with the observed gravity effect after corrections for the height of the satellite or the reference surface. A best fit to the observed gravity is obtained by iterative methods for modeling density differences and the size of the root. Sometimes this approach leads to good coincidence with an isostatic behavior of the Airy type (see Fig. 3.1). For a two-dimensional mountain range such an isostatic behavior results in equal areas of positive (above the mountains) and negative (sideward) Δg deviations as shown in Fig. 3.2. The Gauss integral $\oint \Delta g \, dx$ equals zero across the structure. The negative side lobes originate from the root zone, whereas on top of the mountain the topographic effect dominates. This is the so-called proximity effect described by Woollard (1969), whereby the gravity contribution from mass elements varies with $\cos \theta / r^2$, where θ is the angle from the vertical and r the lateral distance. The iceberg of Fig. 3.1 would also show such gravity behavior. If the width of the mountain range is very large and approaches continental dimensions the proximity effect creates comparable gravity contributions from topography and root in the center of the mountain range, and the gravity undulations concentrate on the edges, a phenomenon known as the "edge effect", as shown in Fig 3.2. In this case $\oint \Delta g \, dx = 0$. Only across mass

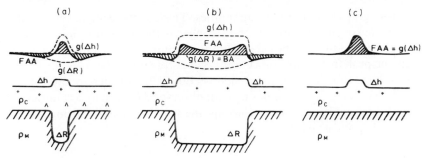

Fig. 3.2 Schematic picture of the gravity behavior across mountain ranges. (a) Narrow structure, compensated ($\oint \Delta g \, dx = 0$); (b) wide structure, compensated ($\oint \Delta g \, dx = 0$); and (c) uncompensated structure ($\oint \Delta g \, dx = 2\pi G \, \Delta M$). Here FAA is the free air anomaly, BA the Bouger anomaly, Δh the excess elevation, and ΔR the crustal root.

anomalies that are not fully compensated does $\oint \Delta g \, dx \neq 0$ and approach $2\pi G \, \Delta M$, where G is the constant of gravity and M the excess mass for a certain area (see, e.g., Grant and West, 1965). In the case of a totally sustained (noncompensated) additional topographic mass the gravity effect of the topography is equal to that of the observed gravity. As mentioned before, the smaller the topographic anomaly, the greater is the possibility that the "elastic" lithosphere and especially the upper crust can sustain the surface load. In this case the equation for Fig. 3.2 must be modified (Brotchie and Sylvester, 1969), i.e.,

$$\rho_C g \, \Delta h = g(\rho_M - \rho_C) \, \Delta R \tag{3.1}$$

is extended to

$$\rho_C g \, \Delta h = g(\rho_M - \rho_C) \, \Delta R^* + D \, \nabla^4(\Delta R^*) \tag{3.2}$$

where ΔR^* is the downward reflection of elastic lithosphere, D the flexural rigidity, and g gravity. ∇ is nabla, a differential operator.

The flexural rigidity D describes the long-term behavior of an elastic plate (i.e., a kind of resistance) against bending moments. It is related to the thickness H of the elastic lithosphere by (Cazenave and Dominh, 1981)

$$D = EH^3/12(1 - \sigma^2) \tag{3.3}$$

where E is Young's modulus, σ the Poisson constant, and H the plate thickness. The flexural rigidity D was successfully introduced into problems of buckling and bending of an oceanic lithosphere near active continental margins. In the more complicated continental lithosphere, with its complex rheological behavior (see Section 4.5, especially Section 4.5.2), the introduction of D seems problematic, but it can provide a first-order assessment of linear topographic and gravity anomalies. There is the general relationship that the thicker the lithosphere, the stronger is the flexural rigidity and the larger is the load that can be carried without compensation. Other terms, like a dynamic push from below or a compressive stress from the sides, might also be introduced on the right-hand side of Eq. (3.2) in order to cause topographic changes such as an uplift with an observed Δh. Such uncertainties will always be present in gravity calculations and will be especially strong in areas of recent tectonic activity.

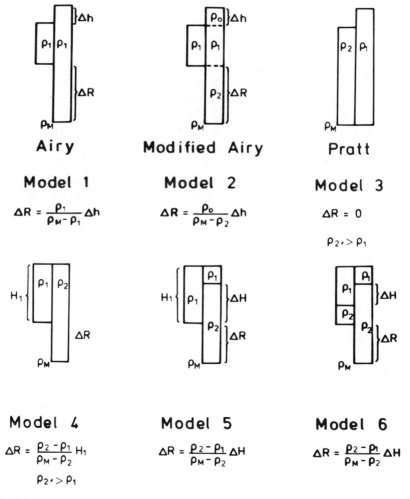

Fig. 3.3 Six models for crustal isostatic compensation. Models 2 and 6 seem to dominate.

In the case of a simple Airy compensation as described by Eq. (3.1), it is often possible to get a good estimate for the expression $\rho_C/(\rho_M - \rho_C)$ once an average Δh is known from topographic data and an average ΔR from seismic data. In most cases $\rho_C/\rho_M - \rho_C$ is between 7 and 8 for young orogenic mountain ranges which are in isostatic equilibrium. Assuming an average density of the crust ρ_C of 2.9 g/cm^3 = 2900 kg/m^3, a mantle density ρ_M about 3.3 g/cm^3 is obtained. A slightly more refined treatment uses the density of the topography separately, as mentioned with Fig. 3.3 (model 2).

3.1.2 CRUSTAL DENSITY VALUES

The densities mentioned above are only average values and not of much use for detailed crustal studies. Table 3.1 gives some density values of representative crustal rocks at zero pressure. Data refer to the "wet bulk density," which is supposed to represent the densities *in situ*. Especially for the first 7 km of depth the density increases rather strongly because of the closing of microcracks under pressure. Values of $\Delta\rho/\rho$ will be about 3 to 6% at 10 km depth. Depending on pressure and temperature, $\Delta\rho/\rho$ may slightly increase or decrease for greater depths. Hence, the values of Table 3.1 can be used only as a guide for a comparison of densities of different rock types. For use in crustal studies a depth correction must be applied as indicated in Table 3.1. A better estimate is generally obtained from the correlation between density and velocity (see Section 3.1.4).

3.1.3 FREE AIR AND BOUGUER ANOMALIES

For the purpose of the following discussion the so-called *free air anomaly* (FAA) and *Bouguer anomaly* (BA) are introduced. All gravity anomalies described in the previous sections refer to the free air anomaly. The FAA is defined as the difference between the observed gravity g (to which the free air correction $H \, \partial g/\partial R$ is applied as a reduction to sea level) and the (latitude-dependent, average) gravity $g_0(\varphi)$ at sea level (exactly at the reference geoid), where H is the elevation, $\partial g/\partial R$ the gravity gradient near the surface, and φ the latitude. For details, see textbooks such as those by Grant and West (1965) and Telford *et al.* (1976). Corrections or "reductions" are necessary in order to compare gravity values obtained at different latitudes of a rotating body and at different elevations. For free air corrections no unknown parameters are required. Therefore, the FAA is widely used for all isostatic problems on the Earth and other planets. As shown in Fig. 3.2, the FAA shows a direct correlation with topography for narrow structures which are more or less isostatically compensated. For a more sophisticated treatment of

Table 3.1

Average Densities of Crustal Rocks at Zero Pressure[a]

Sediments ($\bar{\rho} = 1.6$–2.7)

Unconsolidated sediments (depth = 0–1 km)

Loess	1.6 ± 0.3
Sand	2.0 ± 0.4
Soils	1.9 ± 0.3
Clay	2.2 ± 0.2

Consolidated sediments (sedimentary rocks) (depth = 0–5 km and up to 16 km)

Sandstone	2.5 ± 0.5
Limestone	2.6 ± 0.35
Shale	2.6 ± 0.2
Dolomite	2.7 ± 0.2

Crystalline basement ("normal" continental crust) ($\bar{\rho} = 2.5$–3.1)
(depth = 5–40 km, in some cases 0–80 km)

Igneous rock $\bar{\rho} = 2.78$ ($+4\%$)

Extrusives

Rhyolite	2.5 ± 0.15
Dacite	2.6 ± 0.15
Trachyte	2.6 ± 0.15
Andesite	2.7 ± 0.15
Basalt	2.9 ± 0.10

Intrusives

Granite	2.67 ± 0.1 ($+6\%$)
Granodiorite	2.72 ± 0.05
Diorite	2.85 ± 0.1 ($+3\%$)
Norite	2.92 ± 0.1
Diabase	2.95 ± 0.15
Gabbro	3.00 ± 0.1 ($+3\%$)
Hornblende	3.1 ± 0.1
Gabbro anorthosite	2.75 ± 0.1
Serpentinized peridoite	2.65 ± 0.3

Metamorphic rock $\bar{\rho} = 2.74$ ($+5\%$)

Quartzite	2.6 ± 0.2
Schist	2.6 ± 0.2 ($+5\%$)
Graywacke	2.65 ± 0.2
Granitic gneiss	2.6 ± 0.05
Gneiss	2.75 ± 0.2 ($+6\%$)
Amphibolite	2.95 ± 0.15
Granulite	2.7 ± 0.1 ($+6\%$)
Quartzitic shale	2.8 ± 0.1
Serpentine	2.85 ± 0.2

Upper mantle ($\bar{\rho} = 3.1$–3.4) depth 30 km, in some cases 15 km)

Peridotite	3.2 ± 0.05 ($+4\%$)	Eclogite	3.4 ± 0.05 ($+5\%$)
Dunite	3.25 ± 0.05		
Pyroxenite	3.2 ± 0.1		

---

[a] Density is expressed in grams per cubic centimeter (tons per cubic meter). Data are extracted from Telford et al. (1976), Grant and West (1965), Meissner and Stegena (1977), and Wohlenberg (1982). Values in parentheses refer to $z = 10$ km. For $z > 10$ km only slight changes in ρ occur. Error bars for the normal range are from Wohlenberg (1982).

isostatic behavior, a spherical harmonic expansion for the FAA and the topography is used (see Garland, 1965).

In addition to the corrections mentioned before, the Bouguer anomaly requires another correction, the so-called Bouguer correction, which takes into account the mass of rocks between the gravity station and a reference level. The "simple" BA is defined as the observed gravity g minus the previously mentioned free air and latitudinal corrections minus the Bouguer correction, which is $2\pi\rho GH$, where ρ is the average density between the observation point at elevation H and sea level (or the reference geoid or reference level) and the G gravity constant.

It is often necessary to apply additional corrections. One of these is the topographic correction which takes care of nearby masses (or mass deficits) around the station. Another is the tidal correction, which takes care of the different effects of the attraction of the sun and the Moon on the solid Earth and nearby oceans. These two corrections are generally used to arrive at the "complete" BA, but are sometimes also applied to the simple FAA values.

In general, the BA reflects density anomalies below the reference level. It is widely applied in geophysical prospecting, e.g., for ores. However, the BA is also sometimes used for isostatic problems because of the good negative correlation between the BA and young mountain ranges. A totally uncompensated mountain would show no BA at all if the corrections were correct. In general, however, there is a marked dependence of BAs on elevation, and often a negative BA of 300 mgal is found for a 3000-m elevation (Woollard, 1969). Such correlations have been obtained for many geologic provinces, and often a different state of isostatic compensation and depth of the root zones has been demonstrated. Although a third kind of anomaly, the so-called isostatic anomaly, has been used for isostatic problems, it is felt that too many unknowns regarding density and depth of compensation are required for this treatment.

Another problem of using the FAA or BA for an assessment of crustal densities or isostatic compensation is the "regional gravity" or simply "regional". This originates from long-wavelength contributions which may have their origin somewhere deep in the upper mantle, like the geoid undulations. Many graphic and smoothing techniques and refined filter techniques, such as the use of an upward continuation or the so-called second derivative, are used in practical calculations of shallow crustal structures (see, e.g., Telford et al., 1976). The regional somewhat limits the assessment of isostasy of broad structures like that of Fig. 3.2b, but certainly not so much that, as some critics say, "the regional is a smooth gravity difference, which has to be substracted in order to make the rest look like a good anomaly". For a careful analysis of the possible depth of long-wavelength anomalies filter techniques are essential, because of a possible compensation of crustal structures in the upper mantle. The slow uplift of formerly glaciated areas, indicated by some positive changes of negative gravity values over a large area in Fennoscandia, points toward such asthenospheric processes (Moerner, 1980).

3.1.4 CORRELATION BETWEEN SEISMIC VELOCITIES AND DENSITIES

Based on a great amount of data from boreholes and from high-pressure and even high-temperature studies of varius crustal and mantle rocks, rather reliable velocity–density relations have been obtained (Nafe and Drake, 1959; Birch, 1961; Dortman and Magid, 1968; Woollard, 1969; Christensen, 1965). Birch (1961) suggested a linear relationship, later to be called Birch's law:

$$V_P = a(\bar{m}) + b\rho, \tag{3.4}$$

where V_P is the compressional wave velocity, $a(\bar{m})$ a constant related to the mean atomic weight $\bar{m}$, with $20 < \bar{m} < 21$ for crustal rocks, b a constant, and ρ the density.

A nonlinear relationship was suggested by Christensen and Salisbury (1975):

$$V_{P,S} = a_V(\bar{m}) + b_V\rho^C, \tag{3.5}$$

where a_V and b_V are different constants for P- and S-waves.

Other authors assumed the density to have a more complicated relationship to the seismic velocity V_ϕ, $V_\phi = (V_P^2 - \frac{4}{3}V_S^2)^{1/2}$ (Shankland and Chung, 1974), and others, like Woollard (1969) and Nafe and Drake (1959), who concentrated on a more thorough investigation of crustal rocks, also found the linearity of Eq. (3.4) slightly disturbing. Kern (1982a,b) finds remarkable differences in the V_P–ρ relation between continental and oceanic rocks. Figure 3.4 shows a correlation between V_P, V_S, and ρ for crustal rocks and some relations of the type of Eq. (3.4) for different confining pressure depths (Gebrande, 1982).

Although the scatter of the original data is still considerable, a transformation of velocity–depth to density–depth models is certainly permissible, especially when investigating isostatic or nonisostatic behavior. The derivation of the density from seismic velocities, therefore, provides a good assessment of crustal density structures.

Some caution is necessary in areas of positive geothermal anomalies. Quite a number of observations indicate that the usual V–ρ relation cannot be applied in such zones. Some reasons for this behavior are the existence of

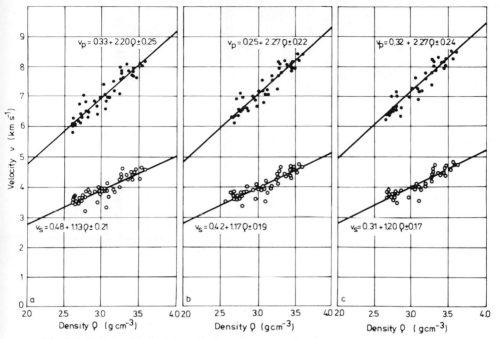

Fig. 3.4   Relationship between velocity and compressional waves V_P, shear waves V_S, and density ρ for confining pressures of (a) 0.2, (b) 0.6, and (c) 1.0 GPa. [From Gebrande (1982).]

partial melt, chemical alterations, and the generation of microcracks by cooling. These effects lead to a stronger decrease of V_P and V_S values than of ρ values (see also Section 6.6.1).

3.1.5   EXAMPLES

The combined use of seismic and gravity surveys is a powerful tool for solving tectonic problems. The use of gravity data is especially helpful if large lateral differences appear, as on profiles across active continental margins. An example is provided by a cross-section from the Pacific Ocean to the mountain ranges of the Colombian Andes (Meissner *et al.*, 1976c) (Fig. 3.5). Here, Bouguer anomalies were used on land and free air anomalies at sea. The conversion from V_P to ρ was performed according to V_P–ρ relations such as those of Fig. 3.4. The initial structural model with V_P layering, converted to a ρ layering, was slightly modified, insofar as still being within the limits of seismic boundary conditions, by the travel times. Such a two-dimensional "linearized inversion" resulted in an iterative approach to the observed gravity, using the method of Talwani *et al.* (1959), which is based on a digital

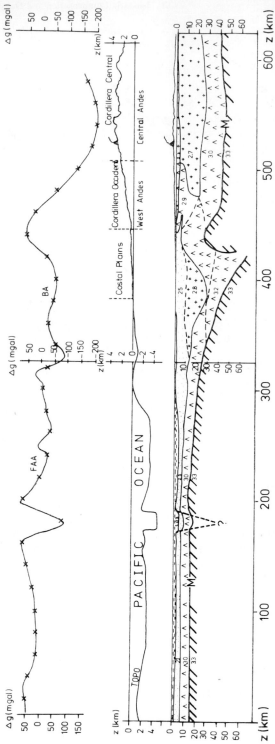

Fig. 3.5 Combined seismic refraction and gravity profile across the active margin in Colombia. Bouger anomalies were used on land, and free air Anomalies were used at sea. Data points are values from model calculations, and curves represent values from measurements.

use of simple gravity formulas. This method, of course, does not give a unique solution, but the seismic boundary conditions limit possible alternatives to a small range. It may be mentioned that only a *two*-layer density distribution —as often used for gravity modeling—for determining the crustal depth with $\bar{\rho}_C$ over a mantle with $\bar{\rho}_M$ in areas without seismic constraints gives a unique solution, although not necessarily the correct one.

In the case of seismic and gravity information of the profiles across the Andes in Colombia, it became evident from the modeling procedure that the West Andes gravity high is very much out of isostatic balance, whereas the Central Andes show the normal negative correlation between the BA and elevation (again, see Fig. 3.5). The West Andes are about 5000 km long and composed of oceanic material. Their weight is probably sustained dynamically by lateral compressional stress related to processes at active continental margins. In the northwestern part of South America a huge obduction of oceanic material has taken place.

An example of very different geologic units adjacent to each other but with no apparent gravity anomalies is provided by the transition from the Paleozoic continental part of western Europe to the old eastern European platform. Although many data from seismology and deep seismic sounding have shown that the crust is generally thin under western Europe and thick under the old platform area, there is no remarkable difference in gravity between these two units of similar elevation. Apparently another kind of large-scale isostasy must be present (Meissner and Vetter, 1976a). The key to this problem is provided by more detailed observations from deep seismic sounding (DSS), which showed that the thin crust in the west has rather low seismic velocities, indicating a sialic composition down to 20–25 km with lower densities, while the crust under the old platform shows higher velocities, indicating a more mafic composition with higher densities from 10–15 km downward. For more details see Section 7.3. The velocity–depth function in both geologic units is now converted to density–depth functions by means of the V_P–ρ relationship of Fig. 3.4. Figure 3.6a shows two velocity–depth diagrams of different areas with similar gravity values, the old Baltic Shield and Paleozoic western Germany, while Fig. 3.6b shows the corresponding density–depth curves. For perfect isostatic behavior the areas $\Delta\rho\,\Delta z$ between the curves should be the same. We can easily see that this holds for the situation depicted in Fig. 3.6b. However, when comparing the Paleozoic curve with that of a very young basin, e.g., the Buchara region (Fergana Basin, Uzbek, USSR) (Godin *et al.*, 1960) (Fig. 3.7), there is nearly no overlapping in the curves and certainly no isostatic equilibrium (Meissner and Vetter, 1976a,b). The Buchara region, a fore deep of the Tien Shan, is influenced by recent dynamic processes. Around such young orogenies—as in deep sea trenches—gravity anomalies are strongly negative.

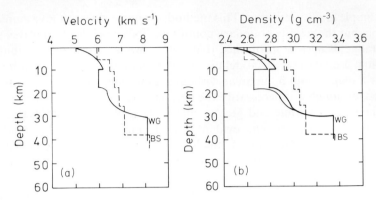

Fig. 3.6 (a) Velocity–depth function and (b) density–depth function of two areas with identical gravity but different crustal structures. Here BS is the Baltic shield, and WG is West Germany. For (b), two extreme V_P–ρ conversions were used. The tendency is for isostatic compensation according to model 6, in Fig. 3.3.

In general, gravity spectra of all wavelengths together with seismic data and V_P–ρ relationships are effective tools for studying the degree of isostatic compensation. The different appearance of isostasy in crustal sections is depicted in Fig 3.3. All these models have been observed in nature; the last one (model 6) is discussed together with Fig. 3.6. The problem of the mechanism for approaching isostatic compensation remains. Sections 4.5 and 7.3 will give some clues with regard to this problem.

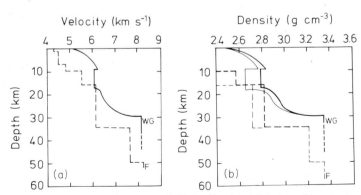

Fig. 3.7 (a) Velocity–depth function and (b) density–depth function of two areas with large differences in gravity. Here WG is West Germany, and F is the Fergana Basin. (Two V_P–ρ conversions, no isostasy.)

3.1.6 GRAVITY MAPS

In some countries there is good coverage of topographic and gravity data. For instance, such a data set for the continental Unites States is available from the National Oceanic and Atmospheric Administration (NOAA) and contains averaged elevation and about 500,000 gravity readings. The data can be homogenized by digital techniques, and a regular grid for Free Air and Bouguer anomalies can be artificially constructed (Arvidson *et al.*, 1982). Such a data set can be "continued" upward (thereby enhancing the long-wavelength anomalies, generally related to deep structures) or downward (enhancing the short-wavelength anomalies, which are necessarily related to density anomalies in the upper crust). These techniques are described in standard textbooks on applied geophysics (e.g., Grant and West, 1965; Telford *et al.*, 1976). Standard image-processing techniques, as developed in various space programs, have been successfully applied to the interpolated and artificially constructed regularly spaced data sets. For technical details see Moik (1980). Two-dimensional filtering with different filter widths, equivalent to an upward continuation of the field to different levels, results in maps which provide an assessment of the depths of density anomalies. These maps can be presented in a color-coding scheme or in the form of shaded relief, with an artificial sun introduced at generally low angles in order to produce the most effective shading.

Figure 3.8 shows shaded relief maps of elevation, free air anomalies, and Bouguer anomalies for the continental United States. A critical comparison of these maps reveals many subsurface features. From west to east, there is the Mesozoic–Cenozoic mountain and basin system of the western third of the United States, containing the Californian Great Valley, a depression in all three maps. There are the Colorado Plateau, the Rio Grande Rift, and the front range, separating the mountainous area from the lowland in the east. While the FAA, by means of its correlation with the topography (see Section 3.1.3), accentuates the mountain and valley structure, the BA shows a large low over the western third of the United States, indicating a strong tendency for isostatic compensation of the high elevations. The circular structure in the south, which is not seen in the topography, reminds one of a huge impact crater about 600 km in diameter. Northward of this structure the Mid-Continent Gravity High begins. This is an elongated narrow structure in both the BA and FAA maps. The FAA map shows the characteristic side lobes, with up to 400 mgal difference between maximum and minimum. The BA is positive (!) and shows about 200 mgal difference in gravity values, indicating that isostatic compensation is totally absent. The structure must be related to an uplift of the Moho or another density anomaly in the middle or lower crust. In the east, the Appalachian system dominates the area with only

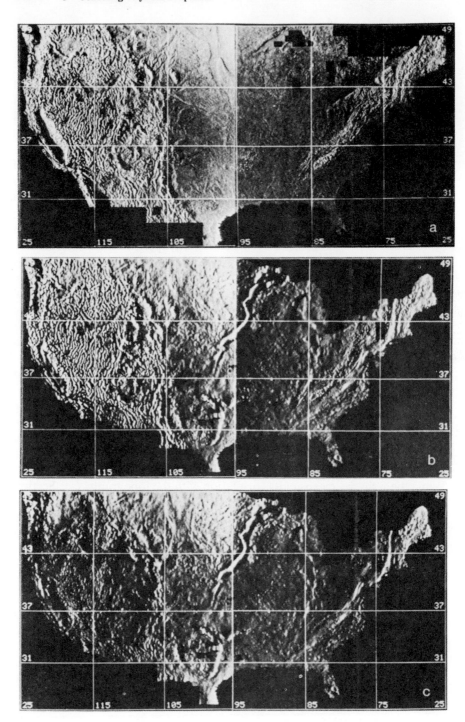

moderate FAA structures and a mild negative BA, which transforms to positive values toward the Atlantic coast, a consequence of the so-called edge effect of the continental–oceanic transition. Many of these areas will be discussed again in Chapter 6. The gravity maps provide important reconnaissance information for later detailed investigations.

3.2 SEISMOLOGICAL METHODS

As mentioned in Section 1.1, it was Mohorovičić at the beginning of this century who defined the Earth's crust in terms of its compressional (P-wave) velocities. Such investigations of body waves from earthquakes are still being carried out, but they lack accuracy compared to the high precision of controlled-source seismology, i.e., the arrangement of special reflection or refraction profiles and the use of explosive (or nonexplosive) sources. Also, *surface waves* produced by shallow earthquakes are widely used in seismology for regional crustal studies. Highly developed inversion techniques provide the tools for a refined interpretation. Historically, these methods have provided the first information on the difference between the continental and the oceanic crust (Gutenberg and Richter, 1936; Rothé, 1947; Caloi *et al.*, 1949; Caloi, 1967).

A major input to studies of the state of stress inside the continental crust has come from investigations of the seismicity. In addition to the alignment of earthquakes along plate boundaries, many faults (but not all of them) show a strong concentration of seismic (and aseismic) events. Fault plane solutions and earthquake mechanisms are other important parameters for the present tectonic stress pattern. Finally, the distribution of seismicity with depth provides information about the crust's brittle and ductile behavior (see also Section 4.5).

The existence of body waves and surface waves was proved in the 1800s. Compressional or P waves (or longitudinal particle displacements) and shear waves or S waves (of transverse displacements) are radiated from all earthquakes. The letter P comes from the Latin *primae undae* and S from *secundae undae*, giving evidence of observations in the past century that P waves arrive earlier and hence travel faster than S waves. Surface waves, stemming from shallow earthquakes, still have smaller propagation velocities than S waves. Seismic waves are recorded by seismometers in a fixed or

---

Fig. 3.8 Shaded relief maps for the conterminous United States (Mercator projections). (a) Elevation averages (simulated sun 20° to the west), (b) free air anomalies (simulated sun 15° to the west), and (c) Bouguer anomalies (simple BA) (simulated sun 15° to the west).

movable configuration. Seismograph stations with three-component seismometers are distributed all over the continents, although with different density.

In modern seismic observatories seismometers are grouped into areal patterns, so-called seismometer arrays, which allow determination of the wave type, azimuth, and dip angle of incoming waves. All kinds of digital recording, bandpass filtering, and inverse filter techniques are used to increase the signal-to-noise ratio. Many seismograph stations are needed for the determination of exact location, focal depth, and focal mechanisms. Great assistance for these calculations is provided by the World-Wide Standard Seismograph Network (WWSSN) (Aki and Richards, 1980).

3.2.1 BODY WAVES FROM EARTHQUAKES FOR CRUSTAL STUDIES

In general, the use of body waves to derive the crustal structure has a more historical character. Only in areas of very high seismicity, such as the Bucaramanga "nest" in Colombia (Pennington *et al.*, 1979), or areas of a very dense seismometer networks, such as CEDAR in California, is the determination of crustal parameters from earthquake body waves comparable in accuracy to explosion seismology. Historically, three body-wave phases related to the thickness and velocities of the crust were derived: the P_g wave, supposed to follow the top of the granitic crystalline basement below the sediments, the P_mP wave, supposed to be a critical or overcritical reflection from the Moho, and the P_n wave, supposed to run along the top of the mantle. While the P_g wave and the P_n wave were first considered to be head waves, the P_mP appeared as a wide-angle reflection, to be observed at distances greater than about 60 km. Figure 3.9 shows this simplified picture of basic travel times and ray paths, which still forms the backbone of explosion seismology too. Often a reduced travel time diagram is used in order to enhance differences between different velocities (and different travel time branches). A pattern similar to that for P waves is obtained for shear waves, called S_g, S_mS, and S_n, although observation of these branches is not so common. As will be shown in Section 3.4.3, some more branches between the P_g and the P_mP have been detected by explosion seismology, among them the refracted (or reflected) waves from the Conrad discontinuity in the middle part of continental crusts, which are supposed to mark the transition from sialic to more mafic or higher-grade metamorphic material with velocities around 6.5 km/s (Giese *et al.*, 1976).

Body-wave data from earthquakes along seismic profiles were obtained by using earthquake "nests" and movable seismometer stations in a number of

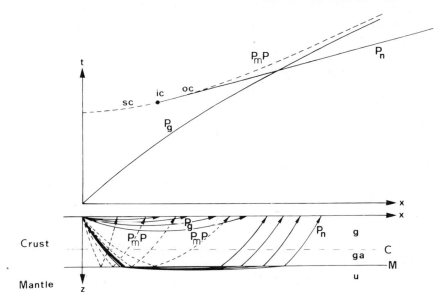

Fig. 3.9 Basic travel time curves of a two-layered crust. Here P_g is the "granitic" (first) event, P_n the refracted (i.e., guided wave in uppermost mantle), P_mP the reflected or diving wave from Moho boundary, sc the subcritical area, ic the critical point, oc the overcritical area, g the gneissic–granitic upper crust, ga the gabbroic–granulitic lower crust, and u the ultramafic mantle.

areas, but, as stated before, their accuracy is limited. Often they are helpful when used together with data from explosion seismology, and they are fitted into the travel time branches by means of their travel time differences $P_g - P_n$ or $P_g - S_g$.

Earthquake body waves really provide information on the lithospheric structure in tectonically interesting areas. They suffer somewhat from uncertainties regarding the focal depth and the rupture process. They have generally low frequencies, and mostly resolve the gross structures in the mantle better than the fine structure in the crust.

Teleseismic body waves are used rather often for the calculation of "travel time residuals" along a profile or along a network (Weinrebe, 1981; Raikes, 1980). In these cases crustal depths or anomalous velocities in the mantle can be delineated. An inversion of travel times of many events measured with many seismometers is used in order to obtain horizontal and vertical velocity patterns of an area. For the Moon, a network of four Apollo seismometers was used to record moonquakes as well as various artificial and natural impacts in order to work out the thickness and velocities of the lunar crust (Toksöz et al., 1974a, 1975; Voss et al., 1976).

Large areal networks of seismometers are often used to monitor the seismicity in tectonically active areas, determining magnitude, seismic moment, frequency relationships, focal depths, and mechanisms. Such arrays, like that of the Central European Seismological Observatory in Graefenberg (Germany), the Norwegian (NORSAR) and Swedish arrays, and the 28-station, 150-km-aperture subarray of the SCARLET network near the central Transverse Ranges in California, are well able to observe areal variations of crustal structure and velocities, although much more attention is focused on the upper mantle. Vetter and Minster (1981) analyzed the P_n velocities of 81 earthquakes and some explosives and found a strong azimuthal dependence, which had been detected previously only in the ocean. Hadley and Kanamori (1977) derived a reasonable three- to four-layer crustal model by the combined use of local earthquakes and explosions. A variation of the thickness of crustal layers in a certain direction was also found from the network analyses—a result that is often difficult to obtain by simple observation along profiles. In general, there is a great deal of redundancy in the investigation of so many events with so many seismographs, and this redundancy is used to select improved model parameters and reduce error bars.

3.2.2   SURFACE WAVES FROM EARTHQUAKES FOR CRUSTAL STUDIES

Surface waves with large amplitudes, a wide frequency range, and generally long wave trains are generated by shallow earthquakes. They differ from body waves in various ways:

(1)   They run along discontinuities with strong impedance contrasts ρV, such as the Earth's surface.

(2)   Their amplitudes decrease exponentially away from the boundary.

(3)   Their geometric amplitude decreases only with $R^{-1/2}$, compared to R^{-1} for body waves. (R is the distance from the source.)

(4)   In practice, all surface waves are dispersive, i.e., different frequencies travel with different velocities.

Rayleigh waves may be guided by a single boundary, e.g., the surface. They are called Stoneley waves if guided along an internal boundary. Their path is generally retrograde elliptic in the direction of propagation; hence, they are recorded by vertical (Z) and horizontal seismographs in the direction of propagation ($H_{\parallel}$). Love waves are horizontally polarized S waves in a waveguide and need at least two boundaries, e.g., the surface and an internal boundary, for their propagation. They are recorded by horizontal seismographs perpendicular to the direction of propagation $(H_\perp)$. In principle, the azimuth can be obtained for any wave from amplitude ratios of two perpendicular horizontal components.

The use of surface waves for crustal studies became known in the 1940s and 1950s through the work of Rothé (1947), Jeffreys (1952), Press and Ewing (1952), and others. A compact review of regional studies is given by Kovach (1978), while the resolving power of surface waves is described by Panza (1980). Evaluation of surface-wave data first requires the determination of the phase and group velocities as a function of frequency. The determination of phase velocities requires at least two seismograph stations or (better) an array. The frequency determination is obtained with a moving window or a similar analysis. Inversion techniques are needed to convert the dispersion curves of Rayleigh (R) and Love (L) waves, i.e., their velocities $V_R(T)$ or $V_R(f)$ and $V_L(T)$ or $V_L(f)$ (T, period; f, frequency), to velocity–depth functions. The "linearized inversion" technique (Backus and Gilbert, 1968, 1970) is based on an iterative procedure employing slight modifications of an initially "known" subsurface structure in order to arrive at a best fit between the modeled and the observed dispersion curves. This technique is especially useful in crustal studies where some basic information is available from DSS and/or gravity data.

In the 1950s it had already been found that surface waves that pass through ocean basins travel faster than waves of the same period that pass through continental shields (although this turned out to be wrong for the Rayleigh group velocity in several frequency ranges). Dispersion curves of Rayleigh waves are influenced by the presence of water and show very small values of the group velocity U between 1.5 and 1.8 km/s for $T = 15$ or 18 s. For larger periods T, standard curves for ocean basins and continents, for group velocities U of Rayleigh and Love waves, and for their phase velocity c were provided by Knopoff and Chang (1977) and Panza (1980) (see Fig. 3.10a,b).

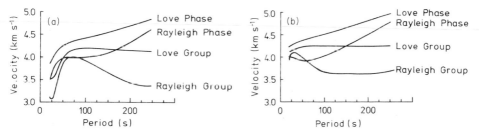

Fig. 3.10 Standard phase and group velocities for (a) continental and (b) oceanic ray paths. [From Panza (1980).]

The curves of Fig. 3.10 refer to the ground modes of surface waves and are based on the crust–upper mantle model described in Table 3.2. While the curves for oceanic paths are all very similar to each other, data from the continents differ considerably, especially for small periods T. As also stated by Calganile and Panza (1979), surface waves at $T = 12$–60 s can be used to resolve the average crustal thickness to within 3–5 km if the elastic parameters are known. Figure 3.11 shows an example of calculated and observed dispersion curves for the Apennines (Calganile *et al.*, 1979). It is based on a

Table 3.2

Model for the Crust–Upper Mantle Structure for an "Average" Continent and Ocean[a]

Continent				
Depth (km)	Thickness (km)	V_S (km/s)	V_P (km/s)	ρ (g/cm^3)
0	10	3.49	6.06	2.75
10	20	3.67	6.35	2.85
30	20	3.85	7.05	3.08
— M ——				
50	65	4.65	8.17	3.45
115	250	4.30	8.35	3.54
365	85	4.75	8.80	3.65
450	200	5.30	9.80	3.98
650	400	6.20	11.15	4.43
1050	240	6.48	11.78	4.63
1290		6.62	12.02	4.71

Ocean				
Depth (km)	Thickness (km)	V_S (km/s)	V_P (km/s)	ρ (g/cm^3)
0	4	0.00	1.52	1.03
4	1	1.00	2.10	2.10
5	5	3.70	6.41	3.07
— M ——				
10	50	4.65	8.10	3.40
60	150	4.15	7.60	3.40
210	240	4.75	8.80	3.65
450	200	5.30	9.80	3.98
650	400	6.20	11.15	4.43
1050	240	6.48	11.78	4.63
1290		6.62	12.02	4.71

[a] From Panza (1980). The models are the basis for the dispersion curves of Fig. 3.10.

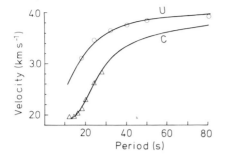

Fig. 3.11 Calculated (———) and observed (○, △) dispersion data for the Apennines. Here U is the group velocity and C the phase velocity. [From Calganile and Panza (1979).]

six-layer crustal and a three-layer upper mantle model with given V_P, V_S, and ρ layering, the Moho being at 36 km. Although the fit between calculated and observed curves seems perfect, the uniqueness of the complicated crustal structure is certainly debatable.

An interesting aspect of surface-wave investigations is their ability to resolve the crustal S-velocity structure in areas where the P structure might be known from DSS. Because of the basic relationship between surface-wave velocities and S-wave velocities, which is especially strong for Love waves, the resolving power for V_P velocities and for ρ is very limited. In most cases the group velocities have better resolution than the phase velocities. The use of higher modes in general does not increase the resolving power. Surface waves usually do not provide a reliable solution for laterally inhomogeneous media or for thin layers. Their wavelength $\lambda = V/f$ generally exceeds 3000 m (for 1 Hz), which is completely inadequate for resolving fine structures of the crust.

Special attention is given to the so-called L_g waves. These are horizontally polarized shear waves (SH) with a period of about 4 s and a velocity of about 3.5 km/s. They are typical for entirely *continental* paths. The crust itself seems to act as a channel with its two boundaries: surface and Moho. Earthquakes within the (upper) crust produce the "channel" waves L_g, being favorably reflected at both boundaries and "guided" with only small energy losses. Recently, coda waves of L_g of local and regional earthquakes were used successfully to determine the Q structure in the crust of the contiguous United States, showing a strong directional variation in attenuation $(1/Q)$ (Singh and Herrmann, 1983). These results are discussed in Section 4.2.

3.3 SEISMICITY

The term "seismicity" or "seismic activity" goes back to Gutenberg and Richter (1954) and describes the geography of earthquakes and their distribution over the Earth's surface. In a second sense it may also mean the

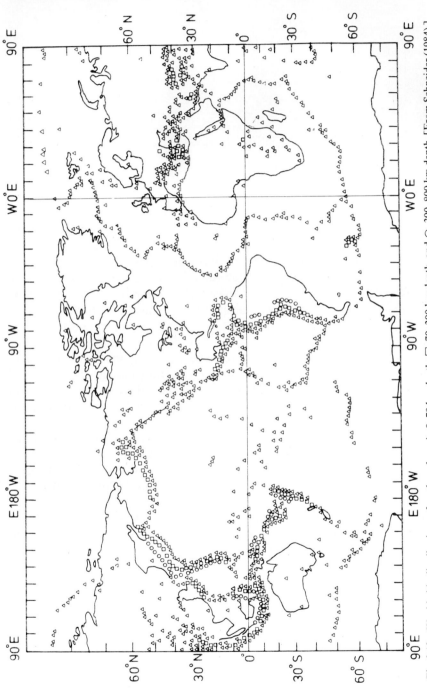

Fig. 3.12 Worldwide distributions of earthquakes; △, 0–70 km depth, □, 70–300 km depth, and ○, 300–800 km depth. [From Schneider (1984).]

magnitude and the focal mechanism of earthquakes in a given area, which are both related to the tectonic stress pattern of a region. The worldwide distribution of earthquakes shows that they tend to occur in belts, for instance, that around the Pacific Ocean, which carries about 75% of the energy release of shallow earthquakes (0–70 km depth). About 23% occur in the Alpine–Himalayan belt, mostly on land but also in the Mediterranean. Figure 3.12 shows the distribution of earthquakes all over the world. They clearly illustrate the plate boundaries (Barazangi and Dorman, 1969). It is now well known (see also Isacks *et al.*, 1968) that more earthquakes originate in the vicinity of plate boundaries than elsewhere. In the context of this book only shallow earthquakes and only those in continental environments are of interest. The worldwide distribution of shallow earthquakes is included in the schematic plot of Fig. 3.12. In general, shallow earthquakes are associated with three types of faults:

(1)   overthrust faults in areas of lateral compression,
(2)   normal faults or dip–slip faults in areas where lateral extension dominates,
(3)   strike–slip faults or transform faults where rigid crustal sections slide past each other.

A combination of types (3) and (1) or types (3) and (2) is very usual.

Figure 3.13 shows a schemtic picture of these three types of faults with their fault planes. Although most earthquakes occur along plate boundaries, a small but significant number are recorded inside oceanic and continental plates. These quakes bear important evidence on critical accumulation of stress inside the plates. In general, the level of seismicity is much larger in tectonically young (and warm) areas than in old shields.

For the intracontinental seismicity, several mechanisms for the concentration of earthquakes in certain fault zones have been proposed (Zoback and Zoback, 1981). The only generally accepted relation is that of zones of earthquakes to favorably oriented preexisting crustal zones of weakness. The preference of earthquakes to occur along a specific crustal lineament often seems arbitrary, as seen in the map of Fig. 3.14 (Illies and Greiner, 1978, and Illies and Baumann, 1982). The observed maximum horizontal shear stress has also been included in this figure. In this area of southern Germany some earthquakes concentrate on the eastern Rhinegraben fault, a long and prominent tertiary dip–slip fault zone, which developed in the Middle Eocene but apparently later transformed into a major left-lateral strike–slip fault. Some focal plane solutions are shown in Fig. 3.15, indicating the strike–slip character of these earthquakes. However, a major source of activity is concentrated on the comparatively small and limited Hohenzollerngraben, outside the Rhine area seen by the easternmost quakes. Certainly, a favorable

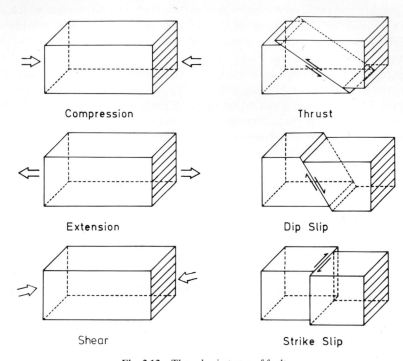

Compression   Thrust

Extension   Dip Slip

Shear   Strike Slip

Fig. 3.13   Three basic types of faults.

Fig. 3.14   Occurrence of earthquakes and maximum horizontal stress in southwest Germany. [From Illies and Greiner (1978). Reprinted with permission of the Geological Society of America.]

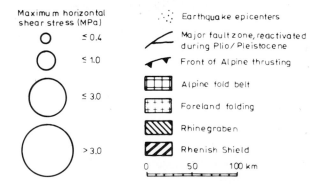

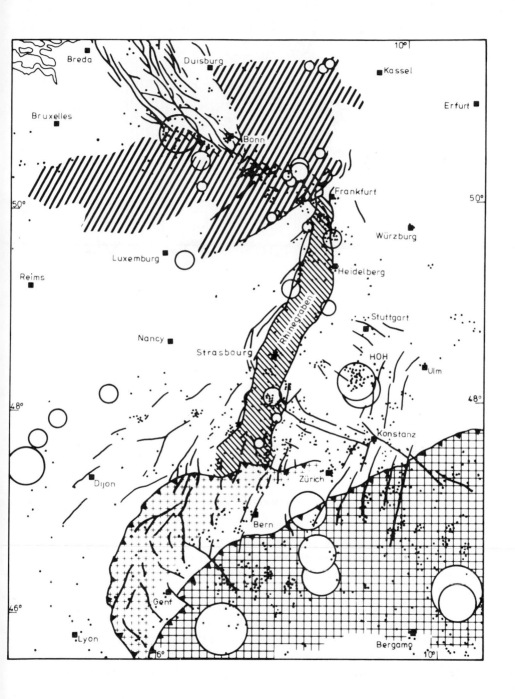

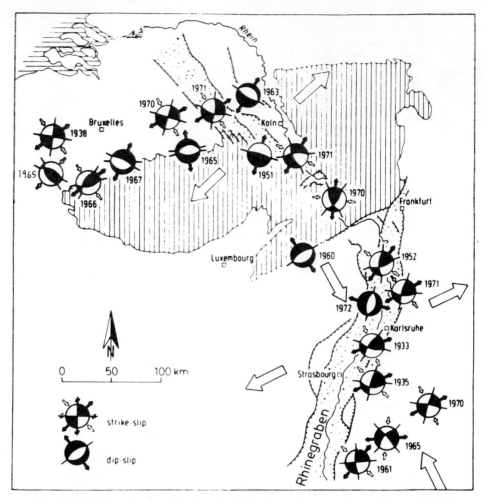

Fig. 3.15 Focal mechanism of earthquakes in southwest Germany. [From Ahorner (1975).]

orientation of the graben and the whole area east of the Rhinegraben with respect to the direction of maximum principal stress, originating from the Alpine region, plays a major role. Even inside the southern Rhinegraben most of the quakes today have a left-lateral strike–slip character with only a few having a (tensional) dip–slip pattern (Ahorner, 1970, 1975). For an explanation of earthquake characteristics, see also Section 3.3.2.

A regional compressive stress system also plays a major role in the occurrence of earthquakes in the northern Mississippi embayment in Missouri. Seismicity occurs far from any plate boundary. It is concentrated in

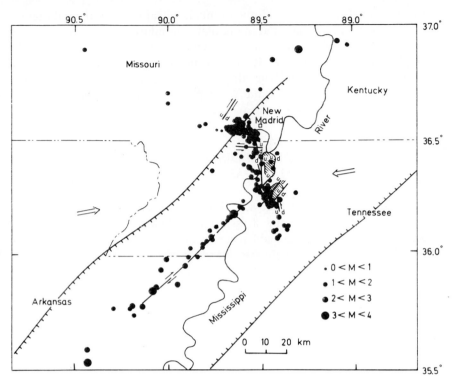

Fig. 3.16 Intracontinental earthquakes in the Mississippi area and their alignment along certain old fault zones. Recent uplift is indicated by (u) and subsidence by (d). The outer lines represent the Precambrian rift zone. [From Zoback and Zoback (1981). Copyright 1981 by the American Association for the Advancement of Science.]

several left-lateral NE–SW (and some minor right-lateral E–W) trending fault zones (see Fig. 3.16). The reason for the strong concentration of seismicity in this area might be the regional ENE–WSW-trending compressive stress connected with zones of weakness in an old Precambrian–Early Paleozoic rift system. The activated (or reactivated) seismic rupture zones are also marked by some uplift movements in the center of seismicity.

In general, there seems to be some correlation between zones of seismicity and long-wavelength gravity anomalies on a large scale and with igneous intrusive rocks on a smaller scale. In the latter case, the intrusives might have lower or higher strength than their surroundings, or they may just be a consequence of regions containing major fault zones, i.e., zones of crustal weakness where the shear modulus and the effective viscosity are low. Stress may concentrate along such zones or transmit a basal shear stress in the asthenosphere or the ductile lower crust upward into the brittle region of the crust. The repetition rate of seismic events in intraplate continental areas is generally much larger than that along plate boundaries, suggesting a slower creep process at depth. Stress measurements by direct observations and by fault plane solutions suggest a rather uniform stress field where no large stress concentrations are evident. Hence the large, deep-reaching crustal fault zones with their potential to transmit a deep-seated (basal?) shear stress to localized zones of weakness near the surface must receive our special attention.

3.3.1   EARTHQUAKE–DEPTH DISTRIBUTIONS IN CONTINENTAL AREAS

It is generally accepted that a shallow earthquake can occur only in a brittle region where strain and stress accumulate until some critical value is reached after which a rather sudden coseismic slip releases some of the stored

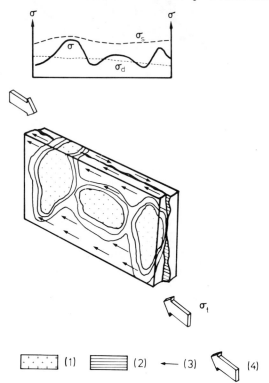

Fig. 3.17   Cross section along a strike–slip fault zone. (1) Asperities, which are locations of difficult slip, (2) planes of past slippage, (3) slip by foreshock or creep, and (4) tectonic stress σ_t. Here, the stress along the fault zone $\sigma = \sigma_t - \sigma_f - \sigma_{as}$, where σ_f is the stress released by foreshocks, σ_{as} the stress released by aseismic creep, σ_s the static friction, and σ_d the dynamic friction.

energy. Whether dilatancy, the nonelastic increase of volume with the opening of microcracks (including a movement of water or a coalescence of microcracks) is the dominant factor or whether a "stick–slip" mechanism plays a major role in the origin of an earthquake is still an open question and will not be the subject of the following arguments. Detailed discussions of the earthquake mechanisms may be found in books such as Bolt (1978), Rikitake (1976), and Kasahara (1981). Figure 3.17 provides a schematic example of accumulation of stress along a (vertical) strike–slip fault. In the brittle, upper part of the crust there are so-called asperities (Wesson et al., 1973) or barriers (Aki, 1979), areas which can sustain rather large stresses and cannot be broken easily. Only after asperities are broken can a major earthquake with a long rupture area be initiated. Note also that once the rupture process has started, not the high static frictional resistance but the lower dynamic resistance determines the rupture process. Today, asperities are considered responsible for foreshocks while barriers which are not broken by the main rupture determine the aftershocks. As shown by Meissner and Strehlau (1982), the maximum rock strength will be at a depth between 5 and 12 km in the continental crust, depending on the thermal regime. These depths will have the strongest barriers. On the other hand, they have the maximum occurrence of earthquakes, as revealed by an analysis of many seismically active continental areas. As will be shown in Section 3.6.5, there is an intimate relationship between the curves of maximum possible stress (rock strength)

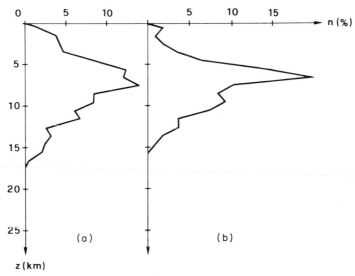

Fig. 3.18 Earthquake–depth distribution for two areas of intermediate heat flow q. The counting interval Δh is 1 km; N represents the number of quakes. (a) Haicheng aftershocks; $N = 1223$ and (b) Greece aftershocks (1978) $N = 109$ and $q = 62$ mW/m^2. [From Meissner and Strehlau (1982).]

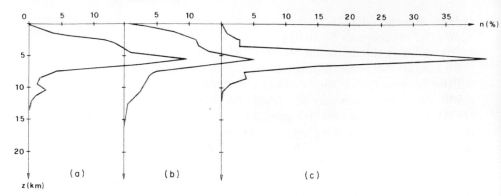

Fig. 3.19 Earthquake–depth distribution for areas with high heat flow; (a) Coso geothermal, $N = 1950$, $\Delta h = 1$ km, $q > 82$ mW/m²; (b) all Coso, $N = 4221$, $\Delta h = 1$ km, $q > 82$ mW/m²; and (c) central California, $N = 425$, $\Delta h = 1$, $q > 82$ mW/m².

and the depth–frequency distributions of continental earthquakes. For the rheological behavior of the crust it is important to note that the number of earthquakes decreases sharply below the peak value (between 5 and 12 km) and dies out at greater depth. Figure 3.18 shows the number of hypocenters for two areas of medium heat flow, and Fig. 3.19 gives examples from areas of high heat flow. The stress drops also seem to be higher in the strong-barrier area, as shown in Fig. 3.20.

The main conclusion from Figs. 3.18–3.20 is that the brittle part of the continental crust ends between 6 and 10 km, followed by a brittle–ductile transition zone of about 5-km thickness. The lower crust is apparently able to respond to stresses by creep. The replacement of stresses released in earth-

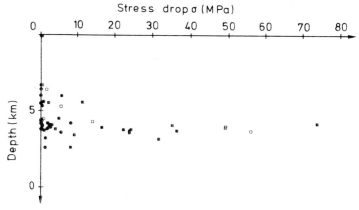

Fig. 3.20 Stress drop–depth distribution of earthquake sequence for 26 events for the San Andreas fault near Bear Valley. (Bakun *et al.*, 1976) Squares represent P-spectra from a Z-seismograph, and circles represent SH-spectra from an H-seismograph. [From Meissner and Strehlau (1982).]

quakes in the strong-barrier zone of the upper crust should definitely take place by creep and by stress accumulation from below, most probably along zones of weakness which might transform into brittle fault zones of the upper crust. This conclusion seems to be valid for true intraplate earthquakes as well as for transform faults on continents, e.g., the San Andreas and the North Anatolian fault zones, both fault patterns with a negligible vertical slip component (Chen and Molnar, 1983).

3.3.2 CRUSTAL STRESSES AS OBTAINED FROM FAULT PLANE SOLUTIONS

In general, crustal stresses are calculated by different methods, including:

(1) fault plane solutions,
(2) *in situ* stress measurements, and
(3) observation of young tectonic lineaments.

Only method (1) will be treated in this section; the other methods are discussed in Section 3.6, especially Section 3.6.2. Focal plane solutions in a way provide the most reliable data on the *present* stress pattern in a certain area. They generally give evidence on the present main stress directions over a depth range between 0 and 20 km, whereas the other methods are more restricted to the near-surface area. The determination of the focal mechanism is based on a number of preconditions. A seismic ray arriving at a seismometer must be traced back into the source region, and it is generally assumed that a compression (or dilatation) arriving at the seismometer as the first P-wave onset also left the source as a compression (or dilatation). The velocity–depth function along the ray path must be known in order to arrive at the right source location. Geometric ray theory is applied to achieve this goal. Many rays of an earthquake—which also means many seismometer stations—must be available around the source region in order to have reliable coverage inside the focal region with regard to the compressional and dilatational first motion.

Regarding the focal region, the rupture area is considered small and may be illustrated by the schematic picture of Fig. 3.21. Please note that the orientation of the fault plane is completely arbitrary, i.e., from vertical to horizontal. Furthermore, there is an "auxiliary plane" perpendicular to the slip movement along the fault plane, which cannot be distinguished from it by means of first motions only. Both planes are called "nodal planes" because they contain no P-wave motion in the far field. In practice, all first-motion displacements are plotted in a projection of a unit sphere around the focus, called the focal sphere. In general, projections to the (horizontal) lower hemisphere are used. The orientation of the nodal planes must be found.

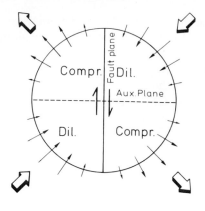

Fig. 3.21 Schematic picture of focal sphere and ray paths with compressional and dilatational first onsets.

Of the four projections used, the Wulff net (stereographic projection) and the Schmidt net (equal area projection) seem to be the most common ones (Kasahara, 1981). For the Wulff net the angle of emergence of the ray, i, is related to the radial distance r by $\tan i/2 = r$, while the Schmidt net has a direct proportionality between i and r. This means that for both nets a vertical plane crosses the center and a horizontal plane is plotted on the margin. All nets are oriented with north pointing upward in the projection on the lower half-plane. Because of the assumed shallow source and the curved path of seismic rays; nearly all seismometers at the surface record seismic waves originating at the lower half-plane, except for a very small area directly on top of the focal sphere. The nodal planes, i.e., the fault plane and the auxiliary plane, form great circles on their intersection with the focal sphere. Certain geometric rules (e.g., the situation of a plane's pole on the other nodal plane) permit the determination of the second nodal plane once the first one has been found by trial and error (see Kasahara, 1981). When both nodal planes have been determined, the sectors of maximum or minimum principal stress may be plotted outside the focal sphere as shown in Fig. 3.22. (For the definition of principal stresses, see Section 3.6.1.) They are generally marked by big arrows and are based on the so-called double couple concept (Vendenskaya, 1956). It should be stressed that the general procedure of plotting a stress arrow pointing to the middle of a compressional (or dilatant) sector is a rough averaging process. It is well known that in practice, i.e., in nature and in the laboratory, the angle between the direction of maximum principal stress and the trace of a fault plane is usually around 30° (Jaeger and Cook, 1976). In general, the direction of maximum principal stress might be *anywhere* within the sector. It is known that any preexisting fault zone, understood as a zone of weakness, can be reactivated if the angle between maximum stress and fault plane trace is anywhere between 5 and 85°. Hence, plotting the stress arrows at 45° is somewhat arbitrary but might be justified

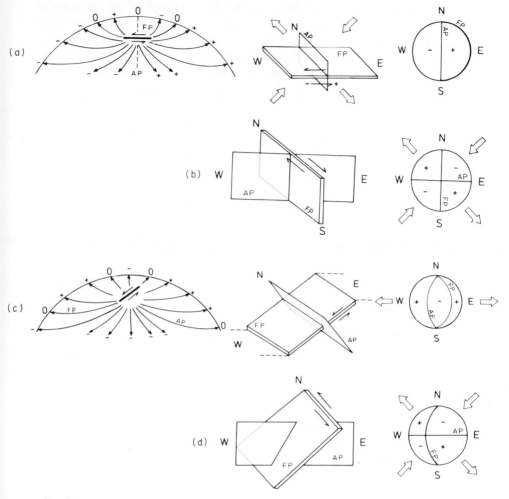

Fig. 3.22 Some fault planes and their picture in the projection on the lower half-plane. Here FP represents the fault plane and AP the auxiliary plane. (a) Horizontal thrust in W–E (or N–S) fault plane, (b) horizontal strike–slip along vertical fault plane (N–S), (c) normal faulting with N–S strike, and (d) horizontal strike–slip along inclined fault plane (N–S).

in determining the *general* stress direction in a certain area based on many focal plane solutions.

Some characteristic examples of fault planes and their picture in the Wulff or Schmidt projection on the lower half-plane are given in Fig. 3.22. In general, the determination of the stress pattern from a tripartite configuration such as Fig. 3.22c is superior to that of a two- or four-sector pattern because the maximum compressional or tensional stresses are often horizontal and dominating near the surface. Whereas the determination of the direction of first movements in the different sectors of the focal sphere and the main stress directions is theoretically unique, the discrimination between the fault plane and the auxiliary plane is ambiguous and not always easy. In practice, several observations may help to overcome this ambiguity; these include:

(1)   a possible lineation of main shocks and aftershocks (along a fault zone),

(2)   a uniform nodal solution of a number of quakes along such lineations,

(3)   geological boundary conditions,

(4)   amplitude determinations of the first arrivals in an area showing variations because of the Doppler effect related to the rupture of the fault zone.

(5)   observations of SH patterns and amplitudes.

Summing up, the fault plane solutions provide important information on the present stress pattern in the rigid upper crust, the general stress release, and the tectonic style of an area.

3.3.3   OTHER USEFUL SEISMOLOGICAL PARAMETERS

One of the most important physical parameters in seismology is the seismic moment M_0. It is defined by

$$M_0 = \mu F U, \tag{3.6}$$

where μ is the shear modulus, F the area of the rupture plane, and U the average displacement. In addition, several empirical relationships have been established between the seismic moment M_0, the spectral behavior of surface waves, and the surface-wave magnitude M_S. The latter is expressed as (Schneider, 1975)

$$\log M_0 = 1.5M_S + [11.8 - \log(\sigma_a/\mu)], \tag{3.7}$$

where σ_a is the apparent stress, to be explained in the next paragraph.

The determination of M_0 is generally performed in three different ways:

(1)   from Eq. (3.6), i.e., from geologic observations and seismic and geodetic estimates of the fault plane area F and the average displacement U,

(2)   from the so-called corner frequency in the spectrum of long-period surface waves,

(3)   from the integral of the seismic amplitudes of P and S waves.

Today, M_0 is the best determinable focal parameter and there is seldom more than a 50% deviation between the different methods.

The apparent stress σ_a is also an important parameter and requires some explanation. It may be defined by the dynamic frictional stress σ_d

$$\sigma_a = \bar{\sigma} - \sigma_d, \quad \text{with} \quad \bar{\sigma} = \tfrac{1}{2}(\sigma_1 + \sigma_2), \tag{3.8}$$

where σ_1 and σ_2 are the stress levels before and after the earthquake. It can be shown (Kasahara, 1981) that σ_a is only slightly larger than $\tfrac{1}{2}\Delta\sigma$, where $\Delta\sigma$ is the stress drop:

$$\sigma_a \geq \tfrac{1}{2}\Delta\sigma \tag{3.9}$$

The stress drop $\Delta\sigma$ can be written in the general form

$$\Delta\sigma = C\mu U/F, \tag{3.10}$$

where C is a geometric shape factor describing the form of the crack.

Figure 3.23 provides a basic explanation of the stress drop during the rupture process. Figures 3.24 and 3.25 show some analyses of a great number of earthquakes. They illustrate that intraplate earthquakes generally have higher stress drops than interplate earthquakes. It must be emphasized that stress drops are generally one to two orders of magnitude lower than the rock strength. This means that during the average rupture process, probably only a small fraction of the accumulated stress is released. The determination of $\Delta\sigma$

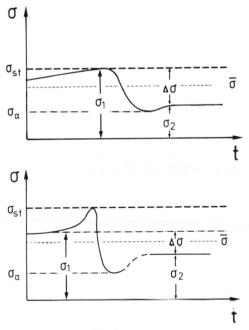

Fig. 3.23 Two models for stress behavior at the time of rupture. Here σ_{st} is the static functional stress, and σ_d is the dynamic fuctional stress.

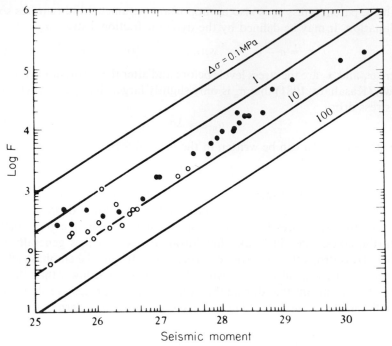

Fig. 3.24   Relationship between the fault surface area F in square kilometers and the seismic moment M_0. Straight lines are from model calculations of circular cracks with constant stress drop; ●, inter-plate; ○, intra-plate. [From Kanamori and Anderson, (1975).]

and/or σ_a, on the other hand, does not give reliable information on the absolute stress level σ_1 or σ_2 (Fig. 3.23). The stress drop apparently is only weakly related to the seismic moment or to the magnitude of the earthquake, as also seen in Figs. 3.24 and 3.25.

Another seismological parameter, strongly debated, is the so-called b value. This is the gradient in a statistical relationship between the frequency of occurrence of earthquakes, N, in a certain area and the observed magnitude M_s:

$$\log N = A - bM_s, \tag{3.11}$$

where N is the number of earthquakes per time unit and A is a constant (the intercept for $M_s = 0$). Generally N is counted per year and ΔM is given in steps of 0.1; $b \sim 1$ and $A = 8.2$ for earthquakes per year, worldwide. Large b values, as shown in Fig. 3.26, mean that the number of small earthquakes is great and might serve as a warning for big ones. Small b values, on the other hand, are related to a comparatively small number of small quakes, and the few large quakes may appear unexpectedly. Large b values are found for the

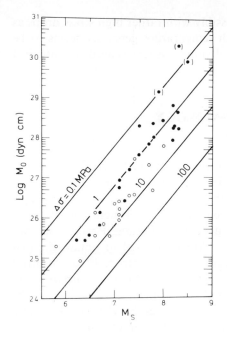

Fig. 3.25 Relationship between seismic moment M_0 and magnitude M_s. Note the difference in stress drop for inter-plate (●) and intra-plate (○) earthquakes.

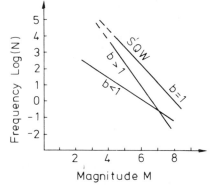

Fig. 3.26 Statistical relationships between the frequency of occurrence N and the magnitude M. Here SQW represents quakes worldwide. $\text{Log } N = A - bM_s$.

Circum-Pacific Belt ($b > 1$), the Moon ($b > 1.3$), and the northern part of Japan ($b \sim 1.06$). Small b values are related to continental rifts ($b \sim 0.7$), the southern part of Japan ($b \sim 0.7$), the Japan Sea ($b \sim 0.66$), and areas with deep foci.

Although there is a general consensus that the b value is a characteristic tectonic parameter for a certain area, there is no general agreement about its meaning. It seems to increase with the number of fractures in an area, which may be related to a mature orogenic phase. Processes involving new as yet unbroken material, such as those of a new rift or trench initiation, seem to be associated with small *b* values, and possibly with rather high stress drops as shown in the previous section. A high temperature generally seems to be correlated with higher *b* values, but much more research is needed to obtain a clear picture.

3.4 CONTROLLED-SOURCE SEISMOLOGY (EXPLOSION SEISMOLOGY)

By far the greatest amount of information on the Earth's crust comes from seismic reflection and refraction investigations. At the time when Mohoro-vičić obtained his first results on the thickness of the crust by seismologic methods in 1909, Mintrop in Göttingen started experiments in which he dropped weights of up to 4 tons from a 14-m scaffolding. Portable seismographs recording seismic waves up to 2 km were used to obtain the first complete seismograms from an artificial source (Mintrop, 1910). Mintrop also introduced seismic refraction methods for detecting subsurface structures, especially salt domes, in Texas and some other areas, mainly by the method of fan shooting. The fact that seismic waves travel faster through a salt body than through most sediments resulted in certain travel time anomalies at and around the salt dome and led to the discovery of most of the shallow salt domes in Texas in the 1920s.

Reflection seismic prospecting also started in the 1920s, but only after the development of appropriate vacuum tube amplifiers made it possible to strengthen the weak, later arrivals by means of feedback circuits and compression of the strong, first arrivals could the reflection method compete with the refraction technique. A general change in seismic prospecting techniques from refraction to reflection in the 1930s was accompanied by the development of electromagnetic receivers, i.e., the transition from large seismographs to much smaller geophones. Also, the possibility of using pattern shooting and multiple-receiver patterns to enhance the signal-to-noise ratio was theoretically examined and tested in this time period.

The introduction of the magnetic tape with analog recording in the 1940s and digital recording in the 1960s was the next big step in the reflection techniques. It was the precondition for the widespread use of digital recording and processing with inverse filter techniques for improvement of the signal-to-noise ratio. Parallel to this development was a renaissance of nonexplosive sources resulting in the widespread use of air gun arrays at sea and Vibroseis-type sources on land.

For a number of reasons, refraction techniques did not profit from these various technical developments in the same way as reflection techniques. The need to compress the first arrivals and enhance the later ones does not exist with refraction techniques because all signal amplitudes are comparable.

Filter techniques, pattern shooting and receiving, and digital recording and processing are also widely used in refraction techniques. All these developments, however, cannot overcome the basic ambiguity of interpretation of refraction data, which will be described in Section 3.4.3. Certainly, the near-vertical reflection method is the only geophysical method that relates a sequence of events from the subsurface uniquely to a sequence of arrivals in the records.

The picture evolving from the parallel use of wide-angle and steep-angle observations was still controversial in the early 1970s. A block structure bounded by steep faults down to the upper mantle was found in the old shield areas of the Russian and Siberian Platform (Sollogub *et al.*, 1968; Davydova, 1975), whereas a more horizontal structure emerged from the first reflection data in western Europe and North America (Meissner *et al.*, 1976a; Schilt *et al.*, 1979). Wide-angle studies suggested a pronounced positive velocity gradient on top of the Moho, while steep-angle reflections showed a laminar or lenticular structure in the lower crust, possibly superimposed on the gradient zone (Meissner, 1973).

These and other discrepancies were more or less resolved in later years. Considering the slant or horizontal rays of wide-angle measurements and their large wavelength versus vertical path of steep-angle reflections and their better resolution, refined crustal models using *both* methods have been obtained in problematic areas (Bartelsen *et al.*, 1982). It will be shown that both methods have certain advantages and both should be used to obtain an optimal crustal picture.

3.4.1   THE DEVELOPMENT OF REFRACTION METHODS FOR CRUSTAL STUDIES: DEEP SEISMIC SOUNDING

Seismic investigations of the crystalline basement of the Earth's crust down to the upper mantle are still dominated by refraction methods. After some sporadic measurements in the late 1920s and 1930s, the large explosion on the island of Helgoland in the North Sea in 1947, which was observed up to 300 km in different directions and showed clear arrivals of crustal and mantle phases (Schulze, 1949, 1974), was the beginning of a period of increased observation of signals from large quarry blasts, underwater explosions, and borehole shots. Up to 10 tons of explosive were used for long-range profiles in western Europe, in the USSR, and in the United States to obtain information from the crust and upper mantle along so-called "lithospheric profiles", which sometimes extended over more than 1000 km (Hirn *et al.*, 1973; Bamford et al., 1976; Belousov *et al.*, 1962; Sollogub *et al.*, 1968; Kosminskaya *et al.*, 1969). Distance between seismometers are usually from 300 m to 5 km.

In the late 1960s and early 1970s the importance of wide-angle events in refraction shooting for crustal studies was noted (e.g., Mueller, 1977). Wide-angle events are usually the strongest arrivals in the records. They are found to originate partly from "true", i.e., first-order, boundaries in the form of critical reflections and partly from zones having a strong positive velocity gradient, often situated immediately above a reflector. The strongest wide-angle events generally come from the Moho. In Fig. 3.9 a schematic picture of these events was given. Rather often the lack of correlatable amplitudes below the critical angle reveals these events to be predominantly diving waves from a strong-gradient zone. The resolution of such gradient zones was improved considerably by consideration of the amplitude behavior, i.e., by modeling techniques with synthetic seismograms (Fuchs and Mueller, 1971; Braile and Smith, 1975). The evaluation of wide-angle events developed into the most important branch of refraction interpretation and forms the basis for crustal velocity–depth profiles in DSS surveys. Wide-angle events were also observed along the lithospheric profiles in the uppermost mantle, where they revealed a certain fine structure. For better correlation of different travel time branches, arrival times t are normalized or reduced by means of:

$$t \rightarrow t_r = t - x/V_r, \tag{3.12}$$

where x is the source–receiver distance, V_r the velocity used for reduction (reducing velocity), and t_r the reduced travel time. All seismograms are mounted on a record section where t_r is plotted versus the distance x. Such a reduced record section is usually plotted by using $V_r = 6$ km/s for crustal and $V_r = 8$ km/s for upper-mantle P phases. For shear waves a factor of $1/\sqrt{3}$ is generally applied to the P-wave values. Figure 3.27 shows an example of such a reduced record section. Note that a travel time branch which shows a velocity identical to the reducing velocity V_r appears as a horizontal, i.e., a visually well correlatable, line.

In general, any curved travel time branch in the sections is indicative of a positive velocity gradient zone (excluding the reflection from a first-order boundary and curved interfaces for a moment). It is the particular strength of crustal studies with refraction/wide-angle techniques that they can reveal and analyze these gradient zones, which generally cannot be found by near-vertical reflection techniques. As will be mentioned in Section 3.4.3, strong-gradient zones result in concave and weak-gradient zones in convex curvatures with respect to the x axis. On the other hand, zones with negative velocity gradients cannot be found directly, either by wide-angle or steep-angle reflection techniques. Only if the depth range of gradient zones approaches that of the dominant wavelength can reflections originate from a gradient zone, the zone itself acting as a low-pass frequency filter.

3.4.2 THE DEVELOPMENT OF (NEAR-VERTICAL) REFLECTION
FOR CRUSTAL STUDIES

The first observations of near-vertical reflections from horizons within the crystalline basement were made with long-running tapes during standard reflection work (Junger, 1951; Dohr, 1975a,b; Galfi and Stegena, 1957; Dix, 1965). Dohr in particular emphasized the importance of these observations and steadily increased the proportion of long-running tapes in seismic prospecting (Dohr and Fuchs, 1967). Reflections from the deep crust appeared sporadic at first and gave a rather diffuse picture of the middle and lower crust. Doubts concerning the reality of such deep reflectors (Steinhart and Meyer, 1961) led to a crustal experiment near Augsburg (Germany) in the northern Molasse basin, where some wide-angle and good near-vertical observations of deep crustal reflections had been gathered previously. In 1964, a large symmetrical observation from the near-vertical to the wide-angle range was performed, and the identities of the near-vertical reflectors (so far recorded only in a statistical sense) and the wide-angle reflectors became apparent (Meissner, 1966, 1967). A similar survey in the Taunus–Hunsrueck area (Rhenish Massif) in 1968 showed even more details, based on a comparison of recorded near-vertical reflections with the velocity–depth function obtained from various wide-angle events (Glocke and Meissner, 1976; Meissner et al., 1976b).

Figure 3.28 gives an indication of the good correlation between the velocity–depth function $V(z)$ obtained from three wide-angle branches and the appearance of near-vertical reflections below the center of the wide-angle, common-midpoint profile with a total length of 180 km. It seems that the strong reflectors coincide with a jump in the $V(z)$ function, or at least with the beginning or with a steep slope of gradient zones. Both early common-midpoint profiles showed that the near-vertical reflectors were real and also demonstrated a general lack of multiple reflections.

In the 1970s the number of observations of deep reflections by standard prospection methods increased considerably, and appropriate experiments were started in many countries (Dohr and Meissner, 1975). Short segments of reflectors were recorded in great numbers. In parallel, the first special deep crustal reflection surveys directed toward specific geological targets, e.g., the form and depth range of prominent fault zones, were performed, often in connection with wide-angle observations (Angenheister and Pohl, 1976; Meissner et al., 1980). The lateral extension increased from 2–3 km in the late 1960s to 20 30 km in the late 1970s. Multiple coverage led to a great improvement in quality. More oil companies extended their recording times to 12 s and generally achieved good resolution.

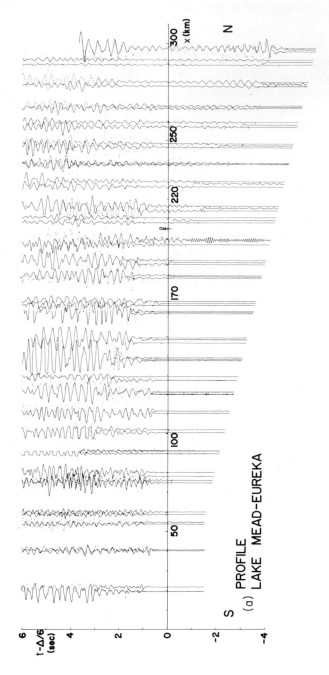

(a) PROFILE LAKE MEAD—EUREKA

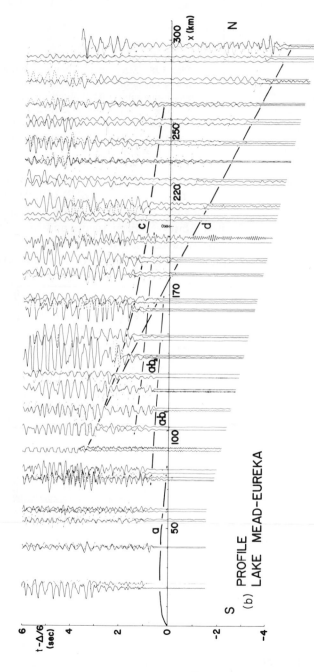

Fig. 3.27 Examples of a reduced record section (a) without correlation and (b) with correlation. Here, *a* represents P_g-wave, ab_1 and ab_2 are diving waves from intermediate boundaries or gradient zones, *c* represents P_mP-waves, and *d* represents P_n-waves. Notice the ending of P_g at about 80 km and the strong arrivals from P_mP in the overcritical range > 90 km. [From Prodehl (1979).]

63

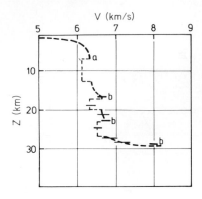

Fig. 3.28 Steep-angle reflections plotted in a V–z curve obtained from a wide-angle, common-midpoint (CMP) observation in the Hunsrueck area. (a) V–z curve from a CMP and (b) reflections from a near-vertical reflection survey near the CMP. [From Glocke and Meissner (1976).]

The year 1975 marked the beginning of nearly continuous operation of the Consortium of Continental Reflection Profiling (COCORP) (Oliver *et al.*, 1976; Brewer and Oliver, 1980). Since then, a concerted effort to study the continental basement in the United States as been made, and many new features of crustal structure have been revealed. Similar groups have been formed, although on a much smaller scale, in Canada, Australia, Great

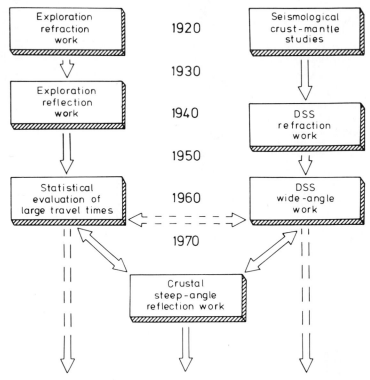

Fig. 3.29 Development of seismic crustal studies.

Britain, France, and Germany. Figure 3.29 gives a short summary of the steps which led to the present-day situation in near-vertical reflection work.

Substantial improvements also took place in seismic sources, recording techniques, and data processing. Besides work by the oil companies, sporadic reflection investigations in Germany have been made using explosives (50–100-kg charges divided over three to five boreholes), whereas COCORP uses conventional vibratory sources (typically five vibrators of 13.5 tons each). Explosives have the advantage that wide-angle observations by portable refraction stations can be performed simultaneously using the same shots. It appears that, until recently, vibratory sources have been too weak to reach lateral distances greater than 80–90 km, even using a sophisticated stacking procedure. The velocity resolution in the present vibratory surveys might also be poorer for the lower crust, because of the comparatively small moveout times, which are a consequence of a spread length of only 10 km. On the other hand, such a huge amount of new structural data has been accumulated in the extensive and comparatively rapid reflection studies of the COCORP group that our knowledge of the structure and development of the continental crust has increased greatly (Oliver, 1980; Brown *et al.*, 1981).

Next to COCORP, the British Institutes Reflection Profiling Syndicate (BIRPS) collected many crustal reflection data, using marine reflection techniques on the continental shelf around the British Isles (Smythe *et al.*, 1982). By means of air gun arrays, long streamers for recording, and multiple coverage, high-quality records north of Scotland, in the Irish sea, and in Southwest Great Britain have been obtained along profiles of some hundred kilometers, taking advantage of the fast progress of marine seismic operations. Today, several thousand kilometers have been accumulated and the structure of Permian and Tertiary basins and their origin have been revealed down to the mantle. Since 1983 ECORS in France and since 1984 DEKORP in Germany have begun field work and have collected several hundred kilometers of reflection lines across the West European continent, extensively supported by wide-angle observations. Among the many advantages of (near-vertical) reflection methods are their ability to follow up to great depths structures which are exposed at the surface, to find the type of tectonic pattern, and generally to provide a three-dimensional picture of the subsurface. Today, reflection investigation of the Earth's crust is on its way to assuming a dominant role in crustal studies similar to its role in oil and gas exploration.

3.4.3 BASIC FEATURES AND STEPS IN REFRACTION/WIDE-ANGLE INTERPRETATION

Most observations take place along straight profiles. After field records have been corrected according to elevation, shot time, etc., they are collected into seismogram sections on a reduced scale, as shown in Fig. 3.27. The individual records are generally normalized; e.g., their amplitudes are plotted relative to the maximum amplitude per trace. Often, a moving-window analysis is applied to obtain frequency information along the trace, and an optimum filter operator is used for filtering of the records.

Such filtered record sections are the basis for a correlation of arrivals. As mentioned in Section 3.2.1, all continental crusts show characteristic travel time patterns. The P_g wave appears as the first arrival, mostly up to distances of 60–120 km. In most areas the P_g wave shows a small convex curvature with respect to the x axis (the geophone distance to the source), its apparent velocities starting with values of about 5.8 km/s near the source (neglecting the sedimentary cover for a while) and reaching final values of more than 6.1 km/s (often about 6.3 km/s) before this branch peters out. Very few areas show a phase correlation of P_g for distance greater than 100 km (Giese, 1972, 1976a,b).

The reason for the curvature of the P_g events and for the end of the P_g branch lies in the increase of velocity with depth in the upper 10 km of most crystalline basements. The increasing lithostatic pressure with depth is responsible for this velocity increase, which is rather strong below about 7 km, equivalent to 2 kbars. Often, dV/dz assumes a constant value for depths greater than about 6–7 km, as found by high-pressure–high-temperature laboratory investigations (Christensen, 1965, 1979; Kern, 1978, 1982b). Gradient zones result in curved ray paths and, hence, in curved travel time branches. In practice, however, positive velocity gradients usually end at about 10 km depth. This is, in general, related to a temperature effect which stops or even reverses dV/dz. Dehydration reactions with an increase of pore pressure and a corresponding decrease in seismic velocity may also play a role at these depths. These effects will be analyzed in Section 6.6.2. An analysis of the generally strong P_mP branch, mentioned in Section 3.2.1, and other crustal waves will be given after some remarks on interpretation.

In general, interpretation of DSS record sections is performed in three stages:

(1)   Assessment of the one-dimensional V-z structure by the application of direct methods and simple formulas, e.g., by solving a two-layer case with constant velocities (intercept time–crossover formulas) or by applying the Wiechert–Herglotz or the Giese algorithm, as shown in the following.

(2)   Ray tracing, i.e., verification of the main travel time branches by model calculations and iterative procedures with the goal of finding a coincidence between observed and calculated travel times by trial and error (see Cerveny et al., 1977).

(3)   Formation of synthetic seismogram sections and their iterative modifications to find a coincidence between observed and calculated amplitudes

and shapes of wavelets along the travel time branches and finally for the whole seismogram section, including multiple events (Fuchs and Mueller, 1971; Braile and Smith, 1975).

A modern treatment of DSS profiles should definitely reach stage (3). The other stages, especially the first one, should always be considered preliminary steps in the interpretation. In the following some of the methods of stage 1 will be mentioned, followed by a more general description of stages 2 and 3.

The simplest assessment within stage 1 is performed with the intercept–time and crossover formulas for a horizontal two-layer case (Telford et al., 1976; Meissner and Stegena, 1977). This gives the first layer's thickness as:

$$Z_1 = V_1 t_i / 2 \cos i_c \tag{3.13}$$

and

$$Z = (x_K/2)[(V_2 - V_1)/(V_2 + V_1)]^{1/2}, \tag{3.14}$$

where t_i is the intercept time, $i_c = \sin^{-1}(V_1/V_2)$, x_K is the crossover distance, and V_1, V_2 are the velocities of the first and second layers respectively.

Standard textbooks of applied geophysics such as those mentioned earlier provide additional formulas for the case of three or more layers, for inclined layering, and for gradient zones (Meissner and Stegena, 1977). An assessment of the dip of any refracted arrival observed by reverse shooting is given by:

$$\delta = \tfrac{1}{2}[\sin^{-1}(V_1/V_{app}^d) - \sin^{-1}(V_1/V_{app}^{up})], \tag{3.15}$$

where V_{app}^d is the apparent down-dip velocity and V_{app}^{up} the apparent up-dip velocity, as observed in the travel time diagrams and in the record section directly.

Similarly, the critical angle $i_c = V_1/V_n$ is given by

$$i_c = \tfrac{1}{2}[\sin^{-1}(V_1/V_{app}^d) + \sin^{-1}(V_1/V_{app}^{up})]. \tag{3.16}$$

Also, the construction of wave fronts is an effective method for controlling $V(z)$ and the refractor velocity, especially for strong-gradient zones and for a set of reversed travel time branches of refracting boundaries (Musgrave, 1961; Meissner, 1961).

Another means of estimating the depth in DSS studies is the use of wide-angle branches, e.g., evaluation of the usually strong and reliable $P_m P$ events for an assessment of the depth of a reflected event from $t^2 - x^2$ diagrams where $\tilde{V}^2$ values present the slope of a (nearly) straight line (see Musgrave, 1961). Giese (1968) has shown that the following two basic equations can be combined:

$$\tilde{V}^2 = (x_2^2 - x_1^2)/(t_2^2 - t_1^2) \sim (x/t)(dx/dt) = (x/t)V \tag{3.17}$$

and

$$z = [\tilde{V}^2(t^2/4) - (x^2/4)]^{1/2} \quad \text{(Pythagorean geometry)} \quad (3.18)$$

($\tilde{V}^2$ from $t^2 - x^2$ diagrams). Introducing Eq. (3.17) into Eq. (3.18) results in (Giese, 1968)

$$z = (x/2)[V(t/x) - 1]^{1/2}, \tag{3.19}$$

where V is V_{app} at the distance x. Assuming that a curved event, e.g., P_mP, represents a true reflection, its reflecting depth is easily calculated from the apparent velocity V_{app} for any t and x values along the travel time branch. For a true reflection the calculated depths should all give the same value of z all along the travel time branch. Deviation of the calculated z values, especially at low apparent velocity (i.e., at the far end of the travel time curve) indicates a deviation from a true reflection. It may show the transition to a diving wave. At the near end of "reflected" events some information on the boundary is also obtained. If apparent velocities are larger than the refractor velocity, e.g., larger than 8–8.4 km/s for the P_mP branch, then a true first-order boundary is present. (A second-order boundary could generate only very weak signals in this range, depending slightly on the ratio of wavelength to the dimension of a V-gradient zone.) These two sets of information from the far end and the near end of a hyperbola-shaped travel time branch are illustrated in the next two figures, based on Eq. (3.19). As seen in Fig. 3.30, the z versus V_{app} curve starts with the (asymptotic) velocity of 6 km/s (at the far end of the wide-angle range) and a fictitious depth of about 28 km which systematically shifts to a "consistent" depth of about

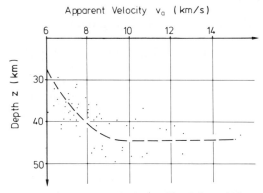

Fig. 3.30 Apparent velocity V_{app} versus depth after Giese's formula for an example from DSS work. $V_{app} > 8$ km/s: subcritical P_mP reflections (from first-order Moho); $V_{app} < 8$ km/s: diving waves from gradient zone on top of Moho. (Moho at 45 km depth.) Dots show scatter of individual V_{app}–z data points.

45 km. This means that we have no true reflection at the far end of the travel time branch, but diving waves. The consistent depth of 45 km for apparent velocities of 9–14 km/s, on the other hand, demonstrates that the beginning of the travel time curve is made of true (subcritical) reflections, situated at the bottom of a gradient zone. The apparent velocities at the far end of the travel time curves often approach a so-called asymptotic velocity V_{asy} (about 6.0 km/s in Fig. 3.30). This minimum apparent velocity corresponds to a nearly horizontal ray path somewhere near the upper part of the gradient zone. It represents the maximum (true) velocity in the overburden (V_{OB}),

$$V_{asy} \geq V_{OB}. \qquad (3.20)$$

Any *true* overburden velocity might be slightly smaller but cannot be larger than the observed V_{asy}.

Figure 3.31 shows a simple model calculation of the travel time and amplitude behavior of some representative velocity–depth structure at the bottom of the continental crust (Vetter and Meissner, 1976). The two models for a 50- and a 30-km-thick crust, which were calculated according to formulas of ray optics (Grant and West, 1965), give a good picture of the relative amplitudes that really occur in most observations.

A good old direct way to interpret P_g and other phases in cases where no low-velocity layers are present, i.e., $V(z)$ is nowhere decreasing, is the Wiechert–Herglotz (WH) method (Meissner and Stegena, 1977). It is based on the Abel integral equation and requires a continuous travel time pattern which is equivalent to a nondecreasing velocity, i.e., $dV/dz \geq 0$. No lateral gradients dV/dx are allowed. For practical calculations the Wiechert–Herglotz integral is approximated by sums. For the P_g branch it can be written as

$$z(x^*) = \frac{\Delta x}{\pi} \sum_{x=0}^{A} \cosh^{-1} \frac{V(x^*)}{V(x)}, \qquad (3.21)$$

where $V(x^*)$ is the final value of the (apparent) velocity for which $z(x^*)$ is calculated, and $V(x)$ is the velocity along the P_g travel time branch at any distance interval Δx. Here $z(x^*)$ is iteratively solved from $x = 0$ to the largest value of x. In this way also $z = z(V)$ is obtained for $V = V(z)$ only.

If the travel time curve is continuous, i.e., the velocity does not decrease anywhere in the crust, the algorithm may be extended to all travel time branches. Such a case is demonstrated in Fig. 3.32. For the "reversed diving wave" B–A from the strong-gradient zone and/or the reflection branch (e.g., P_mP, the concave section of the travel time diagram) a negative counting method must be introduced, i.e.,

$$z(x^*) = -\frac{\Delta x}{\pi} \left[\sum_{x=A}^{x^*} \cosh^{-1} \frac{V(x^*)}{V(x)} - \sum_{0}^{A} \cosh^{-1} \frac{V(A)}{V(x)} \right], \qquad (3.22)$$

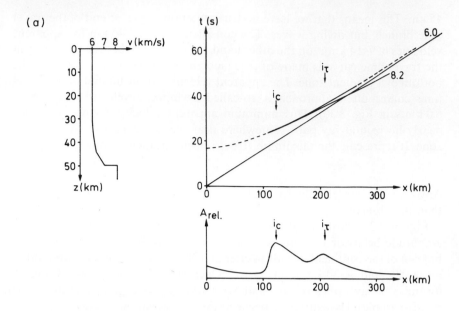

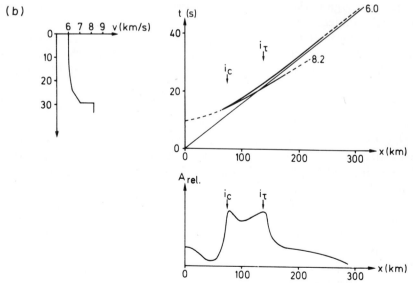

Fig. 3.31 Travel time and amplitude behavior for two selected crustal models. Amplitudes are related to the $P_m P$ events. (a) Shield area and (b) Paleozoic structure; i_c is the critical angle and i_τ the beginning of diving waves. Note the *two* amplitude peaks of $P_m P$ and their shift to smaller distances for smaller Moho depths.

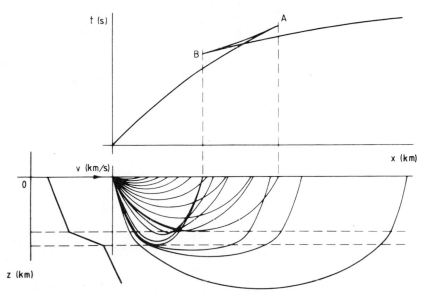

Fig. 3.32   V–z model, travel time branches, and ray paths applicable for a Wiechert–Herglotz solution.

while for the third (convex) branch a positive sign must again be introduced [first term of final Eq. (3.23)], i.e.,

$$z(x^*) = \frac{\Delta x}{\pi} \left[\sum_{x=B}^{x^*} \cosh^{-1} \frac{V(x^*)}{V(x)} - \sum_{x=A}^{B} \cosh^{-1} \frac{V(B)}{V(x)} + \sum_{x=0}^{A} \cosh^{-1} \frac{V(A)}{V(x)} \right].$$

(3.23)

The WH calculation is a fast and direct method. Results may also be used as a starting model for more complicated systems of interpretation. For practical applications all values $V(x)$ *must* be smoothed, e.g., in a V–x diagram, in order to avoid excessive scatter of data points. Shadow zones, i.e., zones with velocity reversals, where gaps in the travel time curves occur, may be attacked by calculating the average interval velocity in these zones, by methods of Dix (1965) if reflections are involved, or by those of Meissner and Stegena (1977) if diving waves from the boundaries of the low-velocity zone are involved. The WH method (or other direct methods) may then be used for velocity–depth calculations.

The cumbersome task of following the normal and reversed travel time branches necessary for the Wiechert–Herglotz algorithm is avoided when using the so-called p–τ *method* (Bessanova *et al.*, 1974; McMechan and Ottolini, 1980). This method converts t–x directly into V–z data. By use of diagrams of intercept time along the travel time curves (t) versus the ray parameter $p = V^{-1}$ a *continuous* presentation of the data is achieved (for all cases where $dV/dz \geq 0$). Figure 3.33 shows the principle of this method. It is interesting to note that the p-τ method is well suited for automatic or semiautomatic interpretation in the slant stack version (Milkereit, 1984). It may even be extended to cases with moderate or small dV/dx and to dipping layers. It also seems to be the best method for the transition to interpretation stage 2, the comparison between calculated and observed travel time branches, and even to stage 3, the amplitude and signal matching procedure, transforming the whole x-t into a p-τ wave field. As mentioned already in this section, all direct methods suffer from their inability to deal with low-velocity zones in the crust. Ray tracing (stage 2) and synthetic seismograms (stage 3) help to decrease the ambiguity inherent in all direct refraction interpretation methods. Systematic iteration of the starting model and careful observation of the amplitude and the shape of the signal (e.g., on a screen) result in a final match between the synthetic and observed seismogram sections. Amplitudes, phases, and their onsets, i.e., the whole seismograms of the two sections, should coincide as closely as possible in the final stage of interpretation.

Among such programs for synthetic seismograms, there is the "reflectivity method" of Fuchs and Mueller (1971) based on the wave equation. It takes into account all possible wave types in an elastic medium. Simpler methods like that of Cerveny *et al.* (1977) are based on simple geometric ray optics but can be adapted easily to lateral velocity inhomogenieties. In general, good fits between the observed and synthetic seismograms are obtained after several iteration processes, i.e., by trial and error. Modern developments and interpretations are also aimed at using absorption values (Q^{-1}) and an-isotropic behavior of crustal layers.

The transition from an interpretation along profiles to that of a network of profiles or that of intersecting profiles and fans must also be discussed. The true dip and strike direction must be found at all intersections in order to be

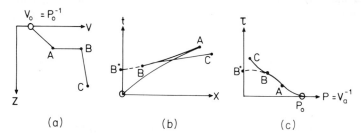

Fig. 3.33   (a) V-z model, (b) travel time branches, and (c) τ-p presentation. Here τ is the intercept time and $p = V_a^{-1}$ = ray parameter = reciprocal apparent velocity.

able to construct depth contour maps. Figure 3.34 shows the geometric situation of a profile AC which has an oblique direction with regard to the strike direction (α) and to the true dip (γ). Figure 3.35 shows a cross section of this profile together with the geometric situation and some formulas for the determination of the strike direction and dip angle. With this method the point of origin of the refracted (or reflected) seismic ray and the depth below the intersection are moved from the vertical plane of the seismic cross section into their true position.

Often it is easier—although usually less accurate—to observe a fan by using the same shot points for a straight profile and an intersection fan. Assuming simple plane layering, the strike and dip angle can also be calculated at the intersection of profile and fan as shown in Fig. 3.36. Fan shooting is still an effective method for estimating an anomalous three-dimensional structure. After corrections to the same shot–geophone distance a ray-tracing method is applied, which often gives fairly good results if several shot points along a profile are available and the anomalous body really is pierced by many rays.

As mentioned before, some reflected or diving-wave phases are often observed from structures inside the continental crust. Figure 3.37 shows a comparison of characteristic travel time curves as recorded in the Buchara depression (Tien Shan Foredeep) and two survey areas in West Germany

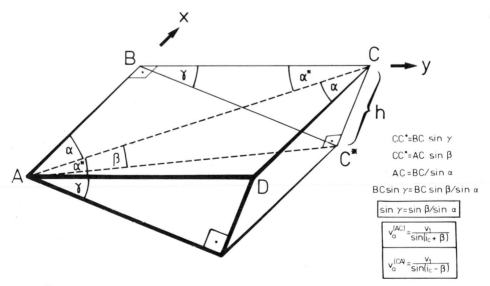

$$CC^* = BC\,\sin\gamma$$
$$CC^* = AC\,\sin\beta$$
$$AC = BC/\sin\alpha$$
$$BC\sin\gamma = BC\,\sin\beta/\sin\alpha$$
$$\boxed{\sin\gamma = \sin\beta/\sin\alpha}$$
$$\boxed{v_a^{(AC)} = \frac{v_1}{\sin(i_c + \beta)}}$$
$$\boxed{v_a^{(CA)} = \frac{v_1}{\sin(i_c - \beta)}}$$

Fig. 3.34 Geometry of a dipping interface along a profile AC. Here γ is the true dip angle; $\tan\gamma = h/\overline{AC}$; α is the azimuth of the profile; and $\alpha^* = \pi/2 - \alpha$.

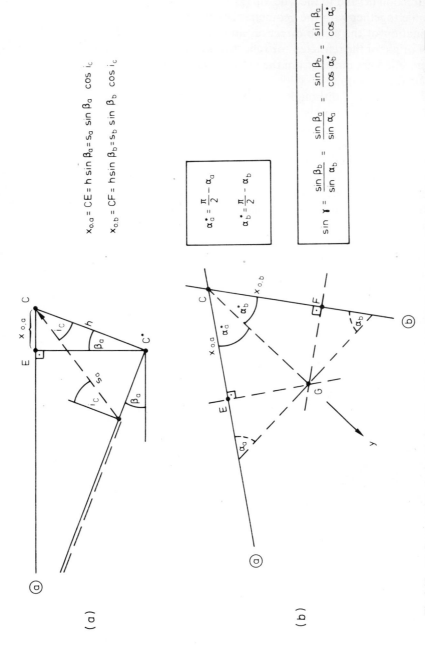

$x_{o,a} = CE = h \sin \beta_a = s_a \sin \beta_a \cos i_c$

$x_{o,b} = CF = h \sin \beta_b = s_b \sin \beta_b \cos i_c$

$$\alpha_a^* = \frac{\pi}{2} - \alpha_a$$

$$\alpha_b^* = \frac{\pi}{2} - \alpha_b$$

$$\sin \gamma = \frac{\sin \beta_b}{\sin \alpha_b} = \frac{\sin \beta_a}{\sin \alpha_a} = \frac{\sin \beta_b}{\cos \alpha_b^*} = \frac{\sin \beta_a}{\cos \alpha_a^*}$$

Fig. 3.35 (a) Cross section along a profile with refracted ray (dashed line) and (b) intersection between two profiles. Rotating $\overline{EC^*}$ around E results in a projection along the surface line EG. The same procedure applied to profile (b) results in surface line FG. Thus G is the projected point of C on the (reflecting or refracting) boundary, and CG shows the (inverse) direction of the true tip.

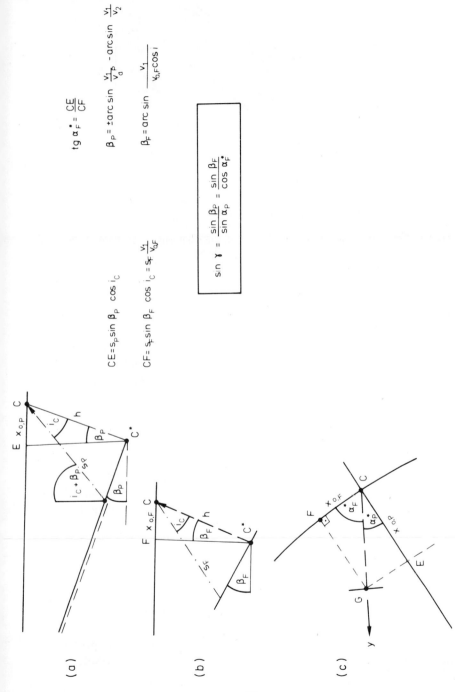

Fig. 3.36 Cross section along (a) a profile, (b) a fan, and (c) intersection between profile and fan. Point G and direction of (inverse) dip CG are determined in a way similar to that in Fig. 3.35.

75

The following equations appear in the figure:

$$tg\ \alpha_F^* = \frac{CE}{CF}$$

$$\beta_P = \pm arc\ sin\ \frac{v_1}{v_a^P} - arc\ sin\ \frac{v_1}{v_2}$$

$$\beta_F = arc\ sin\ \frac{v_1}{v_{a,F} cos\ i}$$

$$CE = s_P sin\ \beta_P\ cos\ i_C$$

$$CF = s_F sin\ \beta_F\ cos\ i_C = s_F \frac{v_1}{v_{a,F}}$$

$$sin\ \gamma = \frac{sin\ \beta_P}{sin\ \alpha_P} = \frac{sin\ \beta_F}{cos\ \alpha_F^*}$$

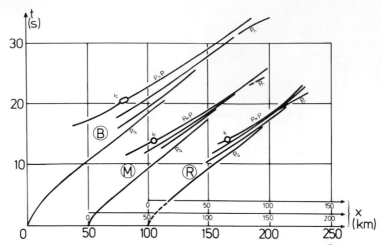

Fig. 3.37   Three travel time diagrams corresponding to the Buchara area Ⓑ (thick crust), the Bavarian Molasse basin Ⓜ, and the Rhenish Massif Ⓡ. Note the similarity of the main travel time branches to P_mP with critical angle i_c, P_n (not fully shown), P_g (ending after roughly 100 km), and the intracrustal (wide-angle) events P_2 and P_3. Data from the Buchara area from Godin *et al.* (1960).

(Variscan Rhenish Shield and Bavarian Molasse). The nature of the intermediate travel time branches varies from area to area (Godin *et al.*, 1960; Meissner, 1976). Like the P_mP phase, these branches often consist of a mixture of a true reflection and diving wave, but generally diving waves dominate. The middle part of continental crusts does not show such characteristic boundaries as the top of the crystalline basement or the crust/mantle transition. The Conrad discontinuity found in many parts of western Europe, where P velocities are supposed to assume values of 6.5–6.8 km/s, cannot safely be correlated with the eastern part of Europe.

Several consequences arise from the observation of $P_2 \ldots P_mP$. From their *late onset* with respect to the x axis we must state that they certainly do not originate from a strong, continuous, and plane first-order boundary. First-order boundaries may be involved as indicated by at least *some* part of the travel time being below the critical angle (see again Figs. 3.31 and 3.35). The problem of V–z data and their correlation with various geologic provinces will be discussed again in Chapter 6.

3.4.4   BASIC FEATURES AND STEPS IN THE INTERPRETATION OF NEAR-VERTICAL REFLECTION SIGNALS

Seismic reflection observations are generally performed along profiles, as demonstrated by the expression "seismic profiling". As the resolution of near-vertical reflection signals is generally higher than that of refracted or wide-angle arrivals (because of their higher frequency and shorter wavelength), more weight must be applied to the *static corrections* of data, using well-known formulas for elevation and weathering corrections (see, e.g., Waters, 1978; Grant and West, 1965; Meissner and Stegena, 1977). Often optimum filters are calculated on the basis of frequency analysis and applied to the field records in order to enhance the signal-to-noise ratio. There are

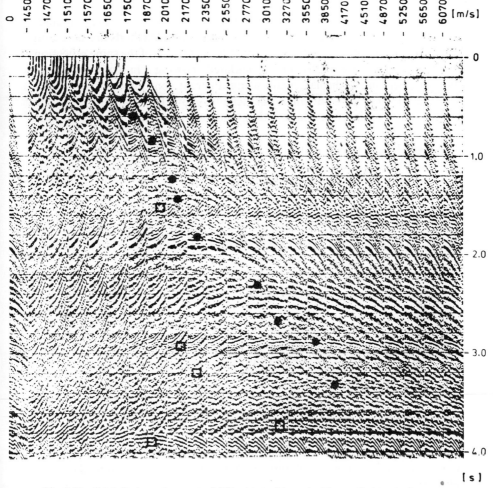

Fig. 3.38 Trial display using a set of different stacking velocities applied to single coverage (dynamically uncorrected record). Best stacking velocities for true reflections (full circles) and multiples (square) are indicated. [From Meissner and Stegena (1977).]

various ways to calculate the dynamic corrections. The most common procedure is to obtain those "stacking" velocities V_{st} which, in a trial display of single records, succeed in converting the reflection hyperbola into a straight line in a plot of t versus x. The "moveout times" (MOs) of the reflection hyperbola must be reduced to zero (Fig. 3.38).

Alternatively, a *trial stack*, i.e., the summation of reflections from the same reflecting element in the subsurface by using various stacking velocities, is performed. Those stacked events which develop the highest amplitudes in a display correspond to the right stacking velocity V_{st} [constant velocity scan display, (VSD)] (Fig. 3.39).

More refined techniques for the determination of the stacking velocities were introduced by Taner and Koehler (1969). These methods are based on the calculation of the coherence between different traces of the same reflection element. A semblance coefficient is calculated showing a measure for coherence. In a semblance display, again using different constant stacking velocities, the maxima of semblance coefficients show the optimum stacking velocity with a higher resolution than the two other methods mentioned before.

The stacking velocity is an important parameter for the calculation of interval velocities. For a plane horizontal layering the stacking velocity is equal to the normal moveout velocity V_{NMO} and to the root-mean-square velocity V_{rms}:

$$V_{rms}^2 = (x_2^2 - x_1^2)/(t_2^2 - t_1^2) \approx (x \, \Delta x)/(t \, \Delta t), \qquad (3.24)$$

where x_2 and x_1, t_2 and t_1 are any set of values along the reflection hyperbola, and

$$\text{for} \quad x_1 = 0 \rightarrow V_{rms}^2 = \frac{x^2}{t_2^2 - t_0^2} \qquad (3.25)$$

or

$$V_{rms}^2 \sim \frac{x^2}{2t \, \Delta t}$$

$$V_{rms}^2 = V_{NMO}^2 = \frac{2}{t_0} \int\limits_0^{t_0/2} V(\tau) \, d\tau, \qquad (3.26)$$

where x represents horizontal distances, t the reflection times along a hyperbola, and t_0 the travel time along a ray path perpendicular to a reflector.

In practice, several factors introduce errors into the simple concept of V_{rms} and V_{st}. First, the reflection hyperbola deviates from a true hyperbola because

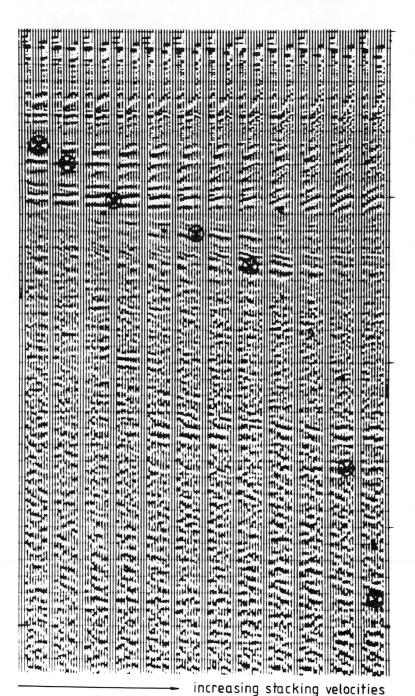

increasing stacking velocities

Fig. 3.39 Trial display using a set of different stacking velocities applied to dynamically corrected and stacked records. Best stacking velocities for true reflections and for multiples are similar to Fig. 3.38. [From Dohr (1981).]

of refracting rays in a more-than-one-layer case. However, lateral inhomo-geneities, anisotrophy, dipping, and curved boundaries also introduce certain errors. As a consequence, V_{rms}^2 will not be a true hyperbola in an $x-t$ diagram or a straight line in an x^2-t^2 diagram. The actual hyperbolas from the upper layers show much stronger deviations than those from deeper layers because of the large $x:z$ ratios and stronger curvature. Often, fitting a straight line in the x^2-t^2 diagram or a hyperbola in the $x-t$ diagram is possible only for small x.

Collecting only small x values for upper reflections and for the stacking, and disregarding larger x values, a process called "muting", often improves the quality of the stacked section. For a more refined treatment of the different velocities V_{st}, V_{rms}, and V_{NMO} the reader is referred to textbooks of applied geophysics (Waters, 1978; Grant and West, 1965). In the case of large deviations between the observed $x-t$ curve and a true hyperbola, model calculations of travel times (and amplitudes) by means of iterative trial-and-error methods must be performed in order to obtain the best interval velocity between reflectors.

For horizontal reflectors and certainly for establishing a first velocity model, the formula of Dix, Dürbaum, and Krey (generally called the Dix formula) is applied (Dix, 1965):

$$V_n^2 = (1/\Delta t_{0,n})(V_{rmsn}^2 t_{0,n} - V_{rmsn-1}^2 t_{0,n-1}), \tag{3.27}$$

where V_n is the interval velocity of the nth layer and $\Delta t_{0,n}$ the difference between t_0 times, i.e., between $t_{0,n}$ and $t_{0,n-1}$. The layer's thickness Δz_n and depth z_n can also be easily obtained:

$$\Delta z_n = V_n \, \Delta t_{0,n}/2 \tag{3.28}$$

and

$$z_n = \sum_1^n \Delta z_n. \tag{3.29}$$

Other formulas for an assessment of depth were given in Eqs. (3.17)–(3.19).

The interval velocity is generally used for a comparison of velocities with laboratory data (see Section 4.1). The formulas also show that the accuracy of the determination of V_{rms} and therefore of V_n increases with increasing x^2, i.e., with the square of the spread length of geophones. If V_n and V_{rms} are to be calculated for the lower crust with an accuracy of 1–2%, a spread length of at least 15 km is required (Bartelsen et al., 1982). The moveout time MO is proportional to the square of the spread length and decreases with increasing V_{rms}. From Eq. (3.24) with $t_2^2 - t_1^2 \approx 2t_{max} \, \Delta t_{max}$ and $x_2^2 \to x_{max}^2$ with $x_1 = 0$ it follows that

$$MO = \Delta t_{max} \sim x_{max}^2/2V_{rms}^2 t_{max}. \tag{3.30}$$

After applying the right dynamic corrections and eliminating the MO, traces with a common reflecting element are stacked together and a stacked record section is achieved. In exploration-type reflection work a 6- to 100-fold coverage is used for crustal studies. The COCORP fieldwork is done with Vibroseis, e.g., with a 12- to 48-fold coverage per trace (Oliver *et al.*, 1976).

Generally the stacked record sections still contain many disturbing events, such as diffracted arrivals, which mostly originate at small-scale inhomogeneities in the subsurface. Moreover, according to the presentation of record sections in a time distance diagram, reflected events from horizons with opposite dip but similar travel times may form a pattern of intersection, an interference, resulting in a complex and misleading part in the record sections. The disentangling of such arrivals and the reduction of diffracted arrivals to a diffraction point are objectives of the *migration process*.

Several *migration methods* exist for the conversion of reflections from the record section x-t into a data set x_{Migr}-t and/or, finally, into x_{Migr}-z sections, where x_{Migr} is the horizontal distance from which a zero-offset reflection in the subsurface really originates, i.e., the x coordinate where the true seismic ray hits the reflector perpendicularly (French, 1974, 1975; Sattlegger, 1964; Claerbout, 1976).

The migration process can still be performed by hand, using wave front constructions and curves of "maximum convexity" from a diffracting point (Grant and West, 1965). Today computer methods are available, e.g., the wave equation method, f-k migration, the Kirchhoff sum of diffraction method, and Claerbout's finite-difference method (Waters, 1978).

The final z-x_{Migr} section should give a true picture of the subsurface along the profile. Figure 3.40 gives an example of the improvement of correlatable events in an unmigrated and a migrated record. In this example, the crisscross patterns of reflections from both flanks of a synclinal feature are disentangled (Fig. 3.40b) and now give a true two-dimensional picture of the structure.

Migration is a well-known process in seismic prospecting. Unfortunately, it requires a good knowledge of the velocity structure, information which often is not available if the same short spread lengths are used for deep crustal studies as for seismic prospecting, with their limited travel times of only about 5 s. It follows that one must either use spread lengths of at least 15 km or perform special expanding spread experiments along a profile. In any case the gathering of precise velocity data should have a high priority, both for providing petrological boundary conditions and for performing the migration process in areas with complicated subsurface structures.

Regarding the seismic structure in migrated records, it should be emphasized that they always contain some energy that is not reflected or scattered from the vertical plane of the seismic cross section. If a network of profiles is available, the two-dimensional migration might be extended to three-dimensional migration, locating diffracting points and reflectors that are laterally

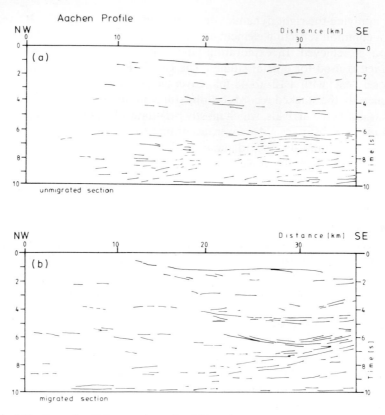

Fig. 3.40 Example of (a) an unmigrated and (b) a migrated record section (Hohes Venn area, Variscan overthrust). [From Meissner and Wever (1985).]

disposed with respect to the vertical plane of the profiles. At the intersection of profiles the lateral offset of depth points might be checked according to the geometric algorithm of Fig. 3.35.

A method which is not used in standard seismic prospecting but has provided a great deal of new information for the understanding of the crust is the statistical evaluation of the frequency of occurrence of reflections per time or per depth interval. Historically, such histograms were the first indications of the existence of quasi-continuous reflecting interfaces in the crust of the Bavarian Molasse basin (Dohr, 1957a,b; Liebscher, 1962, 1964), showing prominent peaks at certain time intervals corresponding to discontinuities such as the Conrad and the Moho. After more continuous deep steep-angle profiles had been observed, the histograms also proved to be useful for comparing the reflectivity patterns of different areas. Generally, the number

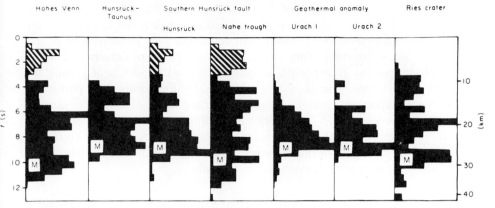

Fig. 3.41 Histograms of reflection density of crustal cross sections through the German Variscides normalized according to maximum number of reflections per 0.5-s two-way travel time. Hatched area represents sediment reflections. [From Meissner and Lueschen (1983). Represented with permission from EAEG.]

of reflections for a 0.5-s two-way travel time interval is calculated. Figure 3.41 shows seven histograms from different areas in West Germany. It is surprising how the reflection density increases with depth. The Moho nearly always marks the base of the crust by showing a strong peak in the histograms followed by a strong decrease in reflection density. The Moho—and sometimes the Conrad—can be determined by such prominent peaks, and often a correlation between different areas is possible. Important conclusions regarding the nature of reflectors and the viscosity of the crust can be inferred from the histograms and will be treated in Chapter 6.

3.4.5   A SEISMIC VIEW OF THE CONTINENTAL CRUST

Today's picture of the Earth's continental crust as obtained by seismic reflection and refraction studies is no longer controversial. Most reflectors detected by near-vertical measurements agree with those from wide-angle investigations and usually also with refracting interfaces. This means that those reflectors are real and not multiple events. A cyclic layering, mostly within an overall positive-gradient zone, is often found in the lower part of warm and young crusts, whereas old shield areas show a more transitional character. Zones with positive velocity gradients dominate in the upper 10 km, as seen from refracted (or diving) waves. Warm crusts often show a velocity inversion below 10 km depth, as also seen from refraction studies. Details of structures in the crust can be observed only by near-vertical reflection studies, e.g., those of the COCORP type. Fault zones also can be detected only by such studies, at least if their dip angle is not too steep. Some fault zones with moderate to steep angles even reach the upper mantle; others end in the lower crust. They mark important zones of detachment along which nappes are transported, sometimes several hundreds of kilometers. Even the determination of seismic velocities, the most important parameter for a correlation with petrology, is no longer a domain of refraction surveys. Reflection seismics with spread lengths of more than 15–20 km are able to reveal velocity anomalies with increased accuracy, laterally and vertically. Later chapters will show that the basic results from seismic measurements are in harmony with those from other investigations.

3.5   THE ELECTRIC CONDUCTIVITY OF THE CRUST

At first sight, the determination of the crustal conductivity and the establishing of conductivity–depth profiles seem to provide a most welcome additional physical parameter for the characterization of a specific area of the Earth's continental crust. Without questioning the importance of obtaining more data on the conductivity, the results obtained after 30 years of electromagnetic, magnetotelluric, and resistivity surveys are disappointing. The reasons for this are the many physical parameters which determine the conductivity or its reciprocal value, the resistivity. A highly conductive rock unit may owe its conductivity to:

(1)   saline water in pores and cracks,
(2)   conductive solid minerals, e.g., of graphitic consistence,
(3)   high temperatures,
(4)   partial melts,

or a complex combination of these four factors (Gough, 1983).

Very low conductivities, on the other hand, seem to be restricted to relatively unfractured crystalline rocks. Generally, moisture (1), chemical composition (2), and temperature (3 and 4) determine the conductivity in laboratory measurements. Field measurements often show shielding or disturbing effects of sedimentary structures, edge effects, and crustal influences.

Some of these effects may lead to a lateral channeling of electric currents over thousands of kilometers and may seriously distort the data obtained from the measurements. In general, the conductivity of the crust will never be calculated with a precision comparable to that of seismically determined parameters, seismic moduli, and density, and even the Moho does not represent a generally occurring electromagnetic discontinuity.

3.5.1   METHODS USED IN FIELD INVESTIGATIONS

There are basically four methods used for determining conductivity in electrical field surveys:

(1)   the four-electrode resistivity–depth probing of the direct-current-conduction method,

(2)   the magnetotelluric method,

(3)   the magnetic array or magnetic gradient method or Wiese (or Parkinson) vector method,

(4)   the electromagnetic or controlled-source induction method.

Methods 2, 3, and 4 are also called induction methods.

Method 1 goes back to Schlumberger and other workers in the field of applied geophysics. For deeper penetration, power transmission lines (not yet in service) or telephone lines have been used (van Zijl, 1969; van Zijl and Joubert, 1975). Dipole arrangements also give satisfactory penetration and are used frequently in the USSR.

In general, there is a wide variety of direct-current techniques as a result of the different possible ways of deploying the current and potential electrodes. The interpretation processes also cover a wide range of methods from resistivity–depth determinations for horizontally stratified media to the qualitative mapping of lateral inhomogeneities (see Grant and West, 1965; Telford et al., 1976; Rikitake, 1966).

Method 2 and method 3 make use of the natural variations and pulsations (high-frequency) of the magnetic field of the Earth originating in the ionosphere. Schmucker's curves are based on the "normal" (quiet) daily variations S_q and variations from "distrubed" periods D_{st}, covering a wide frequency range (Schmucker, 1970, 1974). Any geomagnetic disturbances of magnetospheric origin are modified by induced electric currents. These currents at depth oppose changes in the field (Lenz's law of electromagnetic induction). A good conductor is not penetrated as effectively as a poor one, an effect known as the skin effect (Rikitake, 1966). The magnetic fields of the induced currents overlie and modify the original (primary) ones. They produce amplitude and phase delays of the primary fields. Hence, the observed electric and magnetic components on the Earth's surface are a sum of the two fields, one of external origin and the other internally generated by induction. The phase difference is especially sensitive to any variation of conductivity with depth. In principle, the three components of the electrical field E_x, E_y, E_z and those of the magnetic field H_x, H_y, H_z may be observed along profiles or as point observations. Often Cagniard's method (Cagniard, 1953) is used, estimating the effective impedance $Z(\omega)$ by means of ortho-gonal components of the electric and magnetic fields

$$Z(\omega) = E_x(\omega)/H_y(\omega), \tag{3.31}$$

where ω is the circular frequency. In this approach the apparent resistivity ρ_a (resistivity of an equivalent uniform half-space) is given by [see also Kuckes (1973a,b)]

$$\rho_a(\omega) = \sigma_a^{-1} = (1/\mu_0\omega)|Z(\omega)|^2, \tag{3.32}$$

where σ_a is the apparent conductivity and μ_0 the permeability of free space. The knowledge of $Z(\omega)$ obtained by a frequency analysis of $X_x(\omega)$ and $H_y(\omega)$ is, therefore, sufficient to determine the conductivity changes with depth in a horizontally stratified Earth. For this inversion and more refined techniques, see Cantwell and Madden (1960) or Dowling (1970).

Method 3, the magnetic array (or gradient) method, goes back to Schmucker (1970) and Kuckes (1973a,b). As field measurements of the components of the electric field have traditionally been more difficult because of possible distortions, as mentioned in Section 3.5, the magnetic array method makes use of only magnetic field components. A two-dimensional array is arranged, and the time-varying field is recorded by a number of three-component magnetomers.

The interpretation makes use of the "complex inductive scale length" $C(\omega)$, which is related to the scalar impedance of magnetotellurics $Z(\omega)$ by

$$C(\omega) = Z(\omega)/i\omega\mu_0. \tag{3.33}$$

It can be shown (e.g., Schmucker, 1970) that

$$C(\omega) = \frac{H_z}{(\partial H_x/\partial x) + (\partial H_y/\partial y)}, \tag{3.34}$$

the denominator of Eq. (3.34) showing the two gradients along two perpendicular magnetometer arrays. In an equation similar to Eq. (3.32), the apparent resistivity, as a function of ω, can be expressed by

$$\rho_a(\omega) = \mu_0\omega|C(\omega)|^2. \tag{3.35}$$

Different inversion techniques are then used to convert $\rho_a(\omega)$ to $\rho_a(z)$, i.e., convert the frequency dependence to a depth dependence.

A rather fast and informative variation of method 2 is the method of calculating and plotting so-called geomagnetic induction arrows introduced by Wiese (1965). It has developed into the most frequently used magnetic sounding method in Europe. In deriving the induction arrows a linear relationship between the components of the magnetic variations H_v, H_N, and H_E is assumed such as

$$H_v = aH_N + bH_E, \tag{3.36}$$

where H_v is the vertical component, H_N the north component, and H_E the east component. This equation describes analytically a plane in space. The plane is the strike direction of a conductivity anomaly if we can combine the components of various signals to the same phase (Meyer, 1980), i.e.,

$$H_v(t_0) = a_\omega H_N(t_0) + b_\omega H_E(t_0). \tag{3.37}$$

This is the Wiese relationship (Wiese, 1965). The coefficients a_ω and b_ω determine the geomagnetic induction arrow (or Wiese arrow) in relation to the north direction:

$$\tan \alpha_\omega = b_\omega / \alpha_\omega. \tag{3.38}$$

For t_0 the time of the maximum of the vertical component H_v is usually chosen.

Similar relationships and induction arrows are in use, such as those of Parkinson and Schmucker (Meyer, 1980; Schmucker, 1980), but the Wiese arrow has proved to be a rather stable parameter in field experiments. It is directed toward a high-resistivity anomalous area.

Method 4, the electromagnetic or controlled-source magnetic induction method, is in widespread use by the mineral exploration industry. In contrast to methods 2 and 3, which employ the variations of the natural magnetic fields, man-made primary fields of varying, controlled frequencies are generated by passing an alternating current through a coil or a very long wire. For crustal studies Nekut *et al.* (1977) use a 1.5-km-diameter single-turn excitation loop recording up to distances of 100 km. The recording unit consists of three component μ-metal induction coils recording the incoming magnetic signals. The frequency range is from less than 0.1 to about 400 Hz.

In the interpretation, apparent resistivity (or conductivity) is plotted versus distance. Inversion methods or model calculations with synthetic $\sigma(z)$ models are then used to obtain the best $\sigma(z)$ distribution for a given three- or four-layer case.

3.5.2 SOME EXAMPLES OF CRUSTAL CONDUCTIVITY

As mentioned before, electric conductivity depends on several physical properties. Hence it is no wonder that controversial results are found and there is often no reasonable correlation with data from seismology and seismics. Even the correlation between conductivity and partial melts is not always clear.

In the Afar "hot spot" in Ethiopia there are high-conductivity layers in the upper 50 km where temperatures of predominantly basaltic material must be very near the melting point (Haak, 1977, 1980). Unfortunately, in the African Shield, where the temperature gradient is only half of that in the Afar region, nearly the same conductivity–depth profile is found (Worzyk, 1978; van Zijl and Joubet, 1975). A good correlation has been observed between a prominent magnetic anomaly and a conduction anomaly in the lower crust in South Africa (de Beer *et al.*, 1982; Gough, 1983). The magnetic model implies a rock of very high susceptibility in the lower crust, possibly serpentinite. The explanation for both anomalies is that during hydration iron in the magnesium silicates of olivine or pyroxene is driven out of the serpentine lattice and appears as "stringers" of magnetite (Stesky and Brace, 1973). Such stringers might account for the higher conductivity and the higher magnetic susceptibility.

In Germany, most crustal sections (but certainly not all of them) show high conductivity values in the lower crust. One or two high-conductivity layers are found in the crust of the Variscan mountain ranges, the lower one sometimes crossing the Moho, disregarding any petrological or viscous differences between the lower crust and the mantle (Joedicke, 1981) (see Fig. 3.42). In western Canada there also seems to be a high-conductivity layer in the lower crust or upper mantle (Caner, 1971; Gough *et al.*, 1982).

Many investigations with electromagnetic induction methods have been performed in Hungary and the Carpathian arc. The strong Carpathian anomaly has a linear shape and is explained by a metamorphosis of sediments being transported downward along a kind of subduction zone (Stegena *et al.*, 1975). Partial melting may contribute to the high conductivity values (Rokityansky *et al.*, 1976; Adám, 1980). In the hot Pannonian basin the top of the asthenospheric highly conducting layer (HCL) is at a depth of

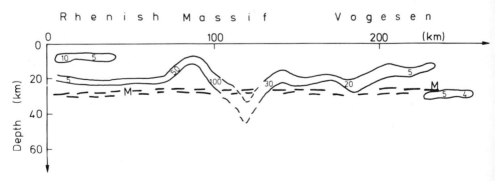

Fig. 3.42 High-conductivity layers along a 250-km-long profile through the German Variscides. Numbers give the conductivity in reciprocal ohm meters. Here *M* represents Moho from DSS studies. [From Joedicke (1981).]

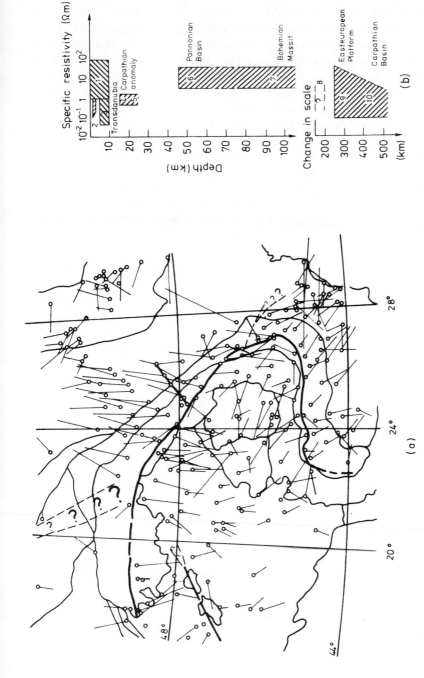

Fig. 3.43 Conductivity data from Hungary, the Carpathian arc, and its surroundings. (a) Wiese's vector. [From Rokityanski *et al.* (1976).] (b) Highly conducting zones of the crust and upper mantle in the Carpathian system and the east European platform. [From Adám *et al.* (1977).]

about 60 km, while in the old eastern European platform area the top of the
HCL is at 200–300 km depth (Stegena *et al.*, 1975). The Carpathian arc
exhibits a well-developed anomaly of the Wiese vectors (Section 3.5.1) with a
pronounced shallow HCL, which is supposed to originate from water release
at the lower part of the flysch complex below a depth of only 10 km
(Rokityansky *et al.*, 1976). Figure 3.43a presents a picture of the induction
arrows and Fig. 3.43b shows an interpretation of resistivity–depth data in the
Carpathian system and the east European platform. Contributions from
metamorphism and dehydration, possibly in conjunction with partial melts
from a subduction zone, definitely play a role in such a hot environment, but
graphitic rocks are also supposed to contribute to the anomalies (Stegena,
1976, 1982b; Adám and Wallner, 1981; Adám *et al.*, 1983).

For the Adirondacks in the northeastern United States Connerney *et al.*
(1980) and Connerney and Kuckes (1980) found a prominent HCL in the
lower crust and attribute this to the existence of water at these deep crustal
levels. In the Basin and Range Province a highly conducting upper mantle
was found by array studies in 1967–1969 (Gough, 1983), but its depth is not
well constrained. The conductive upper-mantle layer correlates rather well
with high heat flow and low seismic velocities. Partial melts may well be
involved in these areas. The Colorado Plateau seems to be less conductive.

This short review of results from field measurements shows that our
knowledge of the reasons for a conductivity anomaly is far from perfect.
Detailed interdisciplinary investigations, especially involving reflection
seismology, are necessary to delineate better correlations between conductivity anomalies and other parameters. Figure 3.44 gives a summary of some
crustal conductivity studies.

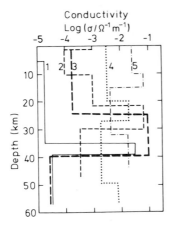

Fig. 3.44 Some crustal–upper mantle conductivity–depth profiles from various electromagnetic induction methods. (1) Kapvaal
craton (Van Zijl, 1977), (2) Adirondacks (Connerney *et al.*, 1980), (3) Limpopo belt (Van
Zijl, 1977), (4) and (5) Variscan structures,
Germany (Joedicke, 1981).

3.6 STRESS MEASUREMENTS AND THE STATE OF STRESS IN THE CONTINENTAL CRUST

Tectonic stress is one of the most important physical parameters for the understanding of tectonism, earthquake patterns, and fault structures. The formation of mountain ranges and grabens and the seismicity in specific areas are consequences of stress patterns which might be rather uniform over an area of several thousand square kilometers.

One of the most debated questions is that of the origin of tectonic stresses. Stress might be created by a convection-induced strain at asthenospheric (or lower crustal) levels, a hypothesis which will be discussed later. It might also originate from the loading effect of the topography, an assumption which seems to be compatible with observed stress patterns around mountain ranges. Accumulated stresses are released from time to time by a rupture process along minor or larger fault zones, i.e., by seismic or coseismic processes. They may also be released by episodic or steady-state creep processes, especially along preexisting zones of weakness, e.g., fault zones. Hence, assessment of the direction and magnitude of stresses in a specific area is of primary importance for understanding tectonism, seismicity and possibly earthquake prediction.

3.6.1 DEFINITION OF STRESS

The purpose of this section is to define stress in a simple and easily understandable way. Exact mathematical definitions may be found in textbooks such as those by Jaeger (1969) and Jaeger and Cook (1976). Stress is the limiting value of force per area, and is generally denoted by the Greek letter σ, i.e.,

$$\sigma = dF/dA, \tag{3.39}$$

where F is the force and A the area. If the force acts perpendicular to the surface area the stress is called *normal* stress. If the normal of dA, for instance, is directed in the x direction (and the force also acts in the x direction), then the normal stress may be denoted by σ_{xx}, the first index describing the plane and the second one the direction of the force. It is evident that for the three-dimensional case there are exactly three normal stress components: σ_{xx}, σ_{yy}, σ_{zz}. Aside from these normal stresses there are *shear* stresses acting *in* the plane dA, denoted by σ_{xy}, σ_{yx}, σ_{zx}, σ_{yz}, σ_{zy}, and σ_{xz}. For reasons of equilibrium $\sigma_{xy} = \sigma_{yx}$ etc.; hence, there are six independent members within the set of nine stress components:

$$\begin{bmatrix} \sigma_{xx} & \sigma_{xy} & \sigma_{xz} \\ \sigma_{yx} & \sigma_{yy} & \sigma_{yz} \\ \sigma_{zx} & \sigma_{zy} & \sigma_{zz} \end{bmatrix}. \tag{3.40a}$$

These nine components constitute a matrix, which sometimes is written using other notations; for instance, in engineering (and often geological) work σ is used for the normal stress and τ for the shear stress. Hence the matrix (3.40a) may also be written as

$$\begin{bmatrix} \sigma_x & \tau_{xy} & \tau_{xz} \\ \tau_{yx} & \sigma_y & \tau_{yz} \\ \tau_{zx} & \tau_{zy} & \sigma_z \end{bmatrix}. \tag{3.40b}$$

It is generally convenient to choose the axes in the direction of *principal stresses*, i.e., σ_1, σ_2, and σ_3. Along these axes the shear stresses vanish, and the matrix assumes the form

$$\begin{bmatrix} \sigma_1 & 0 & 0 \\ 0 & \sigma_2 & 0 \\ 0 & 0 & \sigma_3 \end{bmatrix}, \tag{3.40c}$$

where σ_1 is conventionally chosen to be the largest and σ_3 the smallest principal stress; i.e., the convention used is

$$\sigma_1 \geqslant \sigma_2 \geqslant \sigma_3. \tag{3.41}$$

From some simple geometric arguments for *two* dimensions the basic equations of stress calculations are (see Fig. 3.45)

$$\sigma = \sigma_1 \cos^2 \theta + \sigma_2 \sin^2 \theta = \underbrace{\tfrac{1}{2}(\sigma_1 + \sigma_2)}_{\substack{\text{intermediate} \\ \text{normal stress}}} + \underbrace{\tfrac{1}{2}(\sigma_1 - \sigma_2) \cos 2\theta}_{\substack{\text{greatest} \\ \text{shear stress}}} \tag{3.42}$$

$$\tau = -\tfrac{1}{2}\underbrace{(\sigma_1 - \sigma_2)}_{\tau_{\max}} \sin 2\theta, \tag{3.43}$$

where θ is the angle between σ and the σ_1 axis and, in the two-dimensional case, σ_1 is the largest and σ_2 the smaller principal stress, as seen in Fig. 3.46.

A method for elucidating the relationships mentioned so far is the compilation of *Mohr's diagrams*, which are representations of stress in two or three dimensions. The simplest one is Mohr's circle diagram in two dimensions. As shown in Fig. 3.47, we mark $OP = \sigma_1$ and $OQ = \sigma_2$ and draw

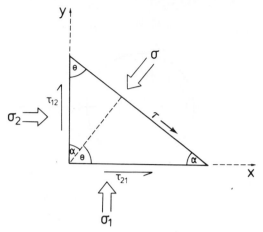

Fig. 3.45 Orientation of stresses in two dimensions; σ represents the normal stresses, τ the shear stresses, and $\alpha = 90° - \theta$.

a circle with diameter PQ and center C. Selecting any point A on the circle, the angle PCA is 2θ, and AB is perpendicular to PQ. The point A has coordinates τ and σ. This representation holds for all values σ_1, σ_2 and all θ. It is evident from Fig. 3.46 and Eq. (3.43) that τ has an extremum for $2\theta = 90°$, i.e., $\theta = 45°$ (Jaeger, 1969).

For three dimensions (and under *in situ* conditions) the greatest shear stress is, in analogy to Eq. (3.42),

$$\tau = \tfrac{1}{2}(\sigma_1 - \sigma_3) = \tau_{max}. \tag{3.44}$$

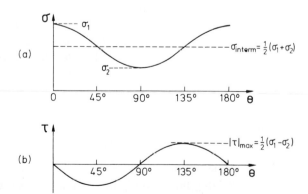

Fig. 3.46 (a) Normal stresses σ and (b) shear stresses τ as a function of the angle θ (see Fig. 3.45); σ_{interm} is the intermediate normal stress and $|\tau|_{max}$ the maximum shear stress.

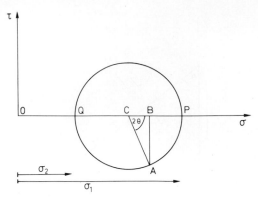

Fig. 3.47 Mohr's circle diagram in two dimensions.

It is across a plane where the normal bisects the angle between the directions of the greatest and the least principal stress σ_1 and σ_3. The values $\tau_2 = \frac{1}{2}(\sigma_1 - \sigma_3)$, $\tau_1 = \frac{1}{2}(\sigma_2 - \sigma_3)$, and $\tau_3 = \frac{1}{2}(\sigma_1 - \sigma_2)$ are often called the *principal shear stresses*, and they correspond to the normal stresses $\frac{1}{2}(\sigma_1 + \sigma_3)$, $\frac{1}{2}(\sigma_2 + \sigma_3)$, and $\frac{1}{2}(\sigma_1 + \sigma_3)$, as may be deduced from the two dimensional case.

It follows from the preceding description that the stress field in a homogeneous medium is given by a triaxial ellipsoid, the semiaxes of which are the principal stresses $\sigma_1 > \sigma_2 > \sigma_3$. If $\sigma_1 = \sigma_2 = \sigma_3$, there is an isostatic equilibrium. If $\sigma_1 > \sigma_2 > \sigma_3$, the medium tries to reach equilibrium by internal movements, i.e., deviations. In this case, σ_1, σ_2, and σ_3 are the principal deviatory stresses and are perpendicular to each other. Here σ_1 always has a compressional and σ_3 an extensional character; the intermediate principal deviatory stress σ_2 can be of compressional or extensional character. For *in situ* stress measurements near the Earth's surface, i.e., in boreholes and tunnels, very often only the *horizontal* principal stresses are considered and the vertical direction is ignored. This is because air cannot accept stress, and near the surface two of the three principal stresses are horizontal; the third one is vertical. There are three cases:

(1) σ_1 and σ_2 lying horizontally: compressional tectonism, reverse faults,
(2) σ_3 and σ_2 lying horizontally: extensional tectonics, normal faults,
(3) σ_1 and σ_3 lying horizontally: horizontal movements, strike–slip faults.

Figure 3.48 gives an example of the direction and magnitude of the two principal deviatory stresses lying horizontally in an area of western Europe dominated by the stress pattern of the Alps. Compare Fig. 3.48 with Fig. 3.14 and 3.15.

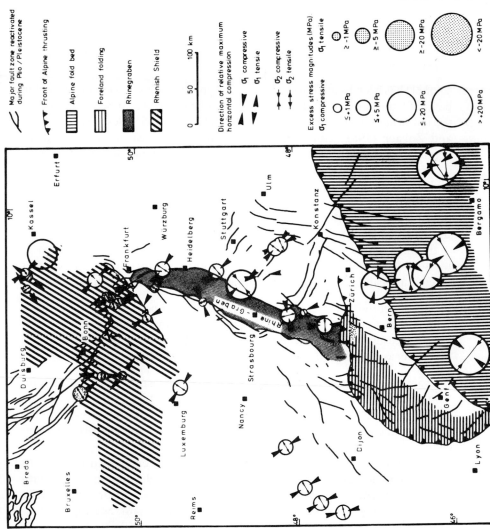

Fig. 3.48 Direction and magnitude of horizontal principal deviatory stresses in western Europe. The shaded area at Bonn has an extensional character; all other areas are characterized mostly by compressional stresses. [From Illies and Baumann (1982).]

Major fault zone reactivated during Plio / Pleistocene

Front of Alpine thrusting

Alpine fold bed

Foreland folding

Rhinegraben

Rhenish Shield

0   50   100 km

Direction of relative maximum horizontal compression

σ_1 compressive
σ_1 tensile
σ_2 compressive
σ_2 tensile

Excess stress magnitudes (MPa)

σ_1 compressive

≤ 1 MPa
≤ 5 MPa
≤ 20 MPa
> 20 MPa

σ_1 tensile

≥ -1 MPa
≥ -5 MPa
≥ -20 MPa
< -20 MPa

For seismic focal plane solutions and all considerations of fault planes the third dimension cannot be ignored. In particular, the increasing compressive stress with depth (i.e., the principal stress in the vertical) plays an important role and may assume all values between σ_1 and σ_3, as will be shown in Section 3.6.4.

3.6.2   METHODS OF MEASURING STRESS *IN SITU*

As mentioned briefly in Section 3.3.2, stress may be assessed basically by three different types of methods, using:

(1)   earthquake focal mechanisms,
(2)   *in situ* stress measurements (i.e., in deep boreholes or tunnels),
(3)   geologic data.

Type (1) has been treated in section 3.3.2. Type (3), providing estimates on the basis of fault pattern, fault offsets, etc., will be described in Section 3.6.4. In the following, the *in situ* stress measurements (type 2) are described. Zoback and Zoback (1981) emphasize the importance of measurements at great depth because large discrepancies might arise between measurements on both a regional and a local scale and because hydrofracture studies often show a high degree of scattering, even in a single borehole, thus indicating a possible decoupled near-surface layer which may be controlled by the nearby topography. In particular, the upper few ten to hundred meters often seem to be disturbed by local factors.

In spite of these limiting statements, *in situ* stress measurements often yield reasonable data, even in shallow holes or tunnels. *In situ* stresses and/or their orientation can be derived by four different methods of measurements:

(1)   direct measurements of rock *stresses* by compensation,
(2)   measurements of *deformations* by "overcoring methods,"
(3)   measurements of *strength* of rocks and *directions* of principal stresses,
(4)   measurements of *directions* of principal stresses by "breakouts."

Examples of (1) are the so-called "flat-jack" measurements (Jaeger and Cook, 1976). A slot is cut into the rock surface of a wall or a tunnel by drilling or sawing, releasing the stress in the slot's surrounding by an expansion of the rock into the slot. This expansion is a strain release, and the displacement changes are measured by previously fixed markers on both sides of the slot. More important, a flat, hydraulic cell (or "jack") is embedded (often with concrete) in the slot. The pressure in the cell is controlled and increased until the displacements are compensated. This "cancellation pressure" is virtually equal to the original stress normal to the plane or the slot. See Fig. 3.49.

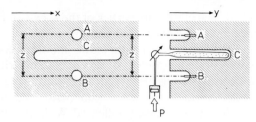

Fig. 3.49 Flat-jack method for *in situ* stress measurements on a vertical rock wall. Here *A*, *B*, and *C* are different drill holes in the tunnel or wall, and *P* is the pressure.

From the walls of tunnels (or the surface of the Earth) such measurements can be performed in various directions; hence, all (or all horizontal) components of the normal stress can be obtained. The flat-jack method is a null method and is one of the few direct measurements of rock stresses.

Extensive studies of possible errors (e.g., Alexander, 1960) include surface effects and creep during the cutting. Also, stress concentrations around the tunnel may contribute to some disturbances, but theoretical assessments and a large number of measurements generally result in reliable values of stresses.

Method 2, the measurements of deformation, includes the popular overcoring methods. In the 1960s several groups (e.g., Leeman, 1964; Merril, 1967) developed borehole deformation gauges for measuring the deformation across one or more borehole diameters. Generally, a borehole is continued centrally with a smaller diameter. After exact determination of this smaller diameter, drilling with the old, larger diameter is continued, thus relieving the stress in the "overcored" central section. See Fig. 3.50.

From the changes of the borehole diameter, i.e., its deformation in all directions, the strain and the corresponding stress can be calculated. The stresses assessed are the principal stresses in the plane perpendicular to the borehole, i.e., generally the horizontal stresses, as shown in Fig. 3.50. But however, with measurements in three inclined boreholes the complete state of

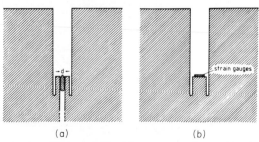

Fig. 3.50 (a) Overcoring-diameter method for *in situ* measurements of strain (and induced stress). (b) Overcoring-strain gauge method.

stress can be obtained. The elastic properties of the rock can be measured by laboratory experiments on the overcored section.

Strain gauge measurements in the center of the smoothed end of a borehole also belong to the overcoring methods. The strain release during and after the overcoring of the central part is observed and transformed into stresses. "Photoelastic" measurements have also been performed to measure the deformation during the strain release (Galle and Wilhoit, 1962).

Measurements of strength, method (3), are the basis of the hydraulic fracturing methods (see Fig. 3.51). In contrast to the methods mentioned so far, they can be used at various depth intervals in a borehole, i.e., quasi-continuously and far from the surface or an excavation. Only the major principal horizontal stress and its direction can be assessed. A portion of the borehole is sealed off by packers; fluid pressure in the sealed portion is increased until fracture takes place. Fluid can now be forced into the hole. The direction of fracture is observed by cameras or by so-called impression packers. This often corresponds to preexisting microcracks, which roughly show the direction of the major principal stress. The method of hydraulic fracturing (hydrofracturing) is widely used for oil and gas reservoirs to produce artificial openings. Also, hot dry rock projects aimed at extracting

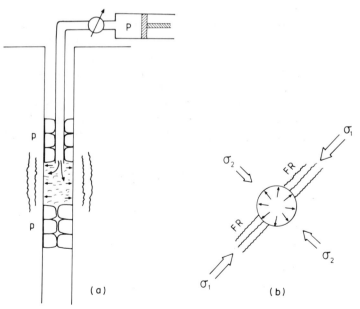

Fig. 3.51 Method of hydraulic fracturing. (a) Cross-section; P, packers; p, pressure. (b) Bird's perspective; FR, fracturing (induced).

geothermal energy from basement rocks use hydraulic fracturing to create openings for water circulation.

In practice, sections of the hole that are free of large cracks are selected for stress determination. Water pressure up to 20 MPa is applied through a remote control valve, first to the packers and then into the injection interval. As seen in the examples of Fig. 3.52, the injection pressure rises quickly and at a rather constant rate to a maximum value P_c. This peak value marks the "breakdown" pressure (or critical pressure). Fracture initiation is indicated by the sudden drop in pressure P. At this point fluid injection is lowered to allow the pressure to settle at a rather constant level P_s, which is needed to keep the new artificial cracks open. This pressure P_s, also called the shut-in pressure, compensates for the natural stress acting on the newly created fracture planes. In general, repeated injections are necessary to establish a reliable level of peak values P_r ("refrac pressure" value of peak pressure at repeated injection). The difference $P_c - P_r$ is considered to be a measure of

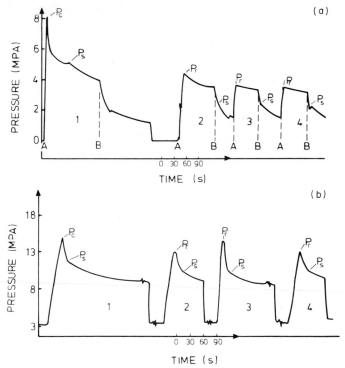

Fig. 3.52 Field records of injection pressure versus time. (a) Pumping between A and B (test at 90.5 m) and (b) pumping until P_c or P_r is reached (test at 312 m). Here P_c is the breakdown pressure, P_s the shut-in pressure, and P_r the refrac pressure. [From Haimson and Rummel (1982).]

the *in situ* tensile strength S_0 of the rock. Extensional *horizontal* fracture takes place if

$$P_c \geq \rho g z + S_0, \tag{3.45}$$

neglecting pore pressure P_0 and assuming that the vertical stress $\rho g z$ represents a principal stress direction.

For the value $\rho = 2.8$ g/cm^3 the vertical stress ρ_v becomes

$$\sigma_v = \rho g z = P_r = 0.027 \quad \text{MPa}/m \text{ times } z \text{ in } m. \tag{3.46}$$

Bredehoeft *et al.* (1976) have shown that in the general case the shut-in pressure P_s is identical to the minimum horizontal principal compressional stress

$$\sigma_{h,min} = P_s, \tag{3.47}$$

while the maximum principal stress is

$$\sigma_{h,max} = 3P_s - P_r - P_0 = S_0 + 3P_s - P_c - P_0 \tag{3.48}$$

with

$$S_0 = P_c - P_r \tag{3.49}$$

and P_0 is the pore pressure.

Many interesting diagrams of horizontal stress, mostly exceeding the vertical stress $\rho g z$, have been plotted as a function of depth. As shown by Rummel *et al.* (1982), the coefficient of internal friction μ^* can be obtained. In general, most tests carried out in central Europe show fair agreement in the amount and direction of stress orientation with the results of overcoring tests as well as with focal plane solutions (Rummel, 1978; Rummel and Alheid, 1979; Rummel *et al.*, 1982).

Method (4), the last method for the study of *in situ* stresses, is the use of borehole breakouts. Such breakouts are elongations of the borehole through natural shear fracturing in the zone of amplified stress differences near the hole. In contrast to the three methods mentioned before, the breakouts only allow determination of the orientations of horizontal principal stresses and only if these are unequal. Nevertheless, such breakouts are frequently observed and very often share a common average orientation (Babcock, 1978), marked by many intervals with noncircular sections oriented in the same azimuth (Gough and Bell, 1981, 1982).

The breakout method differs from the previously discussed borehole method (hydraulic fracturing) in several ways:

(1)　No additional manipulation of the hole is required.

(2)　The breakouts are generally wide, unlike the narrow, tensile fractures in hydraulic fracturing, and are easily observable.

(3)   The breakout azimuth is generally at right angles to the hydraulic fractures, as will be shown later.

(4)   A simple observation of many available wells in an area provides a reliable method for the orientation of principal horizontal stresses.

Figure 3.53 gives an overview of the direction of two different principal horizontal stresses and the situation of the maximum shear stress σ_θ (minimum of the radial stress σ_r) in a circular hole based on formulas derived by Kirsch in 1898.

At the wall, near points P_1 and Q_1, i.e., in the direction of the minimum principal stress σ_2,

$$\sigma_\theta = 3\sigma_1 - \sigma_2 \tag{3.50}$$

and

$$\sigma_r = 0. \tag{3.51}$$

Breakouts, hence, should prefer the σ_2 direction (not σ_1). From a comparison of Figs. 3.53 and 3.51 the different direction of breakouts with regard to the (vertically oriented) hydraulic fractures is evident.

In practice, breakout zones in a borehole can be observed by optical and acoustical imaging. Spectacular pictures were provided by Springer and Thorpe (1982). However, standard calipers, mostly magnetically oriented, also provide reliable mapping of elongations and can be used in the presence of drilling mud. Four-armed dipmeters that give the hole's diameter at four locations have been standard logging tools since the late 1960s. Bell and Gough (1979) and Gough and Bell (1981, 1982) show ample evidence for the consistency of breakouts in various tectonic provinces in Canada and relate their findings to stress directions determined by overcoring measurements,

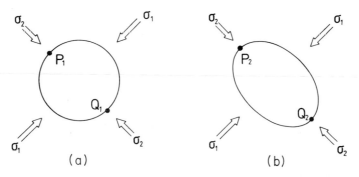

Fig. 3.53   Principle of borehole breakouts. (a) Stresses around circular hole and (b) modified hole diameter. Radial shear stress σ_r is minimum near P_1 and Q_1.

hydrofracturing, earthquake mechanism, and recent movements along fault zones.

As all the *in situ* stress measurement methods are still in an experimental stage, data must always be checked for consistency in a geologically homogeneous area and compared with stresses obtained from focal plane solutions.

3.6.3   STRESS, STRAIN, STRENGTH, AND FRACTURE

This section deals with the background of fracturing and is a kind of introduction to the following section on the estimation of stress from geologic structures. The background is described as briefly as possible. Fracture theory in its simplest form goes back to Eqs. (3.42) and (3.43) in Section 3.6.1 and modifies them according to ideas from Mohr, Coulomb, and Navier (Jaeger, 1969). While according to Eq. (3.43) the maximum shear stress (and hence a possible fracture) occurs across a plane whose normal bisects the angle between σ_1 and σ_2, experimental and *in situ* observations show that the angle of fracture θ_{max} ($= \beta$) is generally larger than 45°; i.e., the angle α between the trace of the fault plane and the σ_1 axis is smaller than 45° (see Fig. 3.54). A reasonable and simple assumption is that fracture takes place if a certain strength σ_0 (the shear strength) plus μ^* times the normal pressure σ across the plane is reached. The parameter μ^* is frequently called the internal friction coefficient, and it can be determined from laboratory and *in situ* observations. Here τ_0 is the "inherent shear strength"; in soil mechanics it is the cohesion. The preceding assumption states that fracture takes place along a plane at which the shear stress τ becomes equal to

$$|\tau| = \tau_0 + \mu^*\sigma \tag{3.52}$$

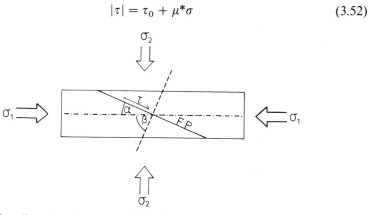

Fig. 3.54   Two-dimensional geometry of fault plane (FP) illustrating normal stresses σ, shear stresses τ, and fault angles (α, $\beta > \alpha$).

(Coulomb criterion, Coulomb, 1773). See also Fig. 3.55. The quantity $|\tau| - \mu^*\sigma$ may be formed from Eqs. (3.42) and (3.43):

$$- \mu^*\sigma + |\tau| = -\tfrac{1}{2}\mu^*(\sigma_1 + \sigma_2) + \tfrac{1}{2}(\sigma_2 - \sigma_1) \qquad (3.53)$$

$$\pm (\sin 2\theta + \mu^* \cos 2\theta).$$

Fracture takes place when $- \mu^*\sigma + |\tau|$ first attains τ_0 [Eq. (3.52)]. Equation (3.53) has a maximum for

$$\tan 2\theta_{\max} = \mp 1/\mu^*. \qquad (3.54)$$

For all considerations of fracture this special angle $\theta_{\max}$ will be called β. It is convenient to use the minus sign in Eq. (3.54). In this case Eq. (3.54) may be written as

$$\beta = \tfrac{1}{2}[\pi - \tan^{-1}(1/\mu^*)] \qquad (3.55)$$

(see Figs. 3.54 and 3.55 again) and the value of the maximum of Eq. (3.53) is

$$\tfrac{1}{2}(\sigma_1 - \sigma_2)\sqrt{\mu^{*2} + 1} - \tfrac{1}{2}(\sigma_1 + \sigma_2). \qquad (3.56)$$

It is easy to apply Mohr's circle diagrams to the study of faults. Equation (3.52), for instance, represents a pair of straight lines in a σ–τ diagram. As mentioned before, any state of stress can be represented by a Mohr circle. No faulting will occur as long as Mohr's circles are totally within the lines $|\tau| = \tau_0 + \mu^*\sigma$. However, if one of Mohr's circles touches the lines, failure will

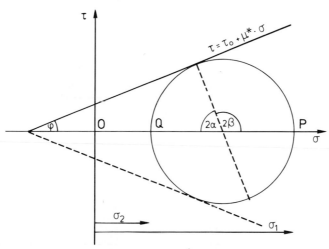

Fig. 3.55 Two-dimensional Mohr's diagram with Mohr's envelopes and fault angles α and β; cf. Fig. 3.47.

take place. The angle α in this case gives the preferred direction of the trace of the fault plane with the σ_1 direction, being between 0 and 45° (usually 22–30°). The two limiting lines are called the Mohr's envelopes.

The value for the internal friction μ^* may also be expressed by tan φ. As seen from Fig. 3.54, the following relations hold:

$$2\beta = 90° + \varphi = 180° - 2\alpha \qquad (3.57)$$

and

$$\varphi = 2\beta - 90° = 90° - 2\alpha; \qquad \alpha = 45° - \tfrac{1}{2}\varphi, \qquad (3.58)$$

where α is the angle between σ_1 and the trace of the fault plane. Table 3.3 shows some values of the parameters α, β, and φ as used in Fig. 3.55.

Any number of Mohr's circles may be fitted between the limiting Mohr's envelopes. The larger σ (or $\sigma_1 + \sigma_2$), the larger is the shear stress that causes fracturing, and the larger is the strength of the material. This means that increasing lithostatic pressure (circles to the right side of Mohr's diagram) results in a greater shear strength of rocks. Small pressure (and especially extension) correlates with a small shear strength.

The two envelopes $|\tau| = \tau_0 + \mu^*\sigma$ are transformed into two planes of fracture for the three-dimensional case, forming angles α to the σ_1 direction. The two possible conjugate planes correspond to the two possible focal plane solutions for a specific stress system $\sigma_1, \sigma_2, \sigma_3$, as mentioned in Section 3.3.2. The maximum shear stress [Eq. (3.44)] and the criterion of failure do not involve σ_2. In all equations [Eqs. (3.41)–(3.43), (3.53), (3.56)] σ_2 is replaced by σ_3. It might be argued that the Coulomb criterion of Eq. (3.52) might be

Table 3.3

Some Values of the Angles α, β, φ (Fig. 3.55) and μ^* = tan φ (Coefficient of Internal Friction) for Angles Given in Degrees[a]

μ^*	φ	β	2β	α	Remarks
0	0	45	90	45	Clays
0.18	10.2	50.1	100.2	39.9	Montmorillonite, vermiculite
0.3	16.7	53.3	106.7	36.7	Sands
0.6	30.5	60.5	121	29.5	Prebroken rocks
0.85	40.4	65.2	130.4	24.8	Prebroken rocks at zero pressure
1	45	67.5	134	22.5	Geologic structures
2	63.4	76.7	153.4	13.3	
	90	90	180	0	

[a] Unbroken rocks show great scatter in experiments. "Remarks" refer to some typical values. Lines 2, 4, and 5 are from Byerlee (1978).

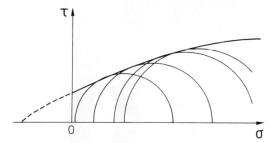

Fig. 3.56 Possible (curved) shape of *in situ* Mohr's envelope.

too simple and could not be representative for very high pressures. Mohr himself in 1900 proposed a more general relationship such as

$$|\bar{\tau}| = f(\sigma), \tag{3.59}$$

i.e., a relationship where $\mu^* \neq$ const. Later workers, including Griffith (1921), Brace and Orange (1968), and Murell (1964), specified Eq. (3.59) for various problems. The name "Mohr envelope" is still used and the envelope is found to be slightly convex downward in most experiments, as shown in Fig. 3.56, so that the traces of the plane of fracture become inclined at an increasing angle to the σ_1 direction. On the other hand, modern friction experiments relevant to geological data of Byerlee (1967, 1978), performed on prebroken samples, clearly show a constant μ^* up to 200 MPa (for moderate temperatures). The reason for performing experiments with prebroken samples is that *in situ* rocks also contain a great number of microcracks and fissures. Their behavior against stresses may be described by a linear relationship over a wide pressure range, i.e.,

$$\tau = 0.5 + 0.6\sigma, \tag{3.60}$$

where $\tau_0 = 0.5$ and $\mu^* = 0.6$ correspond to α and $\varphi \sim 30°$. Equation (3.60) is often called Byerlee's law. It will be discussed in Section 4.3.2.

3.6.4 APPLICATION TO (BRITTLE) FAULTS

It is the objective of this section to apply the previously derived formulas and relations to *in situ* fault zones and to relate their shape and character to the *in situ* stress conditions. From the trace of a fault zone and from its type (i.e., normal, reverse, or strike-slip) the principal stresses may be obtained as shown in Fig. 3.57. Remember that in all cases $\sigma_1 > \sigma_2 > \sigma_3$ and the greatest maximum shear stress is in the $\sigma_1-\sigma_3$ plane for the three-dimensional case.

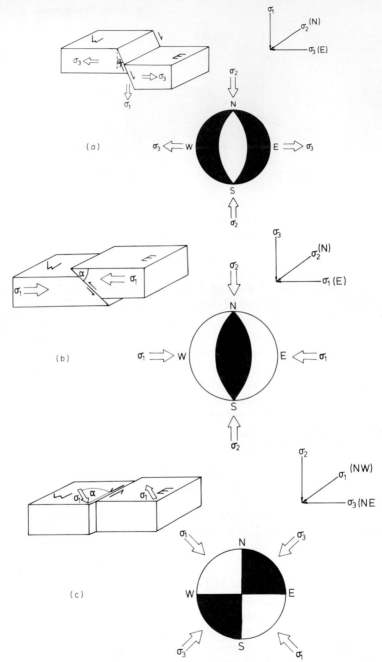

Fig. 3.57 Fault zones, stress pattern, and focal plane solutions. (a) Normal faulting NS, (b) reverse faulting NS, and (c) strike–slip faulting NS. Black arrows represent displacements along faults, and open arrows represent principal stresses.

Figure 3.57 shows basic cases that are common near the surface. These cases represent (a) normal faults, extensional tectonism; (b) thrust = reverse faults, compressional tectonics; and (c) strike–slip faults.

For thrust faulting (Fig. 3.57b) the least principal stress is vertical and the angle α between the trace of the fault plane and σ_1 is generally smaller than 45°, as explained in the last section. For a strike–slip fault (Fig. 3.57c) the intermediate principal stress σ_2 is vertical. For a dip–slip (or normal) fault the greatest principal stress is vertical; hence, α is measured from the vertical and again $\alpha < 45°$. Remember that under the given stress systems *two* possible angles α with the direction of σ_1 are always possible in principle, i.e., α and $180° - \alpha$, as already shown for the focal plane solutions. On the other hand, if the trace of the fault zone and α are known, only one stress system is deduced.

This deduction of the stress system from observations of fault zones by geologic or seismic methods is of the utmost importance for tectonic problems. The method is, of course, similar to earthquake focal mechanism studies, but it integrates over a longer time interval and permits the deduction of stress systems in various geologic epochs. Comparisons with earthquake mechanism studies often show a favorable correlation for the last 5–10 million years. It is tempting to correlate the observed fault angle α with the coefficient of friction $\mu^* = \tan \varphi$ and to find those parts of the fault where small or large μ^* values are present. Two types of faults are plotted in Fig. 3.58. Figure 3.58a shows a thin-skinned nappe such as those of the Appalachian or Hohes Venn area. From the generally observed listric shape of the fault zone we may deduce a weak $\mu^*(\mu^* \to 0)$ zone in the upper 2–3 km and a strong zone ($\alpha \to 0$, $\mu \to \infty$) at 5–6 km depth. This is in good agreement with the stress calculations of Meissner and Strehlau (1982), from which a strong peak in strength (and seismicity) was postulated at these depths. The deeper part, such as that of the Wind River Thrust, dips with about 30° and would show $\mu^* \sim 0.6$, in agreement with Byerlee's experiments. For extensional structures (Fig. 3.58), which preferably show a listric shape, a weakening of the material (decreasing μ^*) is observed for increasing depth.

It is this weakening of the material with depth—strictly speaking, the weakening with temperature—which casts some doubts on the preceding correlation between fault angle α and μ^*. There are listric faults whose deepest parts run horizontally (Bally et al., 1981), and there are extensional faults under very small angles (L. Brown, 1983 private communication). As will be shown Section in 3.6.5, fault zones may really be initiated in the ductile lower crust by processes other than brittle. The widespread observation that the angle α often exceeds 45° in nature, especially for extensional structures and depths greater than 5 km, is a strong indication of the ductile behavior of the lower crust (Meissner and Wever, 1985). Furthermore, an increased ductility (decreasing viscosity and strength) might also be present within the

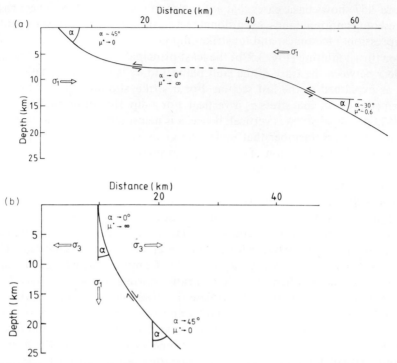

Fig. 3.58   Examples of traces of (a) a thrust fault and (b) a normal (listric) fault, where α is the fault angle to σ_1 and μ^* the coefficient of internal friction.

fault zone itself. We also stress that any fault zone with an increased ductility may be kept active even under a change of stress pattern. Once a zone of weakness has been established in a certain direction, it will be reactivated even under the opposite stress system because it represents a plane of minimum strength, which is a result of its own weakness and smaller μ^* values. All these arguments lead to the conclusion that only young and new faults can be taken for a comparison of stresses with focal plane solutions and *in situ* stress measurements.

3.6.5   CALCULATIONS OF MAXIMUM POSSIBLE STRESSES (SHEAR STRENGTH) WITH DEPTH

As mentioned in Section 3.6.4, laboratory data obtained with prebroken rock samples like those of Byerlee (1967, 1978) seem to be appropriate for the assessment of maximum stresses in the Earth's crust because joints, cracks, fissures, and fractures are found at least in the upper 10 km of sediments and crystalline basements (Brace, 1979). Such fractures limit all ambient stresses. These and other experimental studies are discussed in Section 4.3. From a great number of experiments Eq. (3.52) may be written in the form (Byerlee's law):

$$\tau = 0.5 + 0.6\sigma \tag{3.61}$$

[= Eq. (3.60)]. For very small confining pressure a better approximation is obtained with:

$$\tau = 0.85\sigma. \tag{3.62}$$

Only clayish rocks and minerals like vermiculite and montmorillonite do not obey Byerlee's laws but might be approximated by (Meissner and Strehlau, 1982)

$$\tau = 0.18\sigma. \tag{3.63}$$

Such small maximum stresses as in Eq. (3.63) may be involved in parts of special large fault zones where frictional heating and mylonitization of materials are also involved. In the area of the San Andreas fault Zoback $et\ al.$ (private communication, 1983) found $\mu^* = 0.35$.

In order to calculate the maximum stresses as a function of confining pressure (or depth), Eqs. (3.61)–(3.63) must be introduced in Eq. (3.42), which for the three-dimensional case and for the angle of most probable fracture may be written in the form

$$\tau = -\tfrac{1}{2}(\sigma_1 - \sigma_3)\sin^2 \beta. \tag{3.64}$$

Now one of Eqs. (3.61)–(3.63) is inserted into Eq. (3.64) and solved for $\tau_{max} = \tfrac{1}{2}(\sigma_1 - \sigma_3)$, with σ_1 acting horizontally for the compressional stage and vertically for the extensional condition. The angle β (or $\alpha = 90° - \beta$) may be easily taken from Eqs. (3.57) and (3.58) or from Table 3.3 or Fig. 3.54. It is equal to about 60° ($\alpha = 30°$) for Eq. (3.61) with $\mu^* = 0.6$ and to 65° ($\alpha = 25°$) for Eq. (3.62). The values $\tau_{max} = \tfrac{1}{2}(\sigma_1 - \sigma_3)$ are first solved for σ_1. The σ_1 is taken as vertical for the extensional case and horizontal for the compressional one.

Table 3.4 gives an overview of the calculation of maximum stresses. The first three lines describe the calculation of τ_{max} as a function of the vertical stress as mentioned earlier, the first line referring to Eq. (3.61), the next ones to Eqs. (3.62) and (3.63). Note that the maximum stresses for compression are about three to four times larger than those for extension, a fact which may also be deduced from Mohr's diagrams.

Line 4 shows the substitution of σ_1 (or σ_3) by the lithostatic pressure ρgz, a substitution which is justified only for completely dry rocks without pore

Table 3.4

Overview of Calculation of Maximum Stresses (Shear Strength)[a]

Line	Condition and equation used	Extension	Compression
1	Eq. (3.61)	$\tau_{max} = -0.3975\sigma_1$	$\tau_{max} = 1.3975\sigma_3$
2	Eq. (3.62)	$\tau_{max} = -0.34\sigma_1 - 0.283$	$\tau_{max} = 1.06\sigma_3 + 0.885$
3	Eq. (3.63)	$\tau_{max} = -0.158\sigma_1$ (M + V)	$\tau_{max} = 0.232\sigma_3$ (M + V)
4	dry	$P_L = \sigma_{1,d} = \rho gh = 0.27h$	$P_L = \sigma_{3,d} = \rho gh = 0.27h$
5	wet	$\sigma_1 = gh(\rho - \rho_{H_2O})$ $= 0.63\sigma_{1,d} = 0.17h$	$\sigma_3 = gh(\rho - \rho_{H_2O})$ $= 0.63\sigma_{3,d} = 0.17h$
6	Eq. (3.61) dry	$\tau_{max} = -0.1073h$	$\tau_{max} = 0.523h$
7	Eq.. (3.61) wet	$\tau_{max} = -0.0676h$	$\tau_{max} = 0.330h$
8	Eq. (3.62) dry	$\tau_{max} = -0.09h - 0.283$	$\tau_{max} = 0.2862h + 0.885$
9	Eq. (3.62) wet	$\tau_{max} = -0.06h - 0.283$	$\tau_{max} = 0.1803h + 0.885$
10	Eq. (3.63) dry	$\tau_{max} = -0.043h$	$\tau_{max} = 0.0626h$
11	Eq. (3.63) wet	$\tau_{max} = -0.027h$	$\tau_{max} = 0.0395h$

[a] M + V = montmorillonite and vermiculite, h = depth in kilometers, τ_{max} in kilobars, 1 kbar = 100 MPa.

pressure. Line 5 shows the introduction of the pore pressure, which has been assumed to be hydrostatic. The last six lines of Table 3.4 provide the τ_{max} values for dry and wet conditions, for extension and compression, and for Eqs. (3.61)–(3.63) as a function of the depth. Figure 3.59 shows these relationships. The curves for vermiculite and montmorillonite end in the middle part of the crust because of the known instability of clay minerals. There is growing evidence, however, that frictional sliding along brittle fractures does not extend to large crustal depths. With increasing temperature, ductile behavior starts to play a major role. It drastically reduces the possible maximum shear stresses as a function of temperature and water content. This leaves the curves of maximum possible stress with a sharp peak, which is situated between 5 and 8 km for a warm and water-saturated crust (Meissner and Strehlau, 1982). As mentioned before, this peak coincides with that of the frequency of occurrence of earthquakes. Without having yet explained the calculations, Fig. 3.60 shows two examples of the two peaks mentioned earlier.

The upper part of curve A shows the brittle part, according to the calculations performed so far (line 9 of Table 3.4, i.e., a wet crust), while the lower part represents the ductile behavior, which will be treated in Section 4.4. The shape of deformed grains and the dislocation structure of naturally and experimentally deformed samples provide an additional method for assessment of the differential stress and the temperature during deformation.

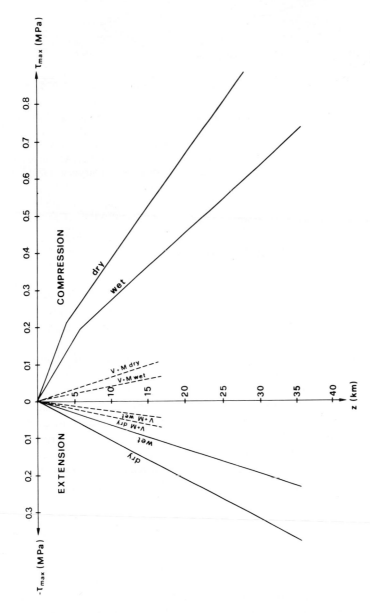

Fig. 3.59 Calculated maximum stresses (STRESSMAX) (i.e., strength) according to Byerlee's law for dry rocks and rocks with hydrostatic pore pressure (wet). Here V + M represents vermiculite and montmorillonite.

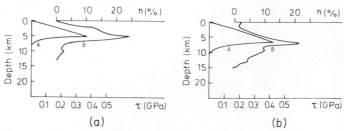

Fig. 3.60   Comparison of STRESSMAX curves (including ductile behavior at depth) and curves of depth-frequency distribution of earthquakes (DEFREQ). (a) High heat flow area Coso; $q = 82$ mW/m^2 and (b) normal heat flow area, northern Greece, $q = 70$ mW/m^2; n is the number of earthquakes per 1-km depth interval, A the calculated STRESSMAX curve for wet conditions (Fig. 3.59), and B the DEFREQ data. [From Meissner and Strehlau (1982).]

3.7   CRUSTAL GEOMAGNETISM

The measurement of the Earth's magnetic field and of its changes in time and space has a long history and constitutes one of the important methods for both general and applied geophysics. Several fields of geomagnetism are related to the Earth's crust.

Continental drift is based first of all on conclusions of *paleomagnetism*. If the terrestrial and other planetary magnetic fields are generated by a self-exciting dynamo action of thermal convection currents within the planet's (outer) core—and no other serious theory is in sight—then it is probable that certain instabilities exist and fields may change considerably in amplitude and even in polarity. From theoretical considerations and from many paleomagnetic measurements and age determinations on rocks, it seems that the Earth's field has always been essentially *bipolar* and—when averaged over some thousand years—even coincident with the axis of rotation (see Fig. 3.61). This is also indicated by a close correlation between paleomagnetic and paleoclimatic data. A prominent example is the clustering of the pieces of Gondwanaland around the South Pole during Permian and Carboniferous times as shown by both a near-vertical inclination of paleomagnetic field lines and an extensive glaciation (Fuller, 1983). One of the spectacular results of paleomagnetism was the discovery that rocks from certain geologic formations, e.g., lava flows or sedimentary layers, show a repeated alternation of magnetic polarity. Simultaneous age determinations on such layers result in the obvious explanation that the geomagnetic field must have undergone repeated reversals.

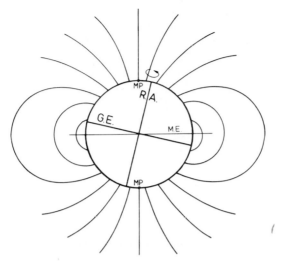

Fig. 3.61. Schematic picture of the Earth's magnetic dipole field; MP is the magnetic pole, R.A. the Rotation axis, M.E. the Magnetic equator, and G.E. the Geographic equator.

Based on this concept, Vine and Matthews (1963) and others interpreted the magnetic stripes of the oceans as an expression of newly created, magnetized basaltic rocks which are carried laterally away from the spreading axes as on a giant magnetic tape recorder. Ocean floor spreading was the obvious explanation for the signature of the magnetic stripes, a signature which had already been established by the correlation of age and magnetic reversals for the continental crust. By knowing the ages and the age intervals of certain magnetic periods and correlating them with the width of oceanic magnetic stripes, the speed of ocean floor spreading, i.e., the speed of the magnetic tape, could be deduced long before drilling into the ocean floor finally proved the spreading. As some continents are fixed to an oceanic plate, the postulation of continental drift was also a consequence of magnetic data. By analyzing the stripes, any direction of movement—at least for the past 200×10^6 years—could be deduced, not only an S–N drift as extractable from the paleomagnetic inclination on land, which can only be related to the paleolatitude. As shown in Fig. 3.62, the (paleo-) latitude of a certain rock unit is directly related to the (paleo-) inclination based on the formulas

$$H = -(m/r^3) \sin \theta = -H_0 \sin \theta, \tag{3.65}$$

$$Z = (2m/r^3) \cos \theta = -2H_0 \cos \theta, \tag{3.66}$$

$$\tan I = \frac{Z}{H} = 2 \cot \theta, \tag{3.67}$$

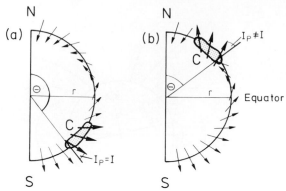

Fig. 3.62   Relationship between latitude and magnetic inclination. (a) Paleoconfiguration of magnetic field at continent (e.g., India); C, continent; θ, colatitude; r, Earth's radius; I, inclination; and the arrows are field lines. (b) Present configuration of magnetic field at continent (e.g., India).

where θ is the colatitude $= 90° - \varphi$ for an axial dipole, m the magnetic moment, r the radius (of the Earth), $H_0 = m/r^3$, I the inclination, H the horizontal and Z the vertical magnetic field strength, and φ the geographic latitude. Equation (3.67), hence, gives a direct relationship between the measured geomagnetic value "inclination", possibly a paleoinclination, and the geographic colatitude, possibly paleolatitude. This means that latitudinal displacements $\Delta\theta$ of continents or continental units may be observed over different age periods by a related ΔI. The declination D is assumed to be always in the direction of the magnetic field when averaged over a couple of thousand years, i.e., $D = 0$. However, the paleodeclination may be quite remarkable ($D \neq 0$) and may provide a direct measure of the angle of rotation of the continental unit.

A new interesting aspect of paleomagnetic studies on land was the discovery that some small areas of the continents were accreted later and came from very different latitudes, as determined by radiometric, stratigraphic, and paleomagnetic observations of I and D. Numerous small continental and oceanic platelets moved northward over periods of millions of years until they are finally attached to one of the big continents. Ben–Avraham et al. (1981) and Nur (1983) have shown that even today about 10% of oceanic areas are covered by such platelets, which move along with the spreading sea floor. However important the impact of paleomagnetism and rock magnetism was for the development of the history of continental drift, their influence on our knowledge of the continental crust is limited to two main contributions. First, magnetic measurements on the ground or from the air can reveal an average depth and give a rough idea of undulations of the (magnetic) basement and, hence, the existence and depth of sediment troughs. Second, such measurements may provide data on the "Curie depth", i.e., the depth were the temperature exceeds the Curie temperature (about 580°C for magnetite). At this temperature the ferrimagnetic properties of rocks and minerals disappear completely. Mapping this depth, hence, provides an additional—though not very precise—tool for determining the temperature. These two topics, the assessment of the magnetic basement and that of the Curie depth, will be subjects of Section 3.7.3. Comprehensive treatments of the principles of geomagnetism are found in textbooks such as those by Stacey (1977b), McElhinny (1973), and Tarling (1983). A review of new data on geomagnetism and paleomagnetism may be found in the U.S. National Report to the IUGG, 1983.

3.7.1 THE PHYSICAL BASIS OF ROCK MAGNETISM AND PALEOMAGNETISM

Like density, the physical properties of ferromagnetic susceptibility (FMS) and natural magnetization (NM) may characterize a certain rock type. Many rocks have a remarkable FMS; when placed in a magnetic field, e.g., the Earth's field, they acquire a certain magnetization of their own, the so-called induced magnetization (IM). Other rocks already have an intrinsic or permanent (remanent) magnetization without being subjected to a (present) outer field; they have so-called natural remanent magnetization (NRM). Both kinds of magnetization, i.e., IM and NRM, are able to produce magnetic anomalies. On a small scale there is generally a large scatter of measured values in both amplitude and direction. The reason for this lies partly in the two different contributions from NRM and IM, which may have different strengths and directions. The other part of the reason is the fact that both kinds of magnetic parameters depend on certain ferromagnetic minerals, such as magnetite (Fe_3O_4), which represent a trace rather than a bulk constituent. A good correlation has been obtained between the susceptibility and the magnetite content of various igneous rocks (Mooney and Bleifuss, 1953; Balsley and Buddington, 1958).

The natural remanent magnetization is the basis for *archeo- and paleomagnetism*. Certain effects during the cooling or compaction of rocks provide a stable memory of the original magnetic field, which is stored in the crystals of the rock as in a computer's memory. The most important mechanism for NRM is the thermoremanent magnetization (TRM). If a magnetic material is cooled below its Curie point (CP), (580°C for magnetite) it acquires a relatively strong and very stable field, even in a weak magnetic field (see Fig. 3.63). The magnetic grains in the rock are still able to reorient their magnetic

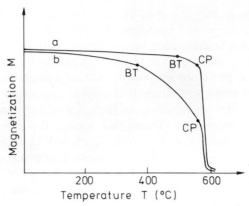

Fig. 3.63 Two examples of thermoremanent magnetization (TRM) below the Curie point (CP). (a) Multidomain grains and (b) monodomain grains; BT, blocking temperature.

moments in a certain temperature interval below the CP down to the so-called blocking temperature, which generally stops reorientation of multidomain grains a few degrees below the CP. Monodomain grains may contribute considerably to the TRM much below the CP. In paleomagnetism we prefer rocks with multidomain grains, which have a high blocking temperature and, hence, exhibit a stable TRM even at temperatures of 400–500°, i.e., at weak to moderate metamorphic events. Rocks that have been cooled in a period of field reversal naturally acquire an inverse magnetization, which provides the basis for the history of field reversals when correlated with radioisotopic ages. In these correlations the time of cooling from a molten stage down to the CP or of a strong metamorphism down to the CP is generally negligible compared to the time scales of the ages involved. For the old controversy between field reversals and self-reversal—a small-scale process based on negative antiferromagnetic coupling between ions or special domains—see Stacey (1977a).

Next to TRM in importance there is the so-called detrital remanent magnetization (DRM), which is a cold and slow process related to the sedimentation of fine-grained particles. Sediments generally contain some iron oxides, e.g., magnetite, whose magnetic moments are partially aligned by the Earth's field during the time of deposition (see Fig. 3.64). Secondary processes may slightly disturb the magnetic orientation, and chemical processes and the influence of the grain size of the magnetic particles play a certain role. Because of these different contributions, a specific sediment type acquires a specific magnetization, which can be detected and recognized with modern techniques. This is the basis for "magnetic stratigraphy" or "polarity stratigraphy," a methodology that has become a major partner of biostrati-

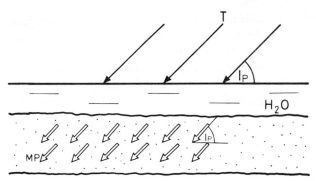

Fig. 3.64 Detrital remanent magnetization (DRM); *T*, magnetic field; *MP*, polarity direction of sedimented particles; and I_p, magnetic inclination.

graphy. For DRM there might be a considerable time lag between sediment formation and the acquisition of remanent magnetization (Henshaw and Merril, 1979; Shive, 1983), although other estimates center on a lag of only about 100 years (Suttil, 1980). For chemical remanent magnetization (CRM), isothermal remanent magnetization (IRM), and other processes of minor importance see the textbooks already mentioned.

3.7.2 MAGNETIC MEASUREMENTS

Magnetic measurements are performed in the field at the surfaces of land and sea, but also from airplanes (aeromagnetic surveys) and satellites, e.g., from POGO and MAGSAT on a low-elevation path around the Earth. In addition, measurements in the laboratory to determine directions (and even intensities) of the old (paleo-) magnetic fields are prerequisites for the whole field of paleo- and archeomagnetism.

The sensitivity of field instruments is now at about 1 nanotesla (nT) or 1γ, i.e., about 10^{-4} to 10^{-5} of the Earth's magnetic field (65,000 nT at the surface of the poles). There is a great variety of magnetometers, one group measuring only the total intensity, i.e., the combined average geomagnetic field plus the anomalous field without any directional information. This type of "scalar" instrument is mostly employed in airplanes and satellites and is based on the nuclear-precession concept. (For details, see textbooks such as Dobrin, 1960.) The other instruments, e.g., those of the balanced torque (torsion) or fluxgate type, are used preferably in ground surveys, although here, too, nuclear-precession instruments with their speed of measurement are beginning to dominate, and fluxgate instruments measuring in perpendicular directions are also used in satellites.

From field measurements on land (and on the sea bottom), the most detailed and small-scale anomalies, like those of ore bodies, fault zones, and intrusions, can be found and translated into models, a well-established method in applied geophysics. Normal marine measurements generally are able to reveal broader (and deeper) anomalies, like those of the oceanic basaltic material with their alternating polarities from one stripe to another. Aeromagnetic surveys, mostly used on land, provide a rough and smoothed picture of magnetic anomalies and are the basis both for applied studies and for the recognition of deeper structures in the Earth. Finally, satellite data, such as those from MAGSAT are able to determine and follow up large-scale anomalies at sea and on land and correlate them with large geologic units and structures.

Although it is theoretically possible to transform field data from the Earth's surface or from one level to any other level, e.g., to that of satellites, by the method of upward continuation, there are still some discrepancies in practice between the predicted and the observed values (Fuller, 1983). Hence, aeromagnetic *and* satellite measurements, especially those covering inaccessible land areas, are certainly of great value today in mapping of the geomagnetic field.

All the geomagnetic data obtained *in situ*, however, could not be properly interpreted without extensive laboratory measurements of the magnetization of minerals and rocks under various conditions. Rock samples marked with their precise orientation are carefully selected from a geologically known rock unit and later cut into cubes along uniform directions. In the laboratory, the old astatic magnetometers for assessing the paleodirections of a sample were replaced by spinner magnetometers during the past 20 years. With these instruments the sample is quickly rotated in three different perpendicular positions. From the induced currents, the paleoinclination I_p and paleodeclination D_p can be easily obtained. A certain number of samples is usually needed to obtain reliable average values of I_p and D_p. Moreover, a stepwise alternating magnetic field up to 0.5 tesla and/or some temperature treatment must generally be applied to get rid of some additional weak magnetization from various sources. A new type of spinner magnetometer has been developed in which no replacement of the sample is required. Cryogenic magnetometers and QUID magnetometers (Fuller, 1983) with the help of microcomputers are able to increase the sensitivity and effectiveness of measurements considerably. Hence, new levels of characterizing weakly magnetized sediment layers have been reached. Within the past 20 years an increasing effort has been under way to determine not only paleodirections but also paleointensities, as based on studies of Thellier and Thellier (1959). In these studies, first performed on baked clays, the samples are heated, and the decrease in their magnetization is observed. A subsequent cooling in (various) artificial magnetic fields induces a TRM, which is compared to the measured values. Several such steps have to be performed, generally on several samples. Careful chemical and radiometric age determinations generally accompany this method. Today, even lunar samples as old as 4×10^9 years have been investigated for their paleodirections and paleointensities (Hood and Cisowski, 1983), and a decrease of the lunar field between 3.8×10^9 and 3.1×10^9 years seems evident.

All magnetic measurements performed so far have shown that the intensity of the Earth's field cannot have changed by more than a factor of two since about 3×10^9 years ago. Moreover, the data clearly support the new mobilistic view of continents as always having been in motion within the Earth's history. A reconstruction of continents for the past 200×10^6 years is shown in Fig. 3.65, which is based mainly on geomagnetic data but also contains geologic information such as dating of strata, structures, and ophiolite belts in the old compressional zones of orogenesis.

3.7.3   ASSESSMENT OF BASEMENT AND CURIE DEPTH

In order to estimate the depth to the (magnetic) basement, sedimentary rocks are generally assumed to be nonmagnetic compared to the magnetized crystalline basement. This reduces the calculation to a two-layer case, which, however, must be considered three-dimensional. In order to assess the shape and depth of a sediment basin a large amount of data must be available, preferably from aeromagnetic surveys, where large-scale anomalies should show up. There are various methods of calculation, starting from the old model calculations of bottomless, vertical-sided prisms of rectangular cross sections (Vacquier *et al.*, 1951), or cylinders of various sizes up to downward continuation of the magnetic fields of the surface (or flight level) (Grant and West, 1965). A modern treatment of field data starts with a two dimensional Fourier analysis to provide the wave-number spectra in x and y necessary for a downward continuation. After several reductions, the most important one being the so-called pole correction (a calculation of the vertical components), and some smoothing, the subsurface relief of the basement can be obtained for various sets of magnetization values. A check for consistency by recalculating the surface field may easily be performed.

The assessment of the *Curie depth* Z_c, i.e., the depth at which the temperature destroys all ferrimagnetic magnetization, is basically the same as the treatment of basement depth. Some large-scale anomalies are considered to originate from undulations of Z_c in an area. Hence, a nonmagnetized undulating half-space is considered to underlie the upper part of the magnetic basement. From two-dimensional Fourier analysis up to the check by model calculations the procedure for assessing the undulations of Z_c is the same as that for basement assessment. Problems arise from stronger magnetizations of dehydrated layers of the lower crust (see next section.)

For most continental areas Z_c is still within the crust. In the thick crust of shield areas, with their lower temperatures, it is deeper than in the warmer and thinner crusts of marginal basements and tectonically younger areas, but in most cases it is in the granulitic lower crust; see Fig. 3.75 in Section 3.8.5, which gives an idea how much Z_c may vary under different temperature

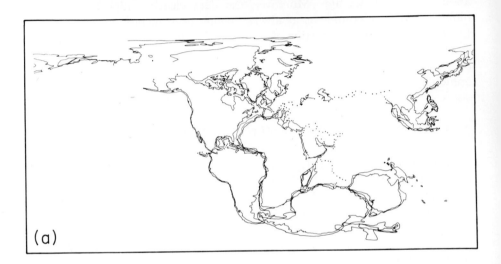

(a)

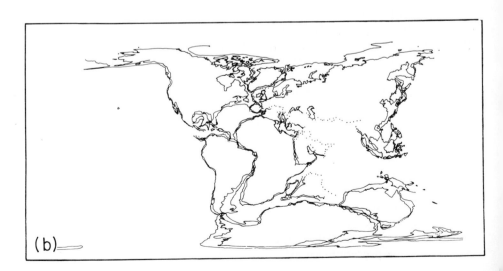

(b)

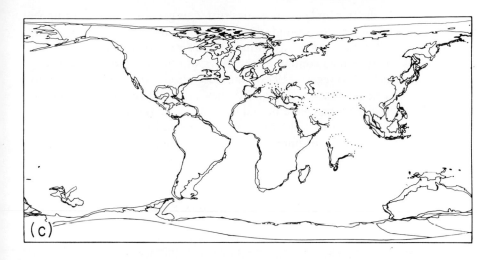

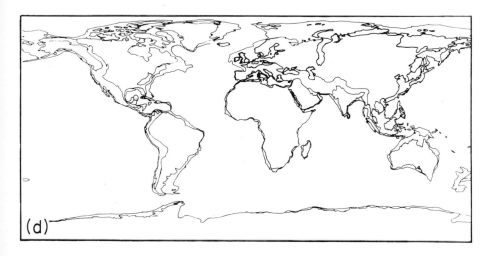

Fig. 3.65 Continental positions for the last 200 Myr from paleomagnetic data from land and magnetic lineaments at sea; (a) 200 Myr, (b)120 Myr,(c)60 Myr, and (d) present day. Redrawn after Smith *et al.* (1982). Maps © Cambridge University Press 1981, from Smith, Hurly, and Briden:Phanerozoic Palaeocontinental World Maps,German Editor: Enke Verlag, Stuttgart. Vedag, Stuttgart.

conditions. In the thin and cool oceanic crust Z_c is generally within the upper mantle, except for areas around rifts, hot spots, and spreading axes. The possibility still remains that serpentinite bodies or serpentinized ultramafic parts of the mantle may contain metal alloys whose Curie temperature may be around 1000°C (Haggerty, 1978), but these observations seem to be only of local importance. Wasilewski and Mayhew (1982) have suggested that the Moho may also be a magnetic boundary, with no magnetic signals coming from mantle sources. Certainly, these and other studies show that the Curie depths and the sources of intermediate and large-scale anomalies are still rather elusive parameters.

3.7.4   MAGNETIC ANOMALY MAPS AND CRUSTAL STRUCTURES

Large-scale crustal anomalies might be obtained from aeromagnetic maps and even from satellite magnetic field data. Actually, the successful isolation of crustal fields with amplitudes as low as 5 to 30 nT from the core fields of 30,000 to 60,000 nT gave impetus to the MAGSAT project. In general, modeling and interpretation of satellite-derived anomaly maps is still in its infancy (Langel, 1982). Figure 3.66 shows a comparison between the results of an upward continuation of aeromagnetic data to the 450-km level, the POGO satellite anomaly map at this level, and the 2° averaged MAGSAT anomaly map at an average elevation of about 350 km; all three maps show the conterminous United States with a contour interval of 2 nT (von Freese *et al.*, 1982). Clearly, the correlation of the two (very similar) satellite anomaly maps with the upward-continued magnetic anomaly map is far from perfect.

Part of the long-wavelength anomalies might come from the lower continental crust, as mentioned in the previous section. Wasilewski and Mayhew (1982) have shown that mafic granulitic rocks, whether found in xenolith suites of Precambrian terrains or outcrops, have the highest magnetization values. From these measurements it follows that a cold lower mafic–granulitic crust with $T < T_c = 560°C$ could be the most strongly magnetized crustal layer. Variations in thickness could provide a part of the long-wavelength anomalies. Also, amphibolites from the granulite–amphibolite facies transition from the exposed lower crust rocks of the Ivrea body showed high values of initial susceptibility and saturation remanence (Wasilewski and Fountain, 1982). These rocks are highly magnetized and have high Curie temperatures T_c. Although the composition of the lower crust, its temperature T, and its T_c value vary considerably from one geologic province to another (see Chapter 6), a correlation between the thickness of the lower crust and magnetic anomalies might be tried in selected areas.

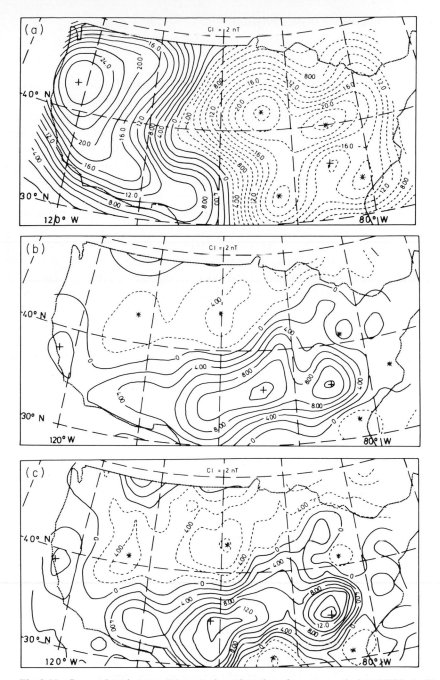

Fig. 3.66 Comparison between (a) upward continuation of aeromagnetic data (450 km), (b) POGO-satellite anomaly map (450 km), and (c) MAGSAT-satellite anomaly map (350 km). [From von Freese *et al.* (1982).]

3.8   CRUSTAL GEOTHERMICS

There is no doubt that thermal processes have played a dominant role in the origin and evolution of the Earth's interior. Assessments of crustal geotherms (temperature–depth relations) for both ancient and present-day conditions are of utmost importance for an understanding of the history and the behavior of the Earth's crust. As mentioned in Section 3.7, temperature determines the Curie depth, the depth of the brittle behavior of the crust, the seismic velocities, the maximum depth of earthquakes, even the stress drop of earthquakes, and—last but not least—volcanic activity.

It is obvious that the accuracy of temperature determinations made from the surface strongly decreases with increasing depth. The deeper we want to go, the stronger is the dependence on model assumptions. Heat conduction in the lithosphere and heat convection in the asthenosphere certainly play a role in the transmission of heat into the crust from below. Heat production is another major contributor to the temperature, and its amount as a function of depth is a critical parameter which is intimately related to the concentration of the radioactive elements uranium-235 (^{235}U), uranium-238 (^{238}U), thorium, and potassium. Fortunately, some geochemical–petrological "geothermometers" exist which provide pressure–temperature (p-T) checkpoints along a geotherm.

Present-day heat flow at the surface has at least five contributions:

(1)   heat from the Earth's accretion,
(2)   heat from core formation,
(3)   heat from tidal friction (?),
(4)   heat from surficial radiogenic sources,
(5)   heat from tectonic processes with enhanced convection.

The influence of solar radiation is negligible.

Points (1) and (2) of the preceding list provide the general global heat level for the Earth. New data from observations of the surfaces of Mercury and other planets clearly show that processes (1) and (2) are so close together in time that they cannot be separated (Meissner and Janle, 1984). From studies by Tozer (1977) and Meissner and Lange (1977) it became evident that after the formation of the core a first period of vigorous fluid convection in the mantle (including the crust) soon transformed into solid-state convection. Solid-state convection may still dictate the thermal regime of the Earth's mantle and that of Venus. Further cooling may have reached a point where the decreasing solid-state convection became so slow that it may be comparable to the heat transport of pure conduction.

While the mantle has solidified from the bottom to the top—because the adiabates are steeper than the solidus curves with depth—the outermost part of the crust/mantle system solidified from the top to the bottom because of strong radiation losses at the surface. This means that the crust and the lithosphere in formation soon cooled down below the solidus. It also means that, except for very hot areas, heat in the crust is transmitted preferably by conduction.

Point (3), tidal friction, could contribute up to 10% to the general thermal level. Certainly more than 35% of today's continental heat flow comes from the decay of radioactive elements in the crust (point 4). It is the upper part of the continental crust with its granitic–gneissic composition in which most of the radioactive elements concentrate. Uranium, thorium, and potassium belong to the "lithophile" elements. They have not only high mobility but also large ionic radii; therefore, they are not easily incorporated into the densely packed structures of the deeper layers of crust and mantle.

Finally, point (5), the heat of tectonic processes, plays a major role in rifting and spreading processes, i.e., in the high temperature of the newly formed oceanic crust. However, compressional tectonic processes like those of subduction regimes or suture zones in continents are also sources of enhanced thermal activity. Subducting slabs of the lithosphere may mobilize the continental asthenosphere, and frictional heating may also contribute to a generally higher heat flow and to all kinds of volcanic activity behind such suture zones. The result may even be the generation of a back-arc basin or at least a strong heating of the crust from below, the convectional or "plume" patterns sometimes modifying the whole crust. A consequence of point (5) is the observation that the highest continental heat flow values are found in tectonically young areas and the lowest ones in old shield areas, as shown in Fig. 3.67 (Chapman et al., 1979). This figure also demonstrates that the continental and oceanic heat flow values show a very different relation to age. See also Section 3.8.4.

A correlation can also be established between lithospheric thickness and heat flow for continental and for oceanic areas. A certain part of the correlation such as that of Fig. 3.67 comes from a continuing erosion of granitic–gneissic material of the upper crust in the old shield areas. Moreover, mountain belts show a large percentage of crustal shortening. Especially in the young mountain ranges a doubling of the upper part of the continental crust (with its high concentration of radioactive elements) is often observed, which certainly contributes to an elevated thermal regime in these young crustal sections of the continents. The coincidence of the average heat flow values of continents and ocean floors will be treated in Section 3.8.4.

An extensive treatment of geothermal topics may be found in textbooks such as those by Smith (1973), Rinehart (1980), Buntebarth (1980), and Cermak and Haenel (1982).

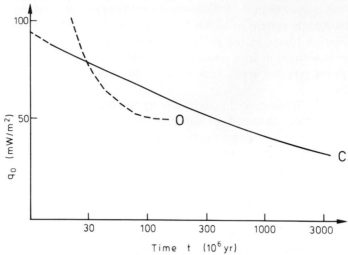

Fig. 3.67   Average relationship between surface heat flow q_0 and age of the last remarkable tectonic event in the area under consideration. Here O is the ocean floor and C the continent. Standard deviation is as large as 25 mW/m². [From Chapman *et al.* (1979).]

3.8.1   DERIVATION OF BASIC FORMULAS FOR THE ASSESSMENT OF CRUSTAL TEMPERATURES

The calculation of temperature in the crust is generally based on the assumption that heat conduction is the dominant process, which is certainly a good estimate for all regions outside active magma chambers. Some solid-state convection can be accounted for by a correction term in the heat conduction equation (Levin, 1962). The general partial differential equation of heat conduction may be written in the form

$$\rho c(\partial T/\partial t) = \nabla K \, \nabla T + K \, \nabla^2 T + A, \tag{3.68}$$

where ρ is the density, c the specific heat, T the absolute temperature, K the thermal conductivity, ∇K is div K, ∇T is grad T, $\nabla^2 T$ is div grad T, and A is the heat production.

If K is independent of the space coordinates, then $\nabla K = 0$ and Eq. (3.68) becomes

$$\rho c(\partial T/\partial t) = K \, \nabla^2 T + A. \tag{3.69}$$

If, in addition, T is stationary, $\partial T/\partial t = 0$, then Eq. (3.69) becomes

$$K\nabla^2 T + A = 0 \tag{3.70}$$

(Poisson's differential equation). If also $A = 0$, we arrive at

$$\nabla^2 T = 0 \tag{3.71}$$

(Laplace's differential equation).

Equation (3.70) is mostly used for temperature calculations in the crust, but with the simplification that T is laterally constant, i.e., that

$$(\partial T/\partial x) = (\partial T/\partial y) = (\partial^2 T/\partial x^2) = (\partial^2 T/\partial y^2) = 0. \tag{3.72}$$

In this case Eq. (3.70) becomes

$$\partial^2 T/\partial z^2 = -A/K. \tag{3.73}$$

Equation (3.73) has various simple solutions, the simplest one being that of only one layer with constant A and K, e.g., an upper part of a horizontally stratified half-space.

After a first integration Eq. (3.73) becomes

$$(\partial T/\partial z) = -(A/K)z + \text{const.} \tag{3.74}$$

The term const is obtained for $z = 0$ and is:

$$\text{const} \left(\frac{\partial T}{\partial Z}\right)_0 = \frac{q_0}{K}. \tag{3.75}$$

This follows from the definition of the surface heat flow or (more exactly) the surface heat flow density, which is:

$$q_0 = K\left(\frac{\partial T}{\partial z}\right)_0. \tag{3.76}$$

This quantity can be determined from two independent measurements at the surface: the thermal conductivity K and the temperature gradient $\partial T/\partial z$. We show in the next section how such measurements are performed and how problematic they are in practice. Returning to Eqs. (3.74) and (3.75), a second integration leads to

$$T = T_0 + (q_0/K)z - (A/2K)z^2, \tag{3.77}$$

the last term showing a deviation from linearity.

An extension of Eq. (3.77) is easily performed through model studies in which K and A for a number of horizontal layers are assumed to be known. At the boundaries of the layers T and q remain smooth. It can easily be shown that

$$T_n(z) = T_{n-1} + \frac{q_{n-1}}{K_n}(z - z_{n-1})^2 - \frac{A_n}{2K_n}(z - z_{n-1})^2 \tag{3.78}$$

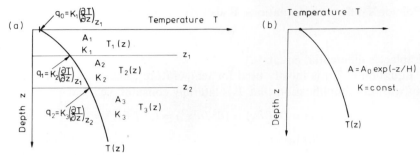

Fig. 3.68   Calculation of the temperature–depth function (geotherm) (a) assuming constant values of heat production A and conductivity K per layer and (b) assuming an exponentially decreasing heat production A with depth.

with

$$q_{n-1} = K_{n-1}\left(\frac{\partial T}{\partial z}\right)_{n-1} = q_{n-2} - A_{n-1}(z - z_{n-1}). \tag{3.79}$$

For a given layered model, T may be calculated iteratively, starting at the surface layer. Figure 3.68a shows a three-layer case.

Another assumption often used in practice is based on an exponentially decreasing heat production (Lachenbruch, 1968) (Figure 3.68b):

$$A(z) = A_0 e^{-z/H}, \tag{3.80}$$

where H is the relaxation depth, i.e., the depth where $A = A_0/e = 0.368 A_0$. Two integrations of Eq. (3.80) finally result in

$$T = T_0 + (q^*/K)z + (A_0 H^2/K)[1 - \exp(-z/H)] \tag{3.81}$$

with

$$q^* = q_0 - A_0 H. \tag{3.82}$$

The term q^* has an important application for distinguishing between different heat sources. It is based on observations that different rocks with different contents of radioactive minerals and, hence, different values of A in a certain area provide a linear relationship between q_0 and A.

Figure 3.69a shows such a relationship with the equation

$$q_0 = q^* + AH \tag{3.83}$$

in which the surface heat flow q_0 is separated into a variable "crustal" contribution AH and a second component q^* which enters the (upper) crust at depth H. The parameter q^* is fairly constant for a particular region and represents the contribution from the mantle (and the lower crust). Calculat-

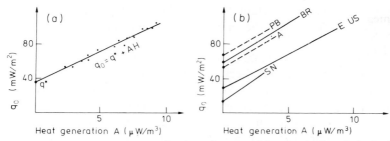

Fig. 3.69   Relationship between surface heat flow q_0 and heat production A obtained from determinations of rock samples in a certain area. (a) Principle of determining reduced heat flow q^* and (b) data from several geologic provinces; PB, Pannonian Basin; BR, Basin and Range Province; E US, eastern United States; and SN, Sierra Nevada. [From Rybach (1973).]

ing H values from the gradient of the straight line of Fig. 3.69a or b, we arrive at values of 7 to 11 km, independent of q^*. This is about the depth of the granitic–gneissic part of the crust, as indicated by seismic P-wave velocities around 6 km/s. The relation of q_0 to q^*, on the other hand, shows how big the contribution of the mantle component really is and that it seldom exceeds 30% of q_0. Figure 3.69b shows relations for various tectonic areas of the Earth's continents (Rybach, 1973, 1976).

3.8.2   METHODS OF MEASURING HEAT FLOW, THERMAL CONDUCTIVITY, AND HEAT PRODUCTION

Heat flow density $q_0 = K\, \partial T/\partial z$ can be measured on land (including lakes) and at sea. As mentioned before, two independent measurements are required: that of the temperature gradient dT/dz, to be obtained by at least two temperature measurements in a borehole or from a needle in sediments, and that of the thermal conductivity K, to be obtained preferably by laboratory data but also by a special observation of heat conduction from a heated needle in the ground (Smith, 1973). For land measurements observations in tunnels have been found to be very useful and more reliable.

The measurement of dT/dz is generally performed with thermistors along a certain depth interval, not too near the surface and not too soon after drilling or injection. The measurements themselves are simple, but in practice many corrections to the measured data must be applied, such as those for topography, possible uplift or subsidence, erosion, glaciation, and other climatic changes. Even more severe is the influence of circulating water in cracks and pores; hence, a detailed investigation of every parameter must be performed, mostly assisted by model calculations. Certainly, a group of heat flow measurements in a certain area will likely average out the more local effects, while for the others the whole tectonic and climatic history must be considered.

The measurement of thermal conductivity K is generally performed in the laboratory using of the Lees–Beck apparatus, especially if samples of solid rocks are to be investigated. These samples from a borehole are compared with a standard material of known thermal conductivity. Instead of boreholes, a thin tube with thermistors and with facilities for collecting samples is released from a ship; the tube penetrates sediments of the ocean or lakes up to 20 m, and also measures the temperatures *in situ* and obtains samples for later measurement of K in the laboratory. A further development of this technique is a needle probe. A hollow needle is inserted into sediments and heated at a known and constant rate, while the rise of its temperature—a measure of the thermal conductivity of the surrounding sediments—is measured with one or several thermistors (Langseth, 1965; Smith, 1973). Efforts are under way to extend this technique of *in situ* determination of K to solid rocks by adjusting a special needle device to very thin holes.

In practice, K does not vary as strongly as dT/dz. But in shales and some other layered rocks it is not independent of direction and must be treated as a tensor. It is generally dependent on temperature and on the confining pressure. Because it consists of two independent terms, namely the conductivity of the lattice and that related to radiation, the actual relation of K to depth Z is not unique. In general, K decreases down to a depth where the temperature is 200–400°C because of increasing confining pressure and more densely packed rock structures, while for temperatures above this range, i.e., in the lowermost continental crust, the contribution from radiation may dominate. Marine sediments generally show low K values.

Measurement of the heat production A is virtually a determination of a rock's content of the radioactive minerals. As mentioned in Section 3.8.1 and shown in Fig. 3.69 determination of the concentration of radioactive elements in collected rock samples is an important basis for assessment of the variety of heat flow values at the surface, q_0, for the "mantle component" q^*, and for the depth H for which q^* is determined. The concentration of U, Th, and K in collected samples is determined by mineralogical and physical laboratory measurements, the latter ones based heavily on mass spectrometry. In modern spectrometers gaseous as well as solid samples can be analyzed (Faure, 1977). Heat production A may then be calculated by taking into account all individual concentrations C of U, Th, and K with an empirical formula (Rybach, 1973)

$$A = (0.718C_U + 0.193C_{Th} + 0.262C_K)0.133\rho \quad \mu W/m^3 \qquad (3.84)$$

where ρ is the density in grams per cubic centimeter.

Table 3.5

Typical Abundances of U, Th, and K for Various Terrestrial Crustal and
Mantle Rocks[a]

	U (ppm)	Th (ppm)	K (%)	Heat production A (μW m^3)
Crustal rocks				
Silicic igneous rocks	4.0	16.0	3.3	2.5
Mafic igneous rocks	0.5	1.5	0.5	0.3
Shales	4.0	12.0	2.7	2.1
Carbonates	2.2	1.7	0.3	0.7
Beach sands	3.0	6.0	0.3	1.2
Possible mantle rocks				
Dunite	0.005	0.02	0.001	0.004
Eclogite	0.04	0.15	0.1	0.04
Oceanic lherzolite	0.02	0.06	0.005	0.01

[a] From Sass (1971).

Table 3.5 gives an overview of the heat production and concentration of U, Th, and K in various continental crustal and mantle rocks (Sass, 1971). In general, there is a correlation between the concentration of U, Th, and K and the amount of quartz in a rock. This observation indicates that the heat production and concentration of heat-producing elements in the lower crust and in the mantle really are small, as suggested by Table 3.5.

Other correlations may be more important for the assessment of heat production by means of geophysical observations. Buntebarth and Rybach (1981) found an empirical relationship between A and the density, and between A and the seismic velocity V_P. The physical reasons are related to the obvious correlation between the cation packing density K^* and the density, ρ. The greater ρ and K^*, the smaller is the space for the large ions of the heat-producing elements, suggesting an inverse relationship between A and ρ. As ρ increases with V_P (see Section 3.1.4), there will also be an inverse relationship between V_P and A.

Figure 3.70, (Rybach and Buntebarth, 1982) shows these relationships, which are certainly a good first-order approximation for an assessment of A, especially in the deeper layers of the crust, where samples are not so easily available. Corrections due to confining pressure may be obtained from Buntebarth (1980, 1982). A summary of data from about 30 areas for which heat flow values and seismic velocity–depth functions are available was

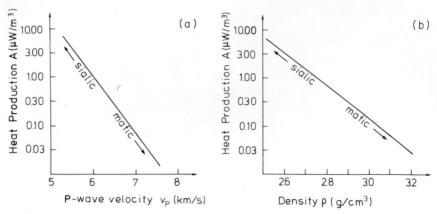

Fig. 3.70 Relationships between (a) heat production and P-wave velocity and (b) heat production and density. [From Rybach and Buntebarth (1982).]

recently provided by Stegena (1984). His relationship between heat production and velocity V_P differs slightly from those of *Fig.* 3.70a and may be written as

$$A = 2.5 \times 10^6 \exp(-2.3V_P) \qquad (3.85)$$

with an estimated error not larger than 20%.

3.8.3   ASSESSMENT OF CRUSTAL TEMPERATURES: GEOCHEMICAL METHODS

As mentioned before, there are a number of methods for calculating geotherms or certain temperature values in the pressure–temperature (p–T) field. The basis of calculations from the surface heat flow values q_0 was given in Section 3.8.1 and the determination of the necessary parameters K and A was discussed in Section 3.8.2; this section describes the basis of the physicochemical "geothermometer," while Section 3.8.4 will give a short summary of some (indirect) geophysical methods for the assessment of temperature.

The principle of geothermometers is based mainly on certain mineralogical equilibrium conditions. Some phases are in equilibrium in a limited temperature range, some in a limited p–T range. In addition, trace elements in certain crystals, the ratios of some isotopes or ions, the solution of SiO_2, and Na-, K-, and Ca-containing minerals provide some clues to the temperature. Finally, the structure of organic particles in sediments changes considerably when they are subjected to certain temperatures. In the following only the most important geothermometers will be mentioned. For detailed information see Buntebarth (1980).

For the *upper crust*, i.e., up to temperatures of about 300°C, the solution of SiO_2 in water from a spring shows a good correlation with temperature, although many corrections must be applied and no mixing of the water of a (strong) spring with other water reservoirs is permitted. The Na–K–Ca thermometer uses the ratio of various concentrations in water and seems to be more reliable than the SiO_2 method alone. Trace elements of certain salts or ores constitute a wide variety of methods for temperature assessments; examples are bromine in NaCl and KCl, and manganese and cadmium in ZnS and PbS, the latter even for $T > 300°C$. A well-established and reliable method for assessing temperatures is the optical determination of the vitrinite reflection coefficient. Its relation to the temperature gradient is based on chemical reactions of organic particles. These reactions are strongly dependent on temperature and are known as coalification, a process which is practically irreversible. During subsidence of sediment basins the organic matter comes under increasing temperatures and changes its structure with temperature and with time. Knowing the age and the history of certain areas, these structural changes of organic matter are represented by the vitrinite reflection coefficient R_m within the wavelength of visible light. Several empirical formulas have been established based on appropriate calibrations between R_m and dT/dz data from boreholes (Buntebarth, 1980).

For deeper layers and higher temperatures the garnet–pyroxene geothermometer plays a dominant role. Based on laboratory investigation at elevated p–T conditions, the equilibrium temperatures are determined and several mineral phases are investigated; some reactions are dependent only on T, others on p and T. The distribution ratio of Fe to Mg in garnet and pyroxene, K_d, strongly decreases with increasing T and weakly increases with increasing P:

$$K_d = (Fe^{2+}/Mg^{2+})^{Gn}/(Fe^{2+}/Mg^{2+})^{Py} \qquad (3.86)$$

where Gn denotes garnet and Py pyroxene. Its relationship to p and T is approximately (Raheim and Green, 1974)

$$K_d = (C_1/T) + (C_2/RT)(p - p_0) + C_3 \qquad (3.87)$$

where C_1, C_2, and C_3 are constants, R is the gas constant, and p is the confining pressure. Boyd (1973) estimates an accuracy of about 5% in T and 20% in p for the garnet–pyroxene thermometer. These and other geochemical methods employing samples of surface rocks and xenoliths provide important p–T checkpoints for the construction of geotherms.

3.8.4 ASSESSMENT OF CRUSTAL TEMPERATURES: GEOPHYSICAL METHODS

Some of the geophysical methods for the assessment of crustal temperatures have already been mentioned; the influence of temperature on the
electric conductivity was discussed in Section 3.5, on the maximum possible
stresses and maximum depth of earthquakes in Section 3.6, and on the Curie
depth in Section 3.7. In addition, density and seismic velocities are strongly
related to temperature. All these relations are not too useful for a reliable
assessment of T and often only provide some limits. In the laboratory,
reliable relationships exist between T and the electric conductivity, between T
and the density, and between T and the velocity for a certain rock unit; but
under *in situ* conditions with unknown rock composition, water saturation,
stress, and crack density, all these relationships seem dubious. Limiting
values such as $T = 450°C \pm 70°$ have been obtained for the maximum depth
of intracontinental crustal strike–slip earthquakes (Strehlau, 1986, in preparation). A good indication of the general temperature level in the crust is the
seismic velocity. From a number of laboratory experiments it is known that
seismic velocities increase with increasing confining pressure and decrease
with increasing temperature (Kern, 1978, 1982b; Meissner and Fakhimi,
1977; Christensen 1965, 1979). Geotherms in warm areas may well show a

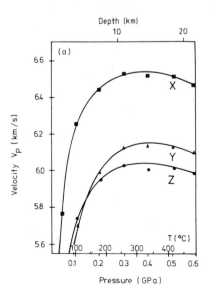

Fig. 3.71 Example of V_p reversals along a geotherm from three-dimensional ultrasonic
experiments. (a) Gneiss and (b) granite. Note also the strong anisotropy of both samples. [From
Meissner and Fakhimi (1977).]

dominance of the dV/dT term over the dV/dp term, and a low-velocity layer is the consequence, provided the rock composition and the pore pressure do not change significantly (see Section 6.6.2). Such a reversal will certainly not take place for the first 200 MPa or down to 7 km depth, where the sensitivity of V versus p and hence dV/dp is very great because of the closing of microcracks in the rocks. However, below these depths low-velocity layers (LVLs) in the crust may appear, and they are observed by seismic refraction (and reflection) studies.

A reversal of velocities with depth will take place if

$$\frac{dV}{dz} = \left(\frac{\partial V}{\partial p}\right)_T \frac{dp}{dz} + \left(\frac{\partial V}{\partial T}\right)_p \frac{dT}{dz} < 0. \tag{3.88}$$

Solving for dT/dz, we arrive at

$$\frac{dT}{dz} > -\frac{(\partial V/\partial p)_T\, dp/dz}{(\partial V/\partial T)_p} \tag{3.89}$$

Values of $\partial V/\partial T$ and $\partial V/\partial p$ are listed in Christensen (1979) and may also be obtained from Kern (1982a). In particular, quartzite and quartz-containing rocks such as most granites and gneisses show large values of $\partial V/\partial T$ and provide a plausible explanation for the large low-velocity zones in the middle

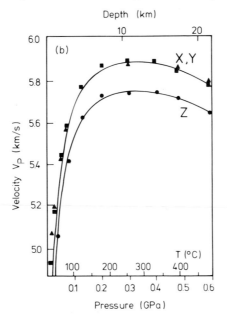

Fig. 3.71 (*Continued*)

and upper crusts of young mountain ranges and even those of still rather warm Mesozoic continents. See also Section 6.6.2.

Hence, the beginning of velocity reversals as indicated by the breakup of the P_g arrivals may be considered as indicative of a certain T value along the geotherm at which dV/dT is beginning to dominate. Figure 3.71 gives an example of such a reversal from laboratory measurements.

In addition to these estimations on a regional scale, an assessment of anomalous temperatures can also be performed by observing seismic velocity changes laterally, e.g., along seismic profiles across a geothermal anomaly. These calculations have the advantage that they may be obtained from levels of constant confining pressure. Based on laboratory measurements by Kern and Richter (1981), the sensitivity of $(dV/dT)_p$ has been plotted as a function of temperature T and temperature deviation ΔT (see Fig. 3.72 and 3.73). These values have been used for the assessment of anomalous temperatures in the Urach geothermal area (Bartelsen *et al.*, 1982). An extensive geophysical survey including S-wave and P-wave velocity studies has been performed in the Yellowstone area. See special issue of the *Journal of Geophysical Research* **87**, B4 (1982) and Section 6.6.1.

Another seismic parameter that has been extensively studied in recent years is the *Q factor* or the specific attenuation Q^{-1}. It also provides a relation to temperature, and an especially strong dependence on T—similar to that of V_S—is observed at temperatures approaching the solidus. Although the absolute value of dQ^{-1}/dT is larger than dV/dT, the determination of

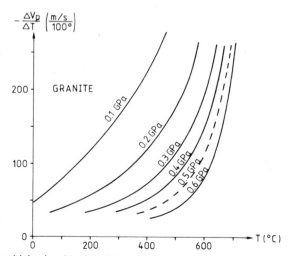

Fig. 3.72   Sensitivity, i.e., $\Delta V_p/\Delta T$, of V_p velocities with temperature at selected pressure levels based on measurements by Kern and Richter (1981).

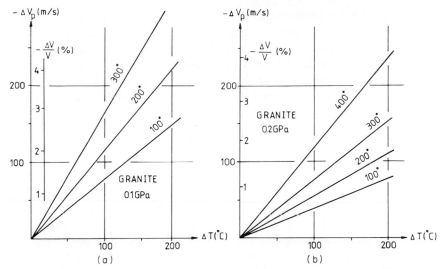

Fig. 3.73 Change in V_P velocity as a function of the (lateral) temperature difference at constant confining pressure (a) at 1 MPa and (b) at 2 MPa. [From Bartelsen *et al.* (1982).]

Q^{-1} cannot yet be performed with the necessary accuracy and may only be used as an additional method for temperature assessment. In general, the combination of surface heat flow values with all geochemical and geophysical data must be performed to obtain the best geotherm possible.

3.8.5 HEAT FLOW VALUES AND GEOTHERMS IN SELECTED GEOLOGIC AREAS

Starting with the largest units, i.e., the continents and the ocean floors, it has puzzled scientists for at least two decades why the average heat flow in the oceans turned out to be nearly identical to that of the continents. If the bulk of the continental heat flow came from the concentration of heat-producing elements in granites and gneisses, then why did oceanic basins without granites and gneisses have the same heat flow?

With the growing understanding of sea floor spreading and the differentiation between crustal (q_0) and mantle (q^*) heat flow, the answer became rather simple: heat flow from the oceanic mantle is much higher than that of continents. Heat below the ridges and in their vicinity is transported by advection and convection from below. The young oceanic ridges, where upwelling of hot material forms the new ocean floor, show the highest heat flow values, although they are not quite as high as predicted by simple thermal models because deeply circulating water leads to strong distortions and large scatter of the data. The spreading oceanic lithosphere cools and assumes the lowest heat flow values in the trenches, especially in those which are covered by thick sediments. The subducting lithosphere takes the low temperatures with it beneath the continents, where it may mobilize the asthenosphere. Table 3.6 gives an overview of heat flow values in continents (Chapman, 1985; Sass, 1971). The table gives an impression of the large deviations and the correlation between q_0 and age, which has already been shown in Fig. 3.67 and shows the average q values and the large deviations.

In the continents a number of heat flow provinces can be defined, and their boundaries can often be correlated with physiographic and tectonic boundaries. Figure 3.74 shows q_0 values in the western United States (Roy *et al.*, 1972). The high heat flow in the Basin and Range Province is connected with extensional features, with a rather shallow Moho and a possible zone of partial melting below a thin lithosphere, while the old Colorado Plateau is much colder and has a deeper Moho.

Table 3.6

Average Heat Flow for Continents[a] and Oceans[b] According to Various Geological Units

Tectonic region	Number of data	Mean heat flow (mW/m^2)	Standard deviation
All continents	1753	61	31
Archean	136	41	10.8
Proterozoic	78	51	20.6
Phanerozoic (non-orogenic)	265	54	20.3
Early Paleozoic	88	52	16.7
Late Paleozoic	514	61	18.3
Mesozoic	85	73	29.1
Tertiary	587	71	36.7
All Oceans	2468	61.4	33
Trenches	78	49	29
Basins	683	53	22
Margins	642	75	39
Mid-oceanic ridges	1065	80	62 (!)

[a] From Chapman (1985).
[b] From Sass (1971).

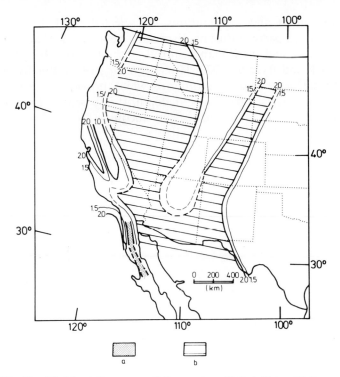

Fig. 3.74 Simplified heat flow map of the western United States. Units are in HFU. (a) Values < HFU = 42 mW/m² and (b) values > 2 HFU = 84 mW/m². [After Roy *et al.* (1972).]

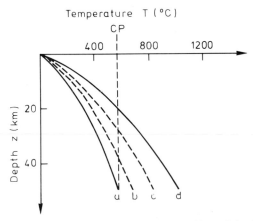

Fig. 3.75 Four average crustal geotherms and the position of the Curie point (CP). (a) q_0 = 1.2 HFU = 50 mW/m², (b) q_0 = 1.5 HFU = 63 mW/m², (c) q_0 = 1.7 HFU = 71 mW/m², and (d) q_0 = 2.0 HFU = 82 mW/m².

Heat flow maps and temperatures for selected depth levels are also available for large parts of Europe. Especially high q^* values are seen in the Pannonian Basin and the Italian peninsula (see Stegena, 1982a,b; Rybach, 1973; Cermak and Rybach, 1979; Horwath *et al.*, 1981). Finally, Fig. 3.75 shows four calculated crustal geotherms which provide some idea of the approximate average temperatures at depth. It is based on Lachenbruch's approximation [Eq. (3.81)], which seems adequate if no additional parameters such as A, H, V_p, and ρ are available.

Chapter 4 | Contributions from Laboratory Experiments

4.0 INTRODUCTION

Laboratory experiments on seismic velocities, densities, resistivities, and fracture and creep have contributed greatly to our understanding of crustal structure and evolution. For the interpretation of *in situ* seismic velocities and possibly of seismic attenuation a knowledge of the parameters in appropriate rock units must be obtained under controlled pressure–temperature (p-T) conditions. The same reasoning can be extended to studies of density, electric resistivity, or thermal conductivity. Without knowing the variability of these parameters for various rock types under controlled conditions, no definite answer as to the state and the range of possible *in situ* values can be given.

Several laboratory experiments, like those on creep and fracture, suffer from limited sample size and time. Creep rate values may differ by a factor of 10^{10} from *in situ* conditions. Large-scale fault zones can hardly be modeled in the laboratory. The migration and pore pressure of H_2O or CO_2 present further problems. Nevertheless, some implications on possible fracture mechanism, creep laws, viscosity, and stresses may be obtained. In Chapter 3 various laboratory measurements were mentioned which were directly related to the interpretation of field measurements. In this chapter a systematic treatment of geophysical laboratory investigations relevant to crustal rocks will be given.

4.1   SEISMIC P- AND S-WAVE EXPERIMENTS

Because seismic velocities as a function of crustal depth can be determined with greater accuracy than any other physical parameter, measurements of seismic velocities on selected rock samples is one of the most important contributions of laboratory experiments. It is the only reliable basis for an assessment of the kind of rocks present in the middle and lower crust, where no drilling results are available. The high frequencies generally used in the pulse transmission technique provide measurements of the necessary accuracy, even at small sample sizes, and do not show a considerable deviation from velocities in the lower frequency range, at least not for solid rocks.

4.1.1   MEASUREMENT TECHNIQUE

The velocities of P and S waves are measured by three different methods (International Society of Rock Mechanics, 1977):

(1)   high-frequency ultrasonic pulse techniques,
(2)   low-frequency ultrasonic pulse techniques,
(3)   the resonant method.

Method 1 works in the frequency range of 100 kHz to 2 MHz with $10-10^3$ repetitions/s. Various piezoelectric ceramics in the form of plates or disks are used for transmission and reception of elastic pulses. Different piezotransducers are used for P and S waves. Various filters, preamplifiers, an oscilloscope, and an electronic counter are prerequisites.

Method 2 works in the frequency range of 2–30 kHz with 10–100 repetitions/s. Transducers may be similar to those of method 1 but may also consist of magnetostrictive elements that are able to generate (and receive) high-amplitude pulses. The other prerequisites are similar to those of method 1.

Method 3 uses a sine wave generator with a frequency range of 1–1000 kHz and pulse voltage as in methods 1 and 2. In contrast to methods 1 and 2, method 3 determines the resonance frequency of longitudinal and torsional vibrations of cylindrical specimens. The bar wave velocity is calculated from

$$V = 2lf_0, \tag{4.1}$$

where l is the length of the bar and f_0 the resonance frequency of the zero mode. The resonant bar method is frequently used for studies of seismic absorption (Section 4.1.5).

Methods 1 and 2 are used mostly in the pulse transmission technique, where the velocities V_P and V_S are calculated from the travel time t:

$$V_P = s/t_P \tag{4.2}$$

$$V_S = s/t_S \tag{4.3}$$

with s the distance traveled. The transmission technique is sometimes modified by using different positions between receiver and transmitter along the sample to imitate "seismic profiling." Such a system is often used for quite large artificial models in order to study details of wave propagation in certain media and at certain boundaries. Instead of the transmission technique, a reflection technique for pulses may be applied for special problems with the receiver near the transmitter. In general, S-wave arrivals are more difficult to observe than P waves because of P-wave noise, instrument noise, etc. In order to measure seismic velocities at high confining pressure and high temperature a considerable effort must be made if pressures up to 1 GPa ($z = 30$ km) and temperatures up to 700°C are to be achieved. The measurement of seismic anisotropy requires a three-dimensional apparatus with six pistons and a very refined heating (and cooling) system. Figure 4.1 shows a cross-section through such a three-dimensional, 200-ton cubic pressure apparatus (Kern, 1982b), which works with 42-mm cubes and transducers of 2-MHz frequency.

4.1.2 DATA FROM V_P AND V_S MEASUREMENTS

In order to determine seismic velocity with carefully selected rock samples, three kinds of measurements may be carried out:

(1) measurements at selected (constant) temperature levels for varying confining pressures p,
(2) measurements at selected (constant) pressure levels for varying temperatures T,
(3) measurements at varying p-T conditions along a given p-T geotherm.

After the pioneering work of Birch (1960, 1961), using P waves and method 1 at room temperature, different schools were established in the United States, e.g., Christensen (1965, 1966, 1974) and Manghnani et al. (1974) and their groups, and in Europe, e.g., Kern (1978, 1979, 1982b), some of them using all three methods mentioned earlier. Some data obtained with the 200-ton cubic apparatus shown in Fig. 4.1 are presented here. Figure 4.2 shows measurements of types 1 and 2 (Kern, 1978) for P waves. The pressure effect is especially strong for pressures below 0.2 GPa, where several kinds of micro-cracks anneal or close and pore space is reduced (Kern and Fakhimi, 1975). At $p \gtrsim 200$ MPa the pressure–velocity relationship tends to acquire a linear slope, although a limited amount of pores might still stay open down to 1 GPa. The effect of temperature is especially strong in quartz-containing rocks like the granites and granulites. The anomalous and nonlinear behavior

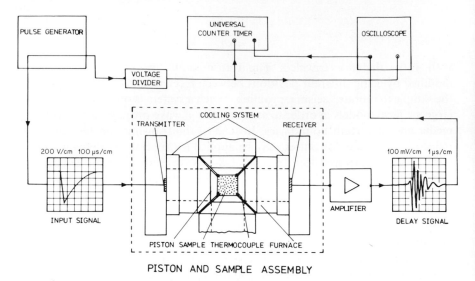

Fig. 4.1   Cross-section through high-p, high-T apparatus. [From Kern (1982a).]

of quartz-containing rocks may be responsible for many crustal low-velocity zones, although the minimum shown in the figures will hardly be reached (with the exception of the vicinity of active volcanoes). Quartz-free rocks like amphibolite and peridotite have a much smaller temperature dependence.

The general decrease of velocity with temperature is a result of thermal expansion of the minerals and the opening of grain boundary cracks. Since quartz and feldspars have the largest thermal expansion coefficients (and a large anisotropy), they show the strongest velocity effects. Comparing P- and S-wave measurements it seems surprising that the temperature effect on the velocity is smaller in S than in P for quartz-containing rocks. Figure 4.3 shows such data that results in very low V_p/V_s ratios and hence low Poisson ratios. This effect may be related to a "dilatant" opening of certain microcracks, which might affect V_P more strongly than V_S. Dilatancy and its impact on V_P and V_S will be discussed in Section 4.3.1.

Very interesting effects which might be of great importance for crustal rocks are dehydration effects, i.e., the release of structurally bound water, mostly connected with an increase of pore pressure. Part or all of the water in basalts, metagraywacke, and metabasalts is released at temperatures between

---

Fig. 4.2   Compressional velocity V_P in granite as a function of (a) pressure p and (b) temperature T with T as parameter in (a) and p as parameter in (b). The dashed lines in (a) were constructed from (b); ●, 0.1 GPa; △, 0.2 GPa; ▲, 0.3 Gpa; ○, 0.4 GPa; ■, 0.5 GPa in (a) and 0.6 GPa in (b). [From Kern (1978).]

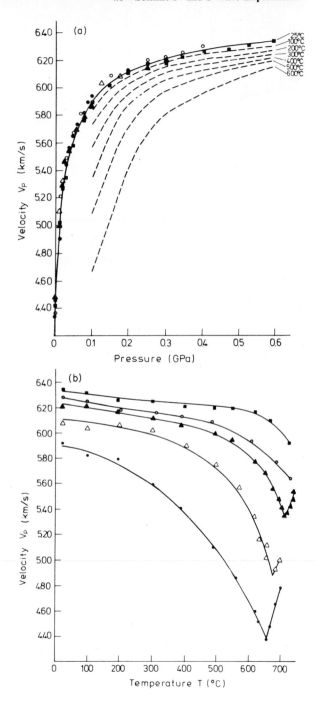

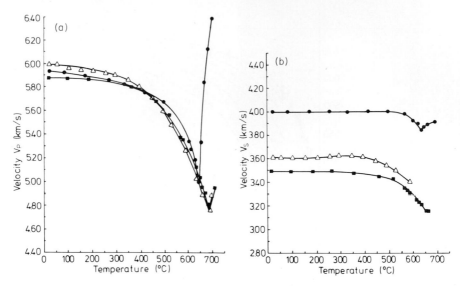

Fig. 4.3   (a) V_P and (b) V_S as a function of T at 0.2 GPa for three rock samples; ●, quartzite; △, granulite; and ■, granite. [From Kern (1978).]

350 and 530°C. Such dehydration reactions produce solid–fluid systems; the effective pressure (lithostatic pressure minus pore pressure) decreases, cracks enlarge or amplify, and V_P and V_S decrease nearly irreversibly, as shown in Fig. 4.4 (Kern, 1982a). Although the process is certainly weaker in nature, where pore pressure may be slowly released during a temperature increase, the dehydration might account for a number of low-velocity layers in V_P and V_S in warm crusts and also in basaltic layers of oceanic crusts. In general, there is no doubt that the additional use of V_S in such experiments and the establishment of a data bank with V_P/V_S and the Poisson constant σ give important new information for the estimation of crustal composition.

Another important topic of investigation is the behavior of V_P at the transition from the solid to the molten stage of rocks. New data on this transition were recently published by Khitarov et al. (1983). Figure 4.5 gives a picture of the V_P behavior with its gradual decrease during the melting. It is interesting to note that for basaltic material (curves 1–7), for the olivine basalt (curve 8), and for peridotite (curves 9 and 10) the strong velocity decrease covers a temperature range of about 200°C during which melting (from the solidus to the liquidus) takes place. The "premelting" behavior of V_P shows a decrease of about 20% for the basaltic samples between 0 and 800°C and a nearly linear slope of about 2% in V_P for $\Delta T = 100°$C. This is much more than the 0.5–1% of V_P for $\Delta T = 100°$C for the granitic sample of Fig. 4.2a and seems excessive compared to similar studies.

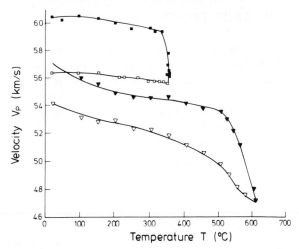

Fig. 4.4 V_P as a function of T for a tholeitic (squares) and a graywacke (triangles) sample during dehydration processes.[From Kern (1982a).]

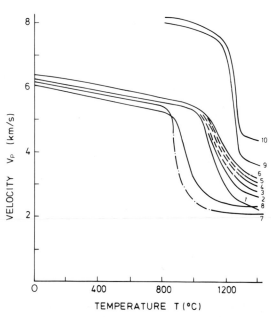

Fig. 4.5 V_P behavior during melting for basaltic (1–7), olivine (8), and peridotitic (9, 10) samples. [From Khitarov *et al.* (1983).]

4.1.3   P-WAVE ANISOTROPY

The anisotropy behavior of seismic waves can also be used for the characterization of rock units. Seismic anisotropy means that seismic velocity (V_P or V_S) depends on the direction of wave propagation; in general $V_x \neq V_y \neq V_z$. Anisotropy should not be mistaken for inhomogeneity, where seismic velocity is a function of the space coordinates x, y, and z; in general, $V(z)$ dominates $V(x)$ and $V(y)$ in crustal studies. In the laboraory, so far only P-wave measurements have been carried out under high p–high T conditions. Mantle rocks like dunites and peridotites have been extensively investigated (Christensen and Ramananantoandro, 1971; Manghnani et al., 1974; Christensen, 1974; Kern, 1978; Kern and Richter, 1981). These investigations were initiated after the detection of a surprisingly strong lateral P-wave anisotropy in the oceanic mantle of the Pacific in 1968 (Morris et al., 1969; Keen and Barett, 1971). Not only the oceanic but also the continental mantle has received greater attention, and (horizontal) P-wave anisotropy has been reported by Bamford (1977). The continental crust should also contain layers and rock units which might be anisotropic. In the early work of Meissner (1967) in the field a weak anisotropy was found for the bulk of the crust while the deepest crustal layers with their strong lamellation exhibited a possible anisotropy of up to 15%. Laboratory investigations of anisotropy of crustal rocks have been reported by Kern and Fakhimi (1975), Meissner and Fakhimi (1977), Kern and Richter (1981), and Kern (1982a). All these measurements were performed with the same 200-ton cubic pressure apparatus shown in Fig. 4.1. Figure 3.75 has already shown an example of such measurements, some of them—as in the example—along a geotherm (method 3), some at constant pressure or temperature. Nearly all crustal rocks showed an anisotropy in these experiments. A three-dimensional anisotropy may be defined by K factors. There is first

$$K_{max} = [(V_{max} - V_{min})/V_{min}]100 \qquad (4.4)$$

in percent, the maximum value of anisotropy, which is nearly always identical to

$$K_M = [(V_{\parallel}^{max} - V_{\perp})/V_{\perp}]100 \qquad (4.5)$$

in percent. Figure 4.6 (Kern, 1982a) shows a kind of summary of K_{max} values as a function of pressure and temperature. It is surprising to see that rocks supposed to occur in at least some continental crusts, like serpentinite, amphibolite, and gneiss, show stronger anisotropies than do peridotites. Granites and granulites, on the other hand, as well as gabbros and metagabbros have very low values and might be below the threshold of seismic detection.

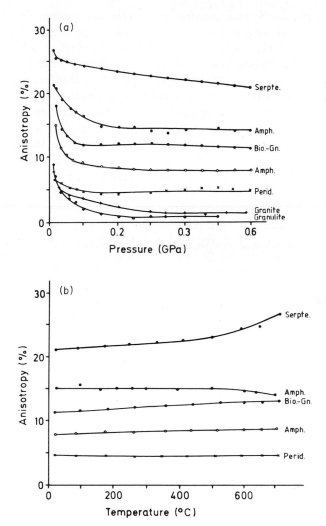

Fig. 4.6   Anisotropy values of V_P in different rock samples (a) as a function of P and (b) as a function of T. [From Kern (1982a).]

Regarding such field observations, it should be stated that the control of velocities for the lowermost crust is certainly poorer than that for the uppermost mantle, where generally clear P_n waves travel. P_n waves can often be followed over more than 200 km and a crossover with profiles of other directions often takes place, whereas $P_m P$ phases are limited to only some tens of kilometers during their penetration into the lower crust. The P_g wave, on the other hand, also offers good control of anisotropy, although generally only for the upper 10 km of the crystalline basin (see Section 3.4.3). So far, however, no anisotropy has been detected, which is possibly not so surprising in view of the small K_{max} values (and the still smaller horizontal anisotropy of granites). Moreover, the anisotropy measured in small samples does not necessarily represent that found in nature, where small-scale and large-scale heterogeneities may be present, especially in the tectonized upper crust.

On the other hand, there are processes which may enlarge the anisotropy values measured in nature. Layers may develop a preferred orientation of "velocity elements" smaller than the predominant seismic wavelength. Such elements may be minerals (to start with the smallest particles), as seems to be the case in the strong anisotropy of the uppermost oceanic mantle. Here, specific processes near the spreading centers with their high-temperature creep processes create a preferred and possibly unique orientation of the highly anisotropic olivine crystals in the peridotitic mantle rocks (Meissner and Flueh, 1979). Such a process may also take place at continental rift zones. In these cases the strong anisotropy of peridotite as well as that of serpentinite and amphibolite as measured in the laboratory is an effect of a preferred lattice orientation of major mineral phases and the anisotropy of mineral constituents.

Other velocity elements are provided by cracks and fissures, especially in the upper part of the crust, often following a preferred direction. A layering of sediments and metasediments where the preferred layer's thickness Δh is smaller than the dominant wavelength λ has also been shown to cause strong anisotropies, which may reach K values up to 20% in some shales and sandstones. Finally, crystallization seams of repeated intrusions almost certainly are strongly anisotropic if $\Delta h < \lambda$. An introduction to problems of anisotropy and a list of references to a rigorous mathematical treatment are given in Meissner and Fakhimi (1977).

The different kinds of anisotropy in nature have not yet been thoroughly investigated. Orientation of three-dimensional anisotropic minerals or mineral constituents will create a three-dimensional anisotropy as found in laboratory measurements (see Fig. 4.5 and 4.6). Hence, velocities will also be different in horizontal directions and could be detected by two, or preferably more, refraction or wide-angle profiles, collecting horizontal velocities of P_g, P_mP, or P_n waves. In addition, this kind of anisotropy usually exhibits the smallest velocity in the vertical direction, which can in principle be measured by near-vertical, i.e., steep-angle reflection work (or by borehole logging or vertical seismic profiling (VSP) in a deep borehole).

Layering will generally create only a "uniaxial" anisotropy. Such a medium is also described as "transversely isotropic", where velocity in only one direction, generally the vertical, perpendicular to the layering, deviates from the others. In sedimentary layering this is always the smallest velocity, because seismic rays have to travel through *all* individual minilayers quasi-perpendicularly. If the layer thickness Δh increases and becomes approximately equal to λ, guided waves in the horizontal will result. A true uniaxial anisotropy can be observed only through a careful analysis of extensive near-vertical and wide-angle reflection work, comparing interval velocities with horizontally traveling waves at the same depth interval. At present such studies are beginning. However, based on laboratory measurements, anisotropy—like attenuation—may well develop into an additional useful parameter for the description of a crustal rock unit.

One group of rocks which are serious candidates for the material in the lower crust are the *granulites*. They may be created from sedimentary sequences being transported down to great crustal depth in collision zones, thereby being transformed into high-density metasediments and finally into dehydrated sialic granulites. Such granulites cannot be distinguished from gabbros on the basis of seismic velocities; both rocks exhibit velocity ranges between 6.7 and 7.4 km/s. There is some hope, however, that discrimination will be possible on the basis of anisotropy, such layered granulites being highly anisotropic compared to gabbros. Layering and an anisotropic lamellation may also be created by the residues of melting processes or large intrusions, and certainly many more studies of anisotropy are essential.

4.2 ATTENUATION EXPERIMENTS

To the seismologist, attenuation of seismic waves is part of the seismic methods. The reason it has not been included in the section on P- and S-wave experiments is the more general use and wider scope of "attenuation" as an anelastic process. It involves different damping mechanisms and might be an even more elusive parameter than anisotropy. Nevertheless, laboratory measurements provide the frame of values which geophysicists should look for in field records. The main problem in the determination of the specific attenuation Q^{-1} lies in the fact that the true loss mechanism is difficult to separate out from other and often stronger damping mechanisms, such as geometric spreading, reflections, transmissions, and scattering. For the near future it might be sufficient to use Q^{-1} only as a rough description of major crustal units, where internal reflection, transmission, and scattering losses may all still be incorporated in a rough "effective Q" or effective Q^{-1}.

4.2.1   DEFINITION OF ATTENUATION

For simplicity, in the following the specific attenuation Q^{-1} or the "attenuation quality factor Q" will always be used. The most important definitions of Q^{-1} are:

$$Q^{-1} = \Delta E/2\pi E_{\max}, \tag{4.6}$$

where $E_{\max}$ is the maximum elastic energy stored during a cycle of loading, e.g., at the passage of a seismic wave, and ΔE is the energy loss during a loading cycle;

$$Q^{-1} = M_I(\omega)/M_R(\omega) = \tan \phi, \tag{4.7}$$

where $M(\omega) = M_R(\omega) + iM_I(\omega)$ is the complex modulus, ϕ is the loss angle or phase difference between stress and strain, and the indices I and R are the imaginary and real parts of the modulus;

$$Q^{-1} = \frac{\alpha V}{\pi f - (\alpha^2 V^2/4\pi f)}, \tag{4.8}$$

which for $Q \geq 100$ gives $Q^{-1} = \alpha V/\pi f$, with α the absorption coefficient, f the frequency, and V the wave velocity; i.e.,

$$\alpha = -\frac{1}{A(x)}\frac{dA(x)}{dx} = -\frac{d}{dx}[\ln A(x)] \tag{4.9}$$

with $A(x)$ the normalized signal amplitude at distance x; and

$$Q^{-1} = \Delta\omega/\omega = \Delta f/f. \tag{4.10}$$

For further definitions see Johnston and Toksöz (1981) and Toksöz and Johnston (1982). It also seems useful to relate Q^{-1} to the behavior of a "standard linear solid." This is a most general combination of linear dashpots and springs and describes the (linear) response of such a system as a function of frequency (Liu *et al.*, 1976):

$$Q^{-1} = \frac{M_I}{M_R} = \tan \phi = \left(\sqrt{\frac{M_u}{M_r}} - \sqrt{\frac{M_r}{M_u}}\right)\frac{\omega\tau}{1 + \omega^2\tau^2} = \tilde{M}\frac{\omega\tau}{1 + \omega^2\tau^2}, \tag{4.11}$$

where M_u is the unrelaxed modulus or modulus in the high-f range and M_r the complex, relaxed modulus. The term $\omega\tau/(1 + \omega^2\tau^2)$ describes a "Debye peak" in the frequency domain with τ the relaxation time.

The "relaxation peak" at $\omega\tau = 1$ is a consequence of linear models and is a well-known physical phenomenon. It has been observed by some seismological methods, by quite a number of laboratory measurements, and recently also by some field observations (Theilen, 1982; Raikes and White, 1984). In general, however, seismological methods and especially those of applied seismics find a frequency-independent Q or a very weak frequency dependence.

In order to account for such observations and in view of the theory that a number of different damping mechanisms and a number of different layers and p–T conditions are all present along a seismic ray path, a kind of bandpass model was developed by Anderson and Minster (1979, 1980) for a certain frequency range which is caused by the superposition of many, slightly different, relaxation peaks. i.e.,

$$Q^{-1} = \tilde{M} \int_{-\infty}^{+\infty} D(\tau) \frac{\omega\tau}{1 + \omega^2\tau^2} d\tau \qquad (4.12)$$

with the distribution function

$$D(\tau) = \beta\tau^{\beta-1}/\tau_2^\beta - \tau_1^\beta \qquad \text{for} \quad \tau_1 \ll \tau \ll \tau_2, \qquad (4.13)$$

for $\beta = 1$ one relaxation peak and for $\beta = 0$ a nearly horizontal plateau between τ_1 and τ_2 will appear. Both cases are shown in Fig. 4.7.

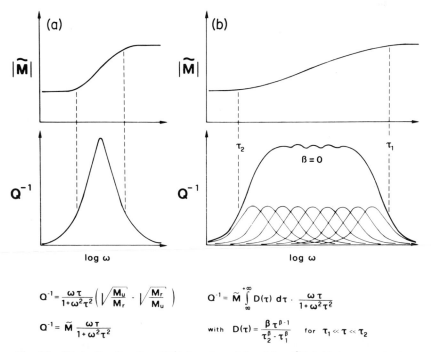

Fig. 4.7 Dissipation function $Q^{-1}(\omega)$ and modulus defect $\tilde{M}(\omega)$ (a) for a standard linear solid and (b) for an absorption band model.

4.2.2   RELATIONSHIP BETWEEN Q_P AND Q_S MEASUREMENTS

From the independent use of P and S waves in laboratory and field measurements additional information can be obtained on the mechanism of Q^{-1}, especially for an assessment of whether the predominant damping mechanism is in shear or compression.

Equation (4.7) may be split into a P- and an S-wave term:

$$Q_P^{-1} = \frac{k_I + \frac{4}{3}\mu_I}{k_R + \frac{4}{3}\mu_R} = \frac{k_I + \frac{4}{3}\mu_I}{\rho V_P^2} \quad \text{for   P waves,} \quad (4.14a)$$

approximating $V_P^2 = (k_R + \frac{4}{3}\mu_R)/\rho$ and with k the compressional and μ the shear modulus, and

$$Q_S^{-1} = \frac{\mu_I}{\mu_R} = Q_\mu^{-1} = \frac{\mu_I}{\rho V_S^2} \quad \text{for   S waves,} \quad (4.14b)$$

approximating $V_S^2 = \mu_R/\rho$. The Q^{-1} for pure losses in compression may be expressed by

$$Q_k^{-1} = \frac{k_I}{k_R} = \frac{k_I}{\rho(V_P^2 - \frac{4}{3}V_S^2)} = \frac{k_I}{\rho V_k^2}, \quad (4.15)$$

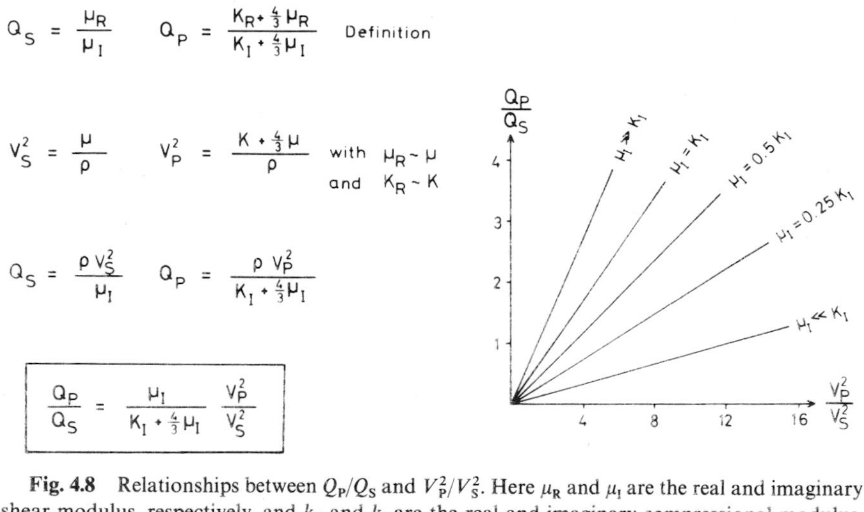

$$Q_S = \frac{\mu_R}{\mu_I} \qquad Q_P = \frac{k_R + \frac{4}{3}\mu_R}{k_I + \frac{4}{3}\mu_I} \qquad \text{Definition}$$

$$V_S^2 = \frac{\mu}{\rho} \qquad V_P^2 = \frac{k + \frac{4}{3}\mu}{\rho} \qquad \text{with } \mu_R \sim \mu \text{ and } k_R \sim k$$

$$Q_S = \frac{\rho V_S^2}{\mu_I} \qquad Q_P = \frac{\rho V_P^2}{k_I + \frac{4}{3}\mu_I}$$

$$\boxed{\frac{Q_P}{Q_S} = \frac{\mu_I}{k_I + \frac{4}{3}\mu_I} \frac{V_P^2}{V_S^2}}$$

Fig. 4.8   Relationships between Q_P/Q_S and V_P^2/V_S^2. Here μ_R and μ_I are the real and imaginary shear modulus, respectively, and k_R and k_I are the real and imaginary compressional modulus, respectively. [Adapted from Winkler and Nur (1982).]

approximating $V_k^2 = V_P^2 - \frac{4}{3}V_S^2 = k_R/P$. Dividing Eq. (4.14b) by Eq. (4.14a) we obtain

$$\frac{Q_P}{Q_S} = \frac{\mu_1}{k_1 + \frac{4}{3}\mu_1}\frac{V_P^2}{V_S^2}. \tag{4.16}$$

The approximations for Eqs. (4.14a) and (4.14b) are correct to within 1% for $Q > 5$ as derived from model calculations. The relation (4.16) goes back to Winkler and Nur (1982) and is illustrated in Fig. 4.8. Fully and partially saturated sediments show Q_P/Q_S values around 1 and a dominance of losses in compression, i.e. $k_1 > \mu_1$. This is an indication of fluid flow processes as the main cause of absorption in porous rocks. Laboratory and field experiments have shown that dry crustal rocks tend to lie in the middle left sector of the diagram, where $\mu_1 < k_1$, while partially saturated sands and sandstones prefer the lower right side, where V_P^2/V_S^2 is high and $Q_S^{-1}/Q_P^{-1} = Q_P/Q_S \leq 1$ and $\mu_1 < k_1$ (Winkler and Nur, 1982; Meissner and Theilen, 1983). Clearly, the information is considerably increased if two quasi-independent additional parameters such as Q_P^{-1} and Q_S^{-1} are available.

4.2.3 LABORATORY TECHNIQUE FOR MEASURING Q^{-1}

According to Zener (1948) and a review by Toksöz and Johnston (1981), laboratory measurements of Q^{-1} may be separated into the methods of

(1) free vibration,
(2) forced vibration,
(3) wave propagation, mostly transmission,
(4) observation of stress–strain behavior.

It is seen from this summary that only method 3 is a purely seismic method.

In method 1 the attenuation is determined by an amplitude decay of successive cycles of free vibrations. Frequencies can be adjusted to match the seismic range of 1–100 Hz by varying the size and attaching different masses to the sample. The determination of a specific loss parameter depends on the mode of vibration.

In method 2 the width of the resonant peak is observed and Q^{-1} is calculated from Eq. (4.10). In general, all resonance methods are comparatively simple and can be adjusted to a wide frequency range. Torsional vibrations for the determination of Q_S^{-1} contain less errors than others.

Method 3, the observation of ultrasonic pulses through a medium, is a most direct method and very similar to *in situ* observations. Unfortunately, the attenuation can be observed only at rather high frequencies, which may not contain important relaxation peaks of (or near) the seismic frequency band. Nevertheless, the basic mechanism affecting Q^{-1} in the dry, partially saturated, and fully saturated states as well as the effect of different pore fluids on Q_P^{-1} and Q_S^{-1} can be separately investigated and the ratio Q_P/Q_S can be determined. Nonconsolidated material can also be investigated with method 3.

Method 4, the observation of stress–strain curves, can again be adjusted to a wide frequency range. The angle between the excitation and the response of bar-shaped samples is measured, and Q^{-1} is given by Eq. (4.7). Torsional modes give Q_E^{-1}, the specific absorption for bar waves.

4.2.4   THE DETERMINATION OF Q^{-1} FROM SEISMIC SIGNALS

So far, most methods for the determination of Q^{-1} from seismic signals of body waves in the laboratory and in the field have assumed a frequency-independent Q^{-1}, at least for a limited frequency interval. According to Eq. (4.11) or (4.12) and Fig. 4.9, Q^{-1} can be determined by

(1)   the modulus defect $\tilde{M}(\omega)$, i.e., observation of dispersion,
(2)   the actual attenuation $Q^{-1}(\omega)$, i.e., observation of absorption.

Method 1 is the main basis of the rise time method of Kjartansson (1979) and Gladwin and Stacey (1974). Extension to frequency-dependent Q^{-1} values is possible. Method 2 is the basis of the well-known spectral division method (Bath, 1974), which determines spectral amplitude ratios as a function of frequency.

It is obvious that for an optimum determination of Q or Q^{-1} the whole seismic signal, i.e., dispersion *and* absorption, should be taken into account. Moreover, with modern techniques such as predictive filters and synthetic seismograms an operator that converts a reference wavelet into the observed wavelet by means of a Q^{-1} factor can easily be calculated (Raikes and White, 1984). Alternatively, the best-fitting Q can be found in a similar way by a trial-and-error analysis by comparing the Q-modified reference wavelet with the observed one. Hence, this wavelet modeling should be superior to the other two established methods. Figure 4.9a shows the principles of all three methods.

It should be stressed that the application of these methods is severely hampered in practice by various obstacles. First, there is the interference by small-scale layering, which creates peg-leg multiples in reflection and transmission, thus disturbing the spectra. Gradient layers may have a similar disturbing effect. Although these and similar sources of disturbance may be attacked by purifying the spectra through homomorphic deconvolution or by

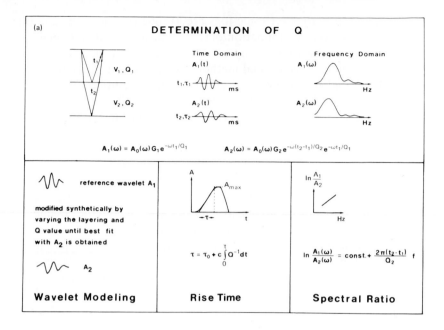

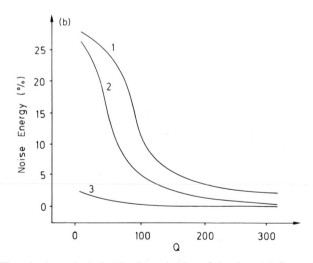

Fig. 4.9 Three basic methods for the determination of Q values. (a) Concept of the three methods [from Meissner and Theilen (1983)] and (b) 25% confidence level of Q determinations of the three methods against noise, i.e., the discrimination power as a function of noise and the Q level. (1) Wavelet modeling, (2) spectral ratio, and (3) rise time [from Jannsen et al. (1985)].

the theoretical calculations if sonic logs are available, serious uncertainties remain. In general, the determination of Q^{-1} is most accurate for small Q^{-1} values and a great thickness of layers. The wavelet modeling technique is superior to the more conventional methods, as seen in Fig. 4.9b. In general, the use of a broad frequency band enhances the resolution. Moreover, it makes it possible to detect a frequency dependence of Q.

For completeness, methods for the assessment of Q from surface waves are also described, although they are hardly used for laboratory investigations. It may be unnecessary to point out that surface waves provide only average Q values for the whole crust and a certain region. In these methods Q^{-1} is obtained from the coda of local earthquakes. The assumption is made that the dominant coda frequency is caused by surface wave scattering and is a function of the source spectrum, the instrument reponse, and the "Q-filter" of the crust (Herrmann, 1980). The effect of the source spectrum may be ignored if its corner frequency is greater than the peak instrument response, which is known. Aki (1969) has proposed an empirical relationship between coda amplitudes, coda frequencies, and Q^{-1}, and Herrmann (1980) provides some master curves for easy assessment. Two complementary methods are based on (1) the shape of the whole coda and a theoretical matching procedure and (2) an estimation of Nuttli (1973) for the vertical component L_g wave where the peak amplitude of the coda a_0 is related to $e^{-\alpha r}/r^{5/6}$, with α the absorption coefficient, related to Q^{-1} by Eq. (4.8), and r the distance from the source.

4.2.5   SOME RESULTS OF LABORATORY MEASUREMENTS

For sedimentary rocks relaxation peaks for Q^{-1} were found in saturated rock samples, while Q^{-1} was very much smaller for dry samples (Spencer, 1981). Relaxation peaks were from 1000 to about 50 Hz for limestone. The influence of small amounts of water has been extensively studied by Tittmann (1977) and Tittmann *et al.* (1980). For lunar rock samples under a very strong vacuum a Q value of nearly 5000 was obtained, in agreement with the Q values from Apollo experiments on the lunar crust and mantle. A very small amount of absorbed volatiles caused the Q values to drop into the range below $Q = 100$. Water has turned out to be the most effective volatile, most likely because of its dipole character and the strong bonding between H_2O and SiO_2.

The other end of the partially saturated regime of rocks also seems to be extremely sensitive. Injections of only $\frac{1}{2}\%$ CO_2 into a fully saturated sand or sandstone caused the Q_P value to drop very sharply while Q_S remained quasi-constant (Muckelmann, 1982) and the Q_P/Q_S ratio dropped to values < 1.

Attenuation generally decreases with increasing confining pressure, in both P and S and for both dry and wet samples; Q^{-1} also decreases with increasing temperature as the relaxation peak is shifted to higher frequencies (Jones and Nur, 1983; Spencer, 1981). It is doubtful whether this last observation can be transferred to *in situ* conditions, where peaks are much smoother and show a plateau-like behavior of $Q^{-1}(\omega)$. In general, increasing temperature will enhance Q^{-1}, but more *in situ* observations are certainly needed.

An interesting new study on the attenuation of rock samples of crust and uppermost mantle was reported by Berckhemer and Hsue (1982) and Berckhemer *et al.*, (1982a,b). Measurements at temperatures near the melting point of rocks were performed by the strain retardation method (method 4) and by forced vibration (method 2) in the torsional mode. The Q_P^{-1} and Q_S^{-1} values were measured or calculated for a broad frequency range. Figure 4.10 shows V_P, V_S and Q_P^{-1}, Q_S^{-1} for different temperatures and frequencies for a dunite. In these experiments Q_P is generally larger than Q_S, especially at the advent of some partial melt. The Q values are between 15 and 1000. Finally, Fig. 4.11 shows the Q values again together with the electric resistivity and the shear modulus as a function of T^{-1}. It is interesting to note that at the beginning of partial melting at about 1200°C the resistivity ρ and the shear modulus drop sharply, while the decrease of Q follows the same linear trend as that at low temperatures.

Investigations of *crustal* rocks show the same trends as those of Fig. 4.10 and 4.11, but Q and μ values are slightly lower because of smaller activation energies. A three-dimensional picture of a gabbro is presented in Fig. 4.12, which clearly shows the dependence of attenuation on the temperature as well as on the frequency.

4.2.6 RELATION OF RESULTS FROM LABORATORY MEASUREMENTS TO THOSE FROM FIELD STUDIES

As mentioned before, Q values between 100 and 1000 were found by laboratory measurements on crustal rocks at normal conditions, but values below 100 were observed at elevated temperatures. With an analysis of deep crustal reflections in the north German plains, Engelhardt (1978, 1979) measured a Q structure of 70 to 150 for sediments, while values over 300 were found for the crystalline crust with a tendency to lower values at the base of the crust. From the coda of surface waves of earthquakes, the average Q_S structure for the United States has been derived for different geologic provinces (Singh and Herrmann, 1983). The Q_S values vary from less than 200 in the thin and rather warm crust of California to more than 1200 in the shield area south of the Great Lakes. It is interesting to note that the cold

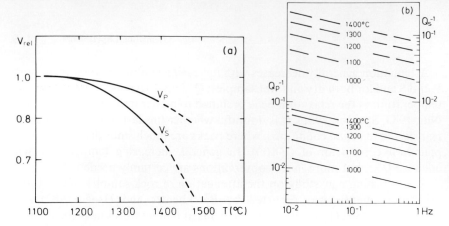

Fig. 4.10 V_P, V_S, Q_P^{-1}, Q_S^{-1} values as a function of temperature T and frequency for a dunitic sample. Note that V_P/V_S as well as Q_P/Q_S increases with T. Note also the *decrease* of attenuation with *increasing* frequency between 10^{-2} and 1 Hz. [From Berckhemer *et al.* (1982a,b).]

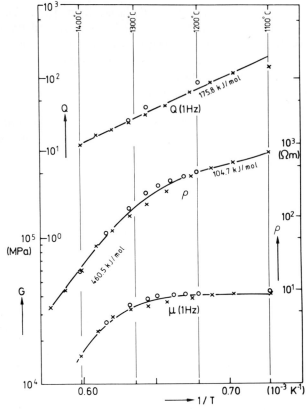

Fig. 4.11 Q values, electric resistivity ρ, and shear modulus μ as a function of T^{-1}. Note that the onset of melt phases between 1200 and 1300°C does not change the slope of the Q curve, while both ρ and μ show a strong "weakening". [From Berckhemer *et al.* (1982a).]

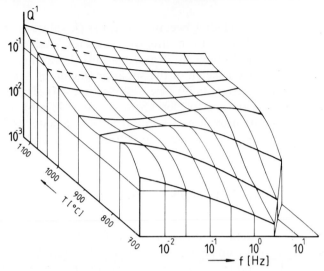

Fig. 4.12 A three-dimensional picture of a gabbro (average grain size = 0.5 mm) showing the attenuation Q^{-1} as a function of frequency and temperature. [From Kamptmann (1984).]

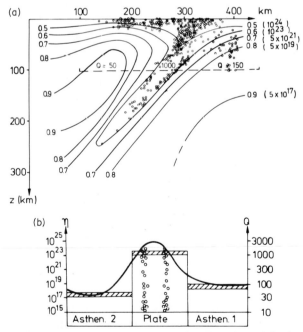

Fig. 4.13 Q and viscosity (η) values in a cross-section through a subduction zone. (a) The slab between the two bands of hypocenters (dots) shows high Q values compared to its surrounding; viscosity isolines are after model calculations. (b) Composition of Q and η values along a cross-section at 100 km depth. [From Meissner and Vetter (1979).]

subducting lithosphere (including oceanic crust) has Q_S values of about 1000, while under the heated back arc basins Q_S values less than 50 have been obtained by seismological body-wave investigations (Barazangi and Isacks, 1971). Figure 4.13 shows a sketch of the situation, including some viscosity values (Meissner and Vetter, 1979) (see also Section 4.4). The Q data are certainly compatible with those of Berckhemer *et al.* (1982a,b). Such sets of field *and* laboratory data provide a real opportunity for a better assessment of the influence of partial melt and temperature from V_P, V_S, and the Q values.

4.3   FRACTURE AND FRICTION EXPERIMENTS

Rock fracture and friction are rather rapid processes and belong to the brittle behavior of rocks under stress. It should be emphasized from the beginning that these processes are often difficult to separate from the contribution of ductile behavior, which is treated in Section 4.4. Most rocks that are brittle under low *p*–low *T* conditions become ductile with increasing *p* and increasing *T*. Figure 4.14 represents a classical experiment by von Kármán (1911) and Griggs *et al.* (1960). Here the transition from brittle to ductile occurs at a confining pressure between 200 and 500 bars (20–50 MPa) at room temperature and between 300 and 500°C at 500 MPa, i.e., at about 15 km depth. Hence, fracture and friction processes must be strongly dominant in the cool upper crust, where ductile behavior is limited to certain minerals and certain areas of stress concentration. Ductile deformation is expected for the lower crust and should be strongly dependent on temperature. Brittle–ductile transition zones must exist over a wide range of temperatures in the middle crust as well as in the uppermost mantle, where materials have higher activation energies and a greater intrinsic rigidity, e.g., peridotitic rocks (Meissner and Strehlau, 1982). Fracture in this context is understood as the process of breaking a rock, including crack propagation; it is a rather sudden failure of initially intact rocks. Frictional sliding, on the other hand, is a sudden motion, often of a stick–slip type, along a preexisting fault. Friction is the resistance toward such a sudden slip along a crack or a prebroken rock.

Experiments on the brittle behavior of rocks are usually performed for a number of purposes. There is a group of experiments directed toward problems of civil and mining engineering, involving stresses up to 1 kbar (100 MPa), equivalent to about 3 km depth. In crustal studies experiments should reach about 1.5 GPa. In such experiments the mechanics of intact rocks—which nevertheless contain microcracks and inhomogeneities of various dimensions—is one topic or research. Precursor phenomena, i.e.,

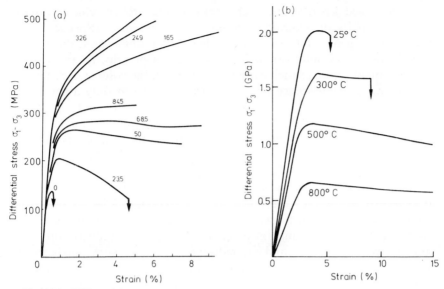

Fig. 4.14 Differential stress versus strain. (a) Confining pressure (in MPa) as parameter and (b) temperature as parameter. [From von Kármánn (1911) and Griggs *et al.* (1960).]

observable changes of physical parameters during crack growth and crack coalescence, are investigated with the final goal of predicting large crustal rupture processes. A second goal of experiments is an assessment of the ultimate strength of the rocks and the maximum stresses that a crustal unit can sustain. In this connection the special roles of volatiles and of gauge material of fault zones as well as the brittle–ductile transition in the continental crust are preferred objectives of investigations.

4.3.1 EXPERIMENTS DURING FRACTURING

The pioneering work of Brace *et al.* (1966) showed that igneous rocks increase in volume before fracturing. Such a volumetric cracking of "intact" rocks is part of the so-called dilatancy phenomenon. It is a partly irreversible opening of pores and microcracks connected with volume increase under increasing stress, with a general weakening of the material and the appearance of precursor phenomena such as decreased seismic velocities and electric resistivity and increased acoustic emissions. With the growth and coalescence of microcracks during the transformation into a real shear fracture, strain hardening transforms to strain weakening. Load-bearing rock bridges (so-called asperities) between cracks must be broken, which produces preseismic activity or acoustic emission in the laboratory (Spetzler *et al.*, 1982). Migrating pore fluids might also have some significance during this stage of microcrack interaction. Precursor values of velocity and electric resistivity might return to their earlier values, before the volume itself shrinks again during the coalescence of microcracks into one or only a few major fracture zones.

Experiments to investigate the development of such fractures might lead to a better understanding of *in situ* observations. It was the field observations in the Garm region (USSR) reported by Semenov (1969) and Nerzerow *et al.* (1969) that showed a marked decrease in V_P and in the V_P/V_S ratio a few years before major earthquakes and a return to the previously observed velocity values shortly before the occurrence of the earthquake. Apparently a large part of the crust in this area must have shown a dilatant behavior. Unfortunately, such a change in precursor activity is not always observed before natural earthquakes, and alternatively the model of stick–slip behavior along preexisting zones and weakness, as postulated by Byerlee and Brace (1938, 1968), might be considered. Such a stick–slip phenomenon is controlled by friction with no (or only small) dilatancy. Stick–slip, however, might also be envisaged as the final sequence of the fracturing process.

In the laboratory (and certainly also in the crust) the nature of rock failure is strongly influenced by the rate of external stress application. In the laboratory the stiffness of the stress-loading system acts as a critical parameter. A servo-controlled electrohydraulic loading system such as that used by Rummel and co-workers (Rummel, 1973, 1974; Rummel and Fairhorst, 1970) is able to perform any self-controlled or strain-controlled application of stress. The development of a fracture, the strain in three directions, the acoustic emission, and the V_P and V_S values can be monitored during the application of stress under a given confining pressure. Basically, the response of a rock sample can be observed at

(1)   constant stress rate $\dot{\sigma}$,
(2)   constant strain rate $\dot{\varepsilon}$,
(3)   constant external stress σ,

as described by Spetzler *et al.* (1982). In recent experiments of Rummel and Frohn (1982) the axial strain rate, according to method (2), was kept constant, while either the intermediate principal stress σ_2 or the corresponding lateral strain ε_2 was also kept constant. The stresses σ_1 and σ_2 as well as the strains ε_1 and ε_2 were monitored, and V_P could be measured in three directions.

Figure 4.15 shows an example of these measurements for a granitic sample. There is first an increase of V_P values in all directions up to a value of about

100 MPa (1 kbar) in stress. This is a well-known effect in the region of elastic deformation and is similar to the increase of V_P with increasing confining pressure. While V_P in the σ_1 direction continues to increase up to failure, the velocities perpendicular to the σ_1 direction decrease. This is a typical example of a microcrack-induced anisotropy. Microcracks develop preferably parallel or subparallel to σ_1. For the real crust we might conclude that a suspicious, i.e., possibly dilatant, area must be monitored in *all* directions in order to be able to observe the V_P dilatancy, which otherwise might be added to the long list of nonprecursor phenomena.

It might be a bit disappointing that in experiments no actual reversal in V_P is seen shortly before failure. When the same experiment was performed on serpentinite samples a small but distinct increase of V_P in the σ_2 direction was observed which was accompanied by a strong increase in acoustic emissions. Figure 4.16 shows two of these experiments in which the velocity recovery of V_P in the σ_2 direction and again the development of an anisotropy are clearly defined. It is speculated by Rummel and Frohn (1982) that rocks which achieve some structural rearrangements by localized ductile deformation, by stress reorientations along crack tips, or by limited frictional sliding along crack surfaces under normal or preferably elevated temperatures may develop a volume decrease or at least an ε_2 decrease as shown in the figure. This would correlate with the observed V_P recovery in the σ_2 direction.

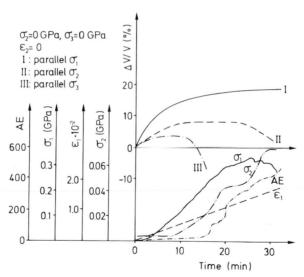

Fig. 4.15 V_P measurements in a granitic sample at constant strain rate $\dot{\varepsilon}_1$ in three dimensions. Stress and strain (σ_1, σ_2 and ε_1, ε_2) are also shown. Here AE represents acoustic emissions. [From Rummel and Frohn (1982).]

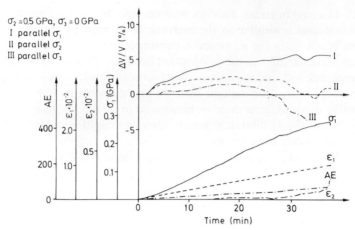

Fig. 4.16 V_P measurement in a serpentine sample. Other symbols are defined in Fig. 4.15. [From Rummel and Frohn (1982).]

Other experiments by Griggs *et al.* (1975) also showed definite decrease in the V_P/V_S ratio for dry samples. A V_P/V_S increase is generally observed during the deformation of fluid-saturated rocks (Hadley, 1975), thus supporting the theory of a pore fluid contribution to the final stage of dilatancy. Experiments on seismic attenuation have also been performed. Byerlee (1978) presented data from an experiment in which P- and S-wave amplitudes were observed in the σ_3 direction. The SV waves, whose oscillations were predominantly parallel to the supposed crack alignment (i.e., in σ_1), showed a very strong attenuation. The SH waves did not show a significant amplitude loss, and the P-wave amplitudes even increased during stress loading. Although in these experiments no quantitative calculations of (anisotropic!) Q^{-1} values could be performed, their application to precursor phenomena of earthquakes and the detection of dilatant areas seems important.

Acoustic emissions seem to be present in the dilatant, prefractured region and also during frictional sliding, and may be associated with the breaking of asperities between cracks in the prefailure stage and the elimination of fault surface asperities during sliding.

A drop in electrical resistivity has been experimentally observed in water-saturated igneous rocks at the beginning of dilatancy (Brace and Orange, 1968), while dry rocks apparently show a resistivity increase. As in applied geophysics, the behavior of electric parameters varies strongly with the ion content and the water saturation of the pores.

For variations of the magnetic susceptibility, the thermoremanent magnetization (TRM), and gas emission as functions of stress loading and dilatancy, see Byerlee (1978). In general, much more controlled experiments in the laboratory and especially in the field are necessary for a close correlation between the two branches of geophysics in order to detect anomalously dilatant regions in which a future rupture will originate.

4.3.2 FRICTION EXPERIMENTS

As has been mentioned, rupture processes usually occur on preexisting faults. Such seismic movements may be considered as a second or a final stage in the development of fault zones and a consequence of fracturing. From the experiments of Brace *et al.* (1966) and Byerlee (1967, 1978) it became clear that the fracturing of intact rocks gives rise to most unsystematic variations of the fracture strength, but experiments on friction along an existing fracture gave a result which was independent of rock type, displacement, and surface condition. This observation seems to be one of the most significant recent discoveries. Hence, a universal rock friction law has been developed, i.e., Byerlee's law, which relates the shear stress τ necessary to overcome static friction on a surface to the normal stress σ across it. This law has been mentioned in Section 3.6.5 and Eq. (3.60) as the basis for stress calculations. Figure 4.17 shows Byerlee's relationship, which is also the foundation for the calculation of frictional strength (maximum possible stress) as a function of effective pressure. Evidence for the existence of fractures in the form of joints, faults, microcracks, and bedding foliation down to 10 km depth has been accumulated in the past 10 years. As mentioned in Section 3.6.5, the lower limit of brittle behavior depends heavily on the temperature and the pore pressure. A transition to ductile behavior may cover a 5–10-km depth interval, a depth of 6–20 km, and a temperature between 400 and 600°C.

The behavior of natural and artificial gouges in friction experiments is of special interest for the simulation of fault zones in the crust. From Byerlee's experiment a strongly reduced coefficient of friction, i.e., $\mu^* = 0.18$ instead of 0.6, was found for the clay minerals vermiculite and montmorillonite. Chlorite may also reduce the friction coefficient. In addition, gouges might develop internal structures inside the gouge zone. From *in situ* experiments near the San Andreas fault, Zoback *et al.* (private communication) found a value of $\mu^* = 0.35$, which is somewhere in the middle between the friction coefficient of clay minerals and that of most other rocks in the laboratory experiments. The effect of pore pressure for calculating the *in situ* effective pressure has already been mentioned in Section 3.6.5. The effect of temperature on μ^* seems to be negligible for dry granite up to 500°C and for gabbro up to 400°C, as measured by Stesky *et al.* (1974). It seems a bit doubtful whether these experimental data can really be extrapolated to *in situ* conditions, where fluid phases may play a major role, especially at higher temperatures and smaller strain rates.

An important consequence of the friction experiments is the implication for the upper limit of possible (maximum) stresses in the upper crust and a comparison of these values with those which could initiate dilatancy. The stress required for friction along preexisting faults is for extension and wet conditions less than 100 MPa, while at least 200 MPa is necessary for the initiation of dilatation phenomena and about 300 MPa for measuring the full

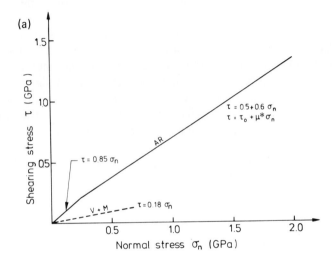

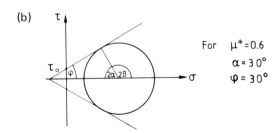

Fig. 4.17   Byerlee's empirical relationship between shear stress and normal stress (a) in friction experiments and (b) related to Mohr's circle; $V + M$, vermiculite and montmorillonite samples; AR, all other rocks; τ, shear stress maximum (STRESSMAX) or shear strength; μ^*, coefficient of internal friction; other symbols are defined in Section 3.6.5.

dilatancy effect. Compare Fig. 4.17 with Figs. 4.16 and 4.15. This reasoning would imply that no dilatancy effects should take place in an extensional tectonic regime, where

$$\tau_{max}^{friction} < \tau_{dil} \tag{4.17}$$

under the usual wet conditions of the upper crust. Even for compression and rather warm areas the frictional stress hardly exceeds 200 MPa. This means that the stress necessary for creating dilatancy effects should be observed only in a cold shield area or an oceanic plate, because here:

$$\tau_{max}^{friction} \geq \tau_{dil} \tag{4.18}$$

even under wet conditions.

Certain strong barriers or especially dry and unbroken rocks might possibly reach such a limiting strength of about 300 MPa, but for a "normal" continental crust it seems unlikely that a maximum stress of about 300 MPa will ever be reached. Certainly, continental transform faults like the San Andreas fault, with its small value of $\mu^* = 0.35$, cannot build up the stresses needed for the dilatancy effect.

This does not mean that controlled seismic experiments in such areas are useless for earthquake prediction research. As seen in Figs. 4.15 and 4.16, the velocity anisotropy—which is still in the elastic range—starts at rather small stress values, and an anisotropy or a change of anisotropy might be measured by an active seismic experiment along various azimuths. A change of attenuation might also be recorded. The strength and the lateral extension of possible crustal barriers, unfortunately, are not yet known and certainly cannot be assessed by laboratory experiments. Even with controlled field data the assessment of such small inhomogeneities might be difficult.

4.4   CREEP EXPERIMENTS

It was mentioned in the previous sections that all rocks become ductile under increasing pressure p and temperature T. Important for the rheology of the crust are the questions of how rocks might differ in their rigid–ductile behavior, what the influence of water, cracks, and pore fluids is, and at what depth and how the transition from rigid to ductile behavior takes place. Other important questions are how much stress can be built up by creep and what the values of viscosity are as a function of creep rate and creep behavior. Many of these questions have been tackled by laboratory studies.

Unfortunately, most creep experiments have been performed with mantle material, i.e., olivine crystals or peridotitic–dunitic rocks. This was a consequence of the increase of our knowledge about the ductile asthenosphere and its implication for sea floor spreading and plate tectonics. Moreover, mantle rocks are generally more homogeneous and simpler to handle than rocks of continental affinity. Convective flow in the asthenosphere could be considered as the flow of olivine. As the importance of fracture and flow during processes of crustal shortening or extension has become better known, some experiments with crustal rocks have been carried out, and their implications for the rheology of the continental crust will be investigated in some of the following sections.

The extrapolation of laboratory data on creep obtained over a period of hours (10^{-4} yr) to geologic times of millions of years (10^{+6} yr) is very problematic. Also, creep rates $\dot{\varepsilon}$, being generally larger than 10^{-7} s^{-1} in the laboratory but smaller than 10^{-14} s^{-1} in nature, pose severe problems to researchers. Only if a correlation between stress and the microstructure could be established for *in situ* and laboratory experiments would it be possible to justify the application of the laboratory data on crustal studies (Kohlstedt, 1979). Figure 4.18 gives an impression of the amount of extrapolation needed in the translation from laboratory data to *in situ* conditions.

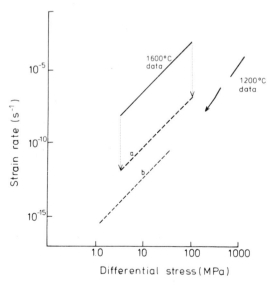

Fig. 4.18 Stress–strain relationships in the laboratory and in possible extrapolation. (a) Extrapolation to *in situ* conditions based on 1600°C data and (b) extrapolation of 1200°C data. [From Kohlstedt (1979).]

4.4.1 EXPERIMENTAL TECHNIQUES

Experimental studies of "plastic" flow or "creep" of solid rocks and crystals concentrate on empirical relationships between stress and incremental strain of rocks under high p–high T conditions without much account of the actual physical processes. In a more recent effort, the atomic structure and inherent small-scale mechanical anisotropy of crystalline solids were investigated. In these experiments much was learned from the deformation of single crystals, before experiments with polycrystalline rocks were interpreted. These data on the nature of dislocations will be discussed in Section 4.4.2.

For the more general investigation of rocks under high p–high T conditions a triaxial testing machine is used. A certain load is applied by at least two pistons corresponding to a stress different from the confining pressure. This load is $\sigma_1(\sigma_1 > \sigma_2 = \sigma_3)$ in compression experiments, and $\sigma_3(\sigma_3 < \sigma_2 = \sigma_1)$ in extension studies. Figure 4.19 gives a schematic view of a triaxial testing machine. The whole setup is similar to that of fracturing experiments, as described in Section 4.3.1.

In all rocks an increase in confining pressure favors the transition from brittle to ductile behavior. Most rocks can be made ductile by pressure alone, although the pressure required to induce ductile deformation exceeds those inside the continental crust. Other experiments at 0.5 GPa, equivalent to about 15 km depth under various temperatures T, show that in almost all polycrystalline materials increasing T suppresses fracture and enhances ductility. In an experiment with marble and constant strain rate, the yield stress was lowered considerably with increasing T and no "strain-hardening" effects occurred for $T \geq 800°C$ (see Fig. 4.13). Strain hardening is an increase in differential stress (and in strength) with increasing strain.

It is interesting to note that some hydrous rocks like serpentinite act rather differently from most other ones. Under similar experimental conditions increasing T first results in a general strain softening at larger strains. Above 500°C a general transition back to brittle behavior occurs, and at 700°C fracture takes place. This particular behavior reflects the presence of a dehydration reaction, causing embrittlement rather than increasing ductility. In these experiments the released water could not escape, and it is doubtful whether such processes really take place under *in situ* conditions, as mentioned already in Section 4.1.2. On the other hand, most natural rocks contain fluid phases in their pores or as a thin film along grain boundaries. It is by now well known that quartz and some other silicates are strong, brittle, and rigid, even under high p–high T conditions, but become weak and ductile in the presence of fluid phases. Data from experiments with dry and wet quartzite were used for the calculation of maximum possible stresses in the

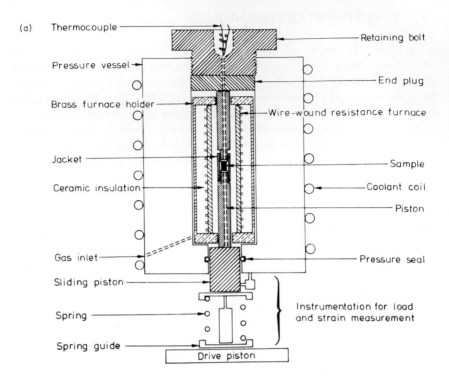

(a)
Thermocouple ——— Retaining bolt

Pressure vessel —— End plug

Brass furnace holder ——— Wire-wound resistance furnace

Jacket ——— Sample

Ceramic insulation ——— Coolant coil

Piston

Gas inlet ——— Pressure seal

Sliding piston ———

Spring ——— Instrumentation for load and strain measurement

Spring guide ———

Drive piston

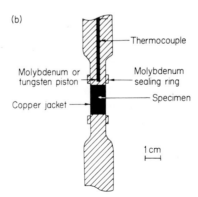

(b)

Thermocouple

Molybdenum or tungsten piston

Molybdenum sealing ring

Copper jacket ——— Specimen

1 cm

Fig. 4.19 Triaxial testing machine. (a) General view and (b) positioning of sample and piston. [From Kohlstedt (1979); Kohlstedt *et al.* (1980).]

crust in Sections 3.6.5 and 4.3.2. The influence of water is often called
"hydrolytic weakening". It involves increased mobility of dislocations as a
result of hydrolysis of Si–O–Si bonds. However, nonchemical effects such as
pore pressure also act as a weakening process, as mentioned in Section 3.6.5.

The major outcome of the conventional, short-term triaxial creep experiments can be summarized as follows.

(1)   Under crustal p-T conditions rocks should be brittle in the upper
crust, transitional in the middle crust, and ductile in the lower crust.

(2)   The most probable mechanisms of failure are faulting (upper crust),
cataclastic flow (middle crust), and ductile creep (lower crust); see Section
6.6.3.

(3)   Differential stresses for these kinds of deformation are high for dry
rocks but are lowered considerably by the presence of high-pressure pore
fluids.

(4)   Dehydration reactions in the middle or lower crust might complicate
the creep behavior.

Figure 4.20 shows a typical experimental strain–time curve which might be
representative for average conditions in the middle and lower continental
crust. It should be emphasized, however, that the duration of loading plays an
important role. Under natural conditions loading is slower by several orders
of magnitude. Consequently, creep processes might be much more dominant
than in the experiments. This also seems to be indicated by the observations
of folded geologic structures in the field, where the grains involved show
recrystallization, annealing, and grain grow, as explained in Section 4.4.2.

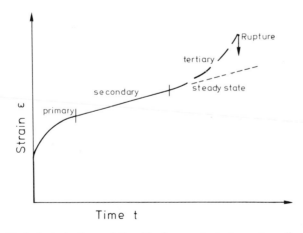

Fig. 4.20 Typical strain–time relationship for constant stress at medium crustal depth
showing two alternatives: tertiary creep and rupture or steady-state creep.

4.4.2   TYPES OF DISLOCATIONS

Experiments with monocrystals reveal some information on the nature of dislocations. The polished surfaces of crystals are no longer smooth after a loading experiment but are crossed by a number of striations, which can be studied with a microscope. In particular, electron and optical transmission microscopy are used to reveal the nature of the striations. They are generally "steps" on the surface of the crystal and represent zones of concentrated deformation between virtually undeformed regions. These zones are commonly parallel to certain planes in the crystal and are controlled by its atomic structure. Along these zones plastic deformation takes place by the sliding of layers of atoms. They are "slip" or "glide" planes. In some crystals there is only one slip plane and direction, also called a slip or glide system; in others several intersecting sets occur. Most rock-forming silicates crystallize in the monoclinic or triclinic system. Hence, only one slip system is to be expected, in accordance with the observation that most silicates are not particularly ductile under dry and medium p-T conditions.

Besides such slip mechanisms there are three other types of plastic deformation, called (1) mechanical twinning, (2) kinking, and (3) shear transformation. For details see Verhoogen *et al.* (1970). All these slip mechanisms, however, when compared with theoretical analysis, could not explain the observations that crystals are, in fact, several orders of magnitude weaker than expected from the simple slip models.

The problem was solved by postulating that the displacement of atoms might not take place simultaneously but by a movement of an imperfection in the crystal structure, a so-called dislocation, along the slip plane, as shown schematically in Fig. 4.21. Much lower shear stresses are required to move a dislocation along the slip plane than would be needed for a simultaneous slip of atoms. Today, dislocations of various types can be directly observed with electron and transmission microscopy. Figure 4.22 gives an impression of the

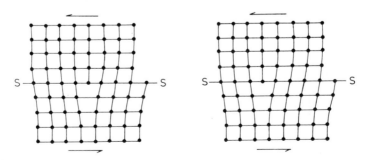

Fig. 4.21   Motion of a dislocation along a slip plane SS.

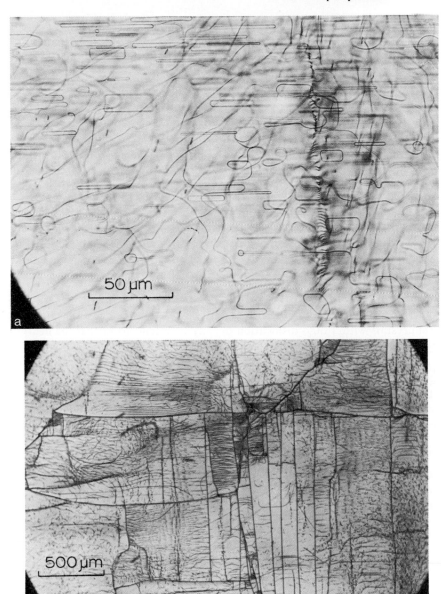

Fig. 4.22 Transmission light micrograph of decorated dislocations in (a) an olivine-rich xenolith from Kilbourne Hole, Texas and (b) an olivine crystal from San Carlos, Arizona. In (b), the darker lines running vertically are low-angle tilt boundaries on planes which are viewed edge-on in this micrograph. Dislocations were decorated by heating a polished thin section for 1 hr at 900°C in air. [From Kohlstedt *et al.* (1976). Copyright 1976 by the American Association for the Advancement of Science.]

density of dislocations in a deformed olivine grain. The two basic types of dislocations are seen, i.e., the edge dislocations shown by the predominantly vertical dark bands and the screw dislocations, represented by the fine lines. Their different nature is explained in Fig. 4.23.

In general, a dislocation can be visualized as a small line separating the slipped from the unslipped part. It can move along any zone of "weakness", e.g., along a line or a curved path or even along a closed loop in a slip plane. In order to describe the type of dislocation two parameters are used: the Burgers vector b, defining the amount and direction of the dislocation's movement, and the local direction of the dislocation line itself, given by the unit vector t. With these two definitions the two basic types of deformations can be recognized:

(1)   In an edge dislocation b is perpendicular to t as shown in Fig. 4.23a.
(2)   In a screw dislocation b is parallel to t as shown in Fig. 4.23b.

In addition to these two basic types of dislocations there are mixed ones composed of edge and screw dislocations.

Dislocations can be moved by stress. They are created at various dislocation sources. Dislocation density increases with the degree of deformation, but crystals with a high dislocation density also tend to anneal (or heal) or recrystallize more readily under appropriate p–T conditions.

The movement of dislocations along slip planes is also called dislocation glide. If a dislocation moves from one slip plane to another, one speaks of dislocation climb, a process which takes place under the influence of high stresses. Dislocations also play a role in the transport of atoms by diffusional processes from one grain boundary to another.

What makes all creep processes in the crust so complicated compared to the olivine rheology is the fact that so many rock-forming minerals exist and most of them vary widely in their capacity to deform plastically. As mentioned in Section 4.4.1, creep properties of quartz and other silicates depend strongly on the presence of water. It is the goal of many experiments

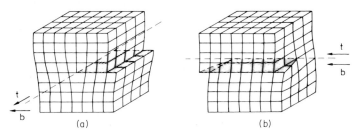

Fig. 4.23   Schematic picture of (a) edge and (b) screw dislocations. Here t is the dislocation line (unit vector) and b the Burgers vector, which defines the dislocation movement.

to relate the microscopic deformation pattern and mechanism observed in the laboratory to those found in nature.

As an example, the dislocation structure of quartz grains found along thrust faults is often severely plastically deformed. Numerous curved dislocations and their boundaries are compared to those of high p–high T plastic flow from laboratory experiments. They provide a measure of the differential stress.

The shapes of the individual quartz grains are also investigated. Some are plastically deformed to 5–10 times their original length. Some grains migrate under high-T conditions. The shape and size of these grains can be correlated with the stress level at which the rocks were deformed, again by comparing the naturally deformed samples with the experimentally deformed samples. Such textural evidence includes elongation, flattening, and bending of grains.

The comparison between natural sample structures and those of artificially deformed crystals is rendered more difficult by the effect of postdeformational grain growth and recrystallization. Such a "healing" is thermally activated and involves diffusion of atoms and migration of grain boundaries. Recrystallization can also proceed synkinematically and drastically affect the strength and creep resistance of a rock or a mineral. Recrystallization reduces the dislocations and grain boundaries and may finally create aggregates completely free of deformation structures, as found in many metamorphic rocks and glacier ice (Verhoogen *et al.*, 1970). However, it has also been found that recrystallized grain size, dislocation density, and the spacing between twin lamellae and slip bands are sensitive mainly to differential stress. Hence, special annealing experiments show by comparison with natural samples that shear stresses may really reach 200 MPa in the crust (Mercier, 1980), while only 2–40 MPa are envisaged for the upper mantle. Stresses estimated from recrystallized grain size are usually lower than those estimated from dislocation density (Kohlstedt and Weathers, 1980; Ord, 1981). At present it is not clear whether this discrepancy arises from the laboratory calibration technique, from a different response of these parameters, from the thermal history, or from the influence of water or other impurities (Kirby, 1983).

4.4.3   EXPERIMENTAL CREEP RELATIONSHIPS

From the previous section it has become evident that dislocation glide and climb as well as diffusional processes can actually deform a crystal and a rock showing different creep mechanisms. These mechanisms are different for different p-T conditions, but they also depend strongly on the individual crystals and the amount of water or other volatiles. Nevertheless, many experiments with metals and olivine lead to a certain generalization of creep laws. One should start with the general creep equation, as mentioned by Weertman (1970) and Weertman and Weertman (1975); i.e.,

$$\dot{\varepsilon} = C_n \tau^n \exp[-(E_c^* + pV)/RT], \tag{4.19}$$

with C_n the creep constant, τ the stress, n the exponent of the creep law $(1, \ldots, 3)$, $\dot{\varepsilon}$ the creep rate, p the pressure, E_c^* the activation energy, R the gas constant, V the activation volume, and T the absolute temperature. Weertman (1970) has shown that a good approximation of Eq. (4.19) is

$$\dot{\varepsilon} = C_n \tau^n \exp(-g^* T_m/T), \tag{4.20}$$

with g^* a constant [about 29 for olivine (Kohlstedt and Goetze, 1974)] and T_m the melting temperature (solidus). Solving Eq. (4.20) for τ we obtain

$$\ln \tau = (1/n)[g^*(T_m/T) + \ln \dot{\varepsilon} - \ln C_n], \tag{4.21}$$

and solving Eq. (4.20) for the effective viscosity η ($\eta = \tau/\dot{\varepsilon}$) we obtain

$$\ln \eta = (1/n)[g^*(T_m/T) + (1 - n)\ln \dot{\varepsilon} - \ln \dot{\varepsilon} - \ln C_n], \tag{4.22}$$

i.e., an equation similar to Eq. (4.21).

At present the choice of Eq. (4.19) or (4.20) as a basis for calculations is debated. Considering the uncertainties in the activation energy and in the activation volume for crustal minerals and rocks (Kohlstedt et al., 1980) it seems that the use of Eq. (4.20) and the related Eqs. (4.21) and (4.22) for the calculation of stresses τ and viscosities η is fully justified. One of the most important parameters in Eqs. (4.21) and (4.22) is the stress exponent n. It is related to the different creep mechanisms mentioned in the previous section. Only for very high and nonnatural shear stress (τ higher than 1 GPa), and in the low-T area, will n be considerably larger than 3. This creep is called *low-T plasticity*. Dislocations move normal to their glide planes; they often annihilate with dislocations of opposite sign. Dislocation density increases with increasing strain, and the flow system is intrinsically transient (Kirby, 1983). It seems that low-T plasticity is of no relevance to *in situ* conditions in the Earth's crust.

A large field of experimental data relevant to *in situ* conditions is described by an exponent equal or similar to 3, i.e., belonging to the much investigated power-law-creep. Dislocations may move normal to their glide planes and may even climb; in addition, bulk crystal diffusion seems to take place. Water may reduce the strength of the material considerably, resulting in a reduction of the activation energy, [Eq. (4.19)], and of the T_m values, [Eq. (4.20)].

A third mechanism that acts under high temperature–low stress conditions and seems to have some long-time relevance for *in situ* conditions is diffusional flow. This involves diffusion of point defects, either through the grains (Nabarro–Herring creep) or along grain boundaries (Coble creep). In contrast to low-temperature and power-law creep, diffusional creep is a linear process; i.e., it shows an approximately Newtonian rheology with $n = 1$ and $\dot{\varepsilon}$ proportional to τ. The three different kinds of creep are illustrated in Fig. 4.24 for olivine with a 1-mm grain size (Ashby and Verall, 1978). We see that for *in situ* stresses between 1 and 10^2 MPa (10 bar to 1 kbar) and for *in situ* creep rates of 10^{-14}–10^{-17} s^{-1}, power-law creep seems to be dominant. Diffusional flow might be important for low creep rates and stresses below 1–10 MPa (10–100 bars). The boundary between these two creep laws plays an important role in the rheology of crust and mantle and will be discussed in more detail by looking at the constants C_1 and C_3 for the two creep laws with $n = 1$ and $n = 3$. First, Eq. (4.20) is rewritten for diffusional creep ($n = 1$)

$$\dot{\varepsilon} = C_1 \tau \exp(-g^* T_m / T) \tag{4.23}$$

and for power-law creep ($n = 3$)

$$\dot{\varepsilon} = C_3 \tau^3 \exp(-g^* T_m / T). \tag{4.24}$$

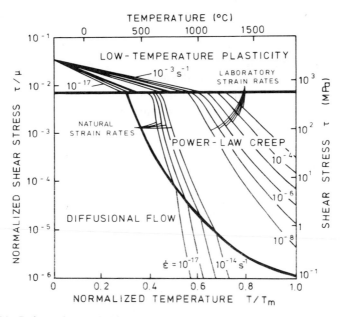

Fig. 4.24 Deformation mechanism map. The material is olivine, and the grain size is 1 mm. The boundary between power-law creep ($n = 3$) and diffusional flow ($n = 1$) (Newtonian) is important for all creep and convection studies, but seems rather elusive. See also Fig. 4.25. [After Kirby (1983).]

According to Weertman (1970), the two constants are

$$C_1 = \alpha_1 \Omega D_0 / R^2 kT, \tag{4.25}$$

$$C_3 = \alpha_3 \Omega D_0 / \mu^2 kT, \tag{4.26}$$

with α_1, α_3, and D_0 constants, k Boltzmann's constant, Ω the atomic volume, R the grain size, and μ the shear modulus. Looking for those values of τ_c for which both creep processes contribute equal amounts we must divide Eqs. (4.23) and (4.24) using Eqs. (4.25) and (4.26); the result is

$$\tau_c = (\alpha_1/\alpha_3)^{1/2} \mu / R. \tag{4.27}$$

Larger values of τ correspond to power-law creep, lower ones to diffusional creep. This is demonstrated in Fig. 4.25 (Meissner and Vetter, 1976a), using values for α_1 and α_3 from Weertman (1970).

Smaller grain sizes require higher shear stresses along the boundary. Small grain sizes and small stresses show diffusional creep with its linear rheology. It seems strange at first that neither the temperature nor the strain rate is involved in this simplified treatment. Comparing Fig. 4.24 and 4.25 for a grain size of 0.1 cm, the value of 1 MPa (10 bars) would be related to $\dot{\varepsilon} = 10^{-14}$ s^{-1} and T/T_m roughly 0.7. Lower T and lower $\dot{\varepsilon}$ would require higher stresses to stay at the boundary. Such a discrepancy may result from a small T dependence of the "constants" α_1 and α_2 and possibly also in slightly different g^* values.

Regarding the rheology of the continental crust, we may state that power-law creep will be dominant, especially in the middle crust and for larger grain sizes. The existence of a major contribution from diffusional flow cannot be ruled out for the lower crust and small grain sizes. In particular, small τ and

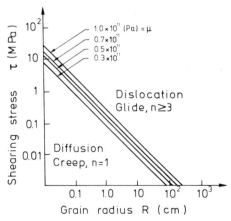

Fig. 4.25 Boundary between dislocation glide and diffusional creep as a function of grain size. Here μ is the shear modulus.

small ε values, which may be found in tectonically inactive areas, e.g., in a period of quiescence after an orogenic event, may show a Newtonian rheology with $n = 1$.

For the asthenosphere, the transition from higher stresses with power-law creep to lower stresses with diffusional creep has been demonstrated by observations of postglacial uplift with its decreasing stress (Meissner and Vetter, 1976a). Uplift velocity and creep rate slow down, and viscosity reaches a quasi-constant value.

For increasing stresses nature seems to provide a self-regulating mechanism. Increasing stress (caused by increasing strain or other sources) does not lead to a catastrophic buildup of stresses but to a transition to the next higher stage of creep mechanism with higher strain rates and easier creep, thereby keeping constant or even reducing the stress.

4.5   SPECIAL IMPLICATIONS OF CREEP DATA FOR THE RHEOLOGY OF THE CONTINENTAL CRUST

Compared to the rather well-known rheology of olivine, our present understanding of the inelastic behavior of materials of the continental crust is very limited. There are at least ten major rock-forming minerals. The weakening processes of water and its pore pressure play a major role. Water promotes crack growth. Hydrogen defects can be introduced by diffusion or crystal growth and further reduce the strength of minerals. The only safe conclusion regarding crustal rheology is provided by the termination of near-surface seismicity in the middle crust, as discussed in Sections 3.3.1 and 3.6.5.

The increasing ductility toward greater crustal depths has also been derived from many laboratory experiments under high p–high T conditions, as mentioned in Section 4.4.1. However, the stresses which are released by rupture processes in the upper crust *must* be replaced by ductile flow from below, thus loading the upper crust continuously.

4.5.1   THE MAXWELL BODY APPROACH

As a starting point for the assessment of the rheological behavior, three simple models are presented for the continental crust, as seen in Fig. 4.26, in which a resistance is combined with a Maxwell element (Jaeger, 1969). As mentioned before, the Newtonian rheology which is included in the Maxwell body might only be valid for small creep rates and small grain sizes, but can serve as a first-order approximation, i.e.,

$$\dot{\varepsilon} = (\dot{\tau}/2\mu) + (\tau/2\eta). \tag{4.28}$$

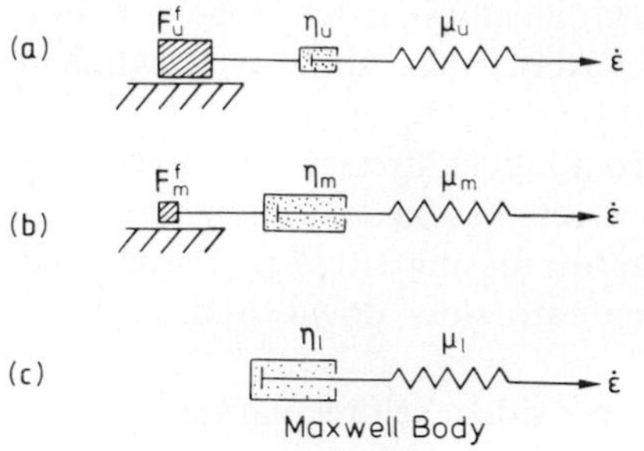

Fig. 4.26 Simple rheological models for different crustal levels. (a) Upper crust, frictional sliding, (b) middle crust, mixed, and (c) lower crust, creep. Here μ is the shear modulus, η the viscosity, and F the frictional resistance.

Its general solution for $\dot{\varepsilon} = \text{const}$ is

$$\tau = \tau_0 \exp\left(-\frac{\mu}{\eta} t\right) + 2\eta\dot{\varepsilon}\left[1 - \exp\left(-\frac{\mu}{\eta} t\right)\right]. \tag{4.29}$$

Assuming complete relaxation for $t = 0$, i.e., $\tau_0 = 0$, Eq. (4.29) transforms into

$$\tau = 2\eta\dot{\varepsilon}\left[1 - \exp\left(-\frac{\mu}{\eta} t\right)\right]. \tag{4.30}$$

As an example, Fig. 4.27 gives solutions of Eq. (4.30) for different viscosities and a constant $\dot{\varepsilon} = 10^{-16}$ s^{-1}. The elastic loading transforms to a constant stress level after some time, if no frictional member is involved in the rheological model.

It is interesting to study the effect of different creep rates of a Maxwell body. In Fig. 4.28 the maximum stresses with $\dot{\varepsilon} = 10^{-16}$ s^{-1} from Eq. (4.30) have been calculated for different creep rates occurring under crustal and upper mantle conditions. Some possible changes of viscosity and stresses from the lower, highly ductile, crust to the stiffer middle crust are indicated in the insert and will be discussed in Section 4.5.2.

For the upper crust, the addition of a frictional resistance, as shown in the model of Fig. 4.26a and b, leads to a sudden stress release once a critical stress τ_c is reached. After the rupture, the stress does not go down to zero (or only does so for an extremely short time during sliding) but reaches a rather stable value τ_0. With $2\eta\dot{\varepsilon} = \tau_\infty$, the long-term average Newtonian creep in the zone of rupture, Eq. (4.30) may be solved for t^*, $\tau_0 \leq t^* \leq \tau_c$:

$$t^* = (\eta/\mu) \ln\left[(\tau_\infty - \tau_0)/(\tau_\infty - \tau_c)\right], \tag{4.31}$$

with t^* the repetition time between ruptures.

Figure 4.29a shows a schematic view of the rupture cycles, and Fig. 4.29b shows a more refined model based on seismological rupture studies (Fitch,

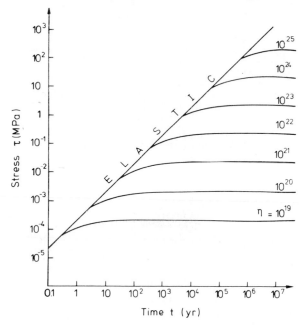

Fig. 4.27 Temporal behavior of stresses for different viscosities (η) and a constant creep rate $\dot{\varepsilon} = 10^{-16}\,\mathrm{s}^{-1}$ (Maxwell body approach).

1979). It is well known that such models can only provide a rough idea about repetition rates of quakes in the upper crust. Nevertheless, the nearly continuous occurrence of earthquakes in a rather small depth interval of the rigid upper crust poses the question of how stresses are transformed into this area from below. This topic will be discussed again at the end of the next section.

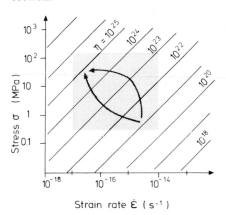

Fig. 4.28 Maximum (i.e., "saturation" stresses) of Fig. 4.27 as a function of creep rate; hatched area represents crustal values of σ, $\dot{\varepsilon}$, and η (Maxwell body approach).

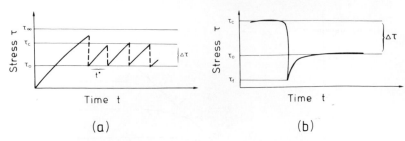

Fig. 4.29 Schematic view of (a) rupture cycle and (b) individual stress behavior at a simple rupture. Compare with Fig. 3.23. [From Fitch (1979).]

4.5.2   AN ASSESSMENT OF STRESSES AND VISCOSITIES AS A FUNCTION OF TEMPERATURE

In the past 10 years some data on the activation energy, the exponent n, and the constant C_n of the general creep equation [(4.19)] have been determined for a number of crustal rocks and minerals. These data are summarized in Table 4.1, which is based on selected values from a recent summary by Kirby (1983) plus some other experimental values. The data clearly show that, in spite of the large scatter of values, power-law creep with a value of n around 3 certainly dominates. This does not mean that for low creep rates $\dot{\varepsilon}$ and low stresses τ, e.g., in the lowermost crust, power-law creep is still the dominant mechanism. However, for the middle crust and probably also for most of the lower crust, especially for tectonically and seismically active areas with a certain stress accumulation, power-law creep must be preferred to diffusional creep.

Hence, for the calculation of τ and η Eqs. (4.22) and (4.23) are used with the constants g^* from Parrish et al. (1976) and Heard and Carter (1968). The equations can easily be solved if the melting point T_m and the temperature T are known as a function of depth z. Figure 4.30 shows some estimates of the temperature–depth function for warm crusts and cold shield areas, and the average crustal structure together with the melting point–depth curves for wet and dry conditions (Ringwood, 1969; Mercier and Carter, 1975). From these data T/T_m can easily be determined for all depths. It should be mentioned that in the last three years so many new values of the activation energy and other constants of crustal rocks have accumulated that the direct approach via Eq. (4.19) has become advantageous (Meissner and Wever, 1985). The outcome of such calculations clearly confirms the existence of a low-η lower crust for all reasonable crustal models.

The calculated data for τ were compared with those for τ_{max} from the calculation of maximum frictional stresses from Byerlee's law, as mentioned

Table 4.1

Various Constants for the Creep Equation $\dot\varepsilon = c_n \tau^n \exp(-E_C^* + pV)/RT$ and $\dot\varepsilon_s \sim c_n \cdot \tau^n \exp(-g^* T_m/T)$

Material	$\log c_n$ (GPa^{-n}/s)	n	E_C^* (kJ/mol)	Reference[a]	T_m(°C) 600 / T_m(K) 873	700 / 973	800 / 1073	900 / 1173	1000 / 1273	1100 / 1373	1200 / 1473	1300 / 1573	1400 / 1673
Albite rock	6.1	3.9	234	(1)	32.2	28.9	26.2	24.0	22.1	20.5	19.1		
Anorthosite	6.1	3.2	238	(1)	32.8	29.4	26.7	24.4	22.5	20.8	19.4		
Quartzite	3.0	2.0	167	(1)	23.0	20.6	18.7	17.1	15.8	14.6	13.6		
	1.9	2.9	149	(2)	20.5	18.4	16.7	15.3	14.1	13.0	12.2		
	1.2	1.9	123	(3)	16.9	15.2	13.8	12.6	11.6	10.8	10.0		
	2.0	2.8	184	(4)	25.3	22.7	20.6	18.9	17.4	16.1	15.0		
	1.8	3.0	223	(9)	30.7	27.6	25.0	22.9	21.1	18.5	18.2		
Quartzite (wet)	3.2	2.4	160	(2)	22.0	19.8	17.9	16.4	15.1	14.0	13.0		
	3.0	2.6	134	(5)	18.5	16.6	15.0	13.7	12.7	11.7	10.9		
	3.7	1.8	167	(3)	23.0	20.6	18.7	17.1	15.8	14.6	13.6		
	3.8	3.0	178	(10)	24.5	22.6	20.0	18.3	16.8	15.6	14.5		
Aplite	2.8	3.1	163	(1)	22.5	20.1	18.3	16.7	15.4	14.3	13.3		
Westerly granite	0.2	2.9	106	(6)	14.6	13.1	11.9	10.9	10.0	9.3	8.7		
	1.6	3.4	139	(3)	19.2	17.2	15.6	14.3	13.1	12.2	11.3		
Westerly granite (wet)	2.0	1.9	137	(3, 7)	18.9	16.9	15.4	14.0	12.9	12.0	11.2		
Quartz diorite	4.3	2.4	219	(3)	30.2	27.1	24.5	22.5	20.7	19.2	17.9		
Clinopyroxenite	9.0	2.6	335	(1)	46.2	41.4	37.6	34.3	31.7	29.3	27.4		
Diabase	6.5	3.4	260	(1)	35.8	32.1	29.1	26.7	24.6	22.8	21.2		
Natural olivine (single crystals)	15.5	3.5	525	(7)			58.8	53.8	49.6	46.0	42.9	40.1	37.7
			535	(7)			37.6	34.3	31.7	29.3	27.4	25.6	24.1
San Carlos, Arizona	17.7	2.5	666	(7)			42.3	38.7	35.6	33.0	30.8	28.8	27.1
Polycrystalline olivine, Aheim (N)	16.1	4.5	498	(8)			55.8	51.1	47.0	43.6	40.7	38.1	35.8
Dunite, Anita	14.0	3.4	444	(8)			49.8	45.5	41.9	38.9	36.3	33.9	31.9

[a] (1) Shelton and Tullis (1981); (2) Koch et al. (1983); (3) Hansen and Carter (1982); (4) Jaoul et al. (1984); (5) Kronenberg and Tullis (1984); (6) Carter et al. (1981); (7) Durham and Goetze (1977); (8) Chopra and Paterson (1981); (9) Heard and Carter (1968); (10) Parrish et al. (1976).

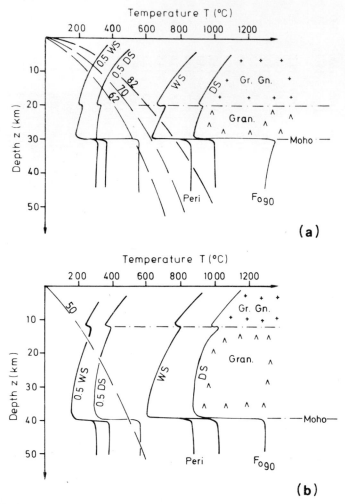

Fig. 4.30 Temperature curves and curves of the dry and wet solidus (DS and WS) as a function of depth for the crustal model of (a) a young continent and (b) a shield area. Here GrGn is granitic–gneissic, Gran is granulitic, Peri is peridotite, Fo_{90} is forsterite 90%, and the numbers along geotherms are the heat flow values in mW/m^2.

in Sections 3.3.1 and 3.6.5. Either τ_{max} from friction (in the upper crust) or the τ from the creep equations (lower crust) will determine the actual maximum stresses in a crustal section. Figure 4.31 shows a combination of all stress data (STRESSMAX). It contains two sets of data for dry and wet conditions and for a given strain rate $\dot{\varepsilon} = 10^{-17}\ s^{-1}$. Within each diagram the curves for increasing temperature result in a strong shifting of the STRESSMAX peak

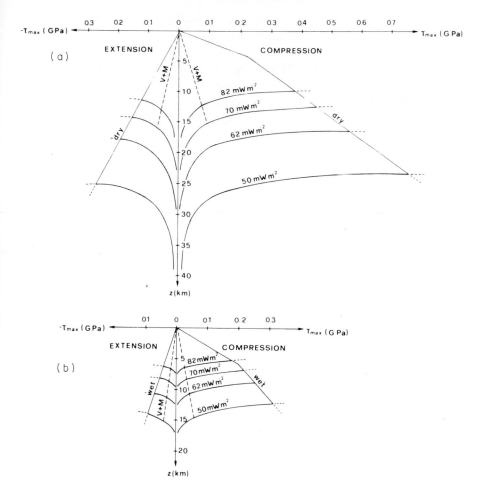

Fig. 4.31 Curves of maximum stresses as a function of depth. Straight lines represent strength (after Byerlee's law). Curved lines represent maximum stresses from creep data; $V + M$ = vermiculite and montmorillonite. (a) $\dot{\varepsilon} = 10^{-17}\,\mathrm{s}^{-1}$, no pore pressure, dry quartz rheology and (b) $\dot{\varepsilon} = 10^{-17}\,\mathrm{s}^{-1}$, hydrostatic pore pressure, wet quartz rheology.

to smaller depth. Using other values for $\dot{\varepsilon}$ results in only small shifts of the peak while the slope of the curve remains the same. They show a slightly stronger stress decrease below the peak for small $\dot{\varepsilon}$ values.

It should be stressed again that only the curves for the hydrostatic pore pressure and wet quartz rheology show a good correlation to the earthquake--depth relation, mentioned in Section 3.3.1. We may conclude from this correlation that at least the upper crust is certainly not dry. For the upper crust the values for the clay minerals vermiculite and montmorillonite (V + M) have also been included in the diagrams. Their weakness leads to rather small possible stresses. Stresses along large fault zones may lie somewhere between the V + M and the ordinary STRESSMAX curves.

For a *dry* lower crust (see upper two diagrams of Fig. 4.31) STRESSMAX might still reach about 50 MPa (500 bars), a value which has been assessed for mylonite zones that were once at great crustal depths (Sibson, 1974; Etheridge and Wilkie, 1979). If the lower crust showed a *wet* quartz rheology it could not sustain such stresses. We must conclude from these data that the lower crust in most areas is dry. If the water content of the crust decreases considerably toward its base, as might be indicated by a number of dehydration reactions under the p–T conditions of the middle crust, other peaks, although not as pointed as the upper one, might occur. The existence of other rock units with a much lower quartz content would also introduce a change in the STRESSMAX curves.

All these possible modifications of the curves, however, are small compared to the upper peak and even compared to the peak at the crust–mantle boundary. Here, at the Moho, the material changes rather abruptly, and olivine rheology with its high activation energies (and higher T_m values) take over. Figure 4.32a provides an overview of the rheological conditions for a moderately warm continent, while a cold continental shield is depicted in Fig. 4.32b. For the calculations Eqs. (4.21) and (4.22) were used. The jump to higher strength and viscosities is particularly strong for dry conditions. Dry forsterite 90 has a high strength under all laboratory conditions. However, for wet conditions and the lherzolite values there is still a considerable jump to high τ–high η values at the Moho. The different rheologies of the lower crust and mantle play an important role in all crust–mantle interactions for fault zones and seismicity. There is growing evidence for a brittle behavior in the high-strength uppermost mantle, both from reflection observations on fault zones and from seismicity.

The viscosity values calculated from the power-law creep equations are in good agreement with the simple assessments of Section 4.5.1. Strain-rate values of 10^{-14} s^{-1}, as estimated for a high-T lower crust, are related to η values of about 10^{20} poise and a maximum shear stress of about 0.1–10 MPa (1–100 bars), as seen from a comparison of Figs. 4.32 and 4.28.

The viscosity minimum in the lower crust has many implications, the most important ones being the possibility for easy access of intruding magmas and the tendency for easy horizontal movements, the formation of multiple reflectors after fractional differentiation along seams, and even the formation of a new, higher Moho from ultramafic cumulates. Clearly, the complex viscosity structure of the continental lithosphere makes it impossible to treat it as a homogeneous plate, as is done in many calculations of bending and

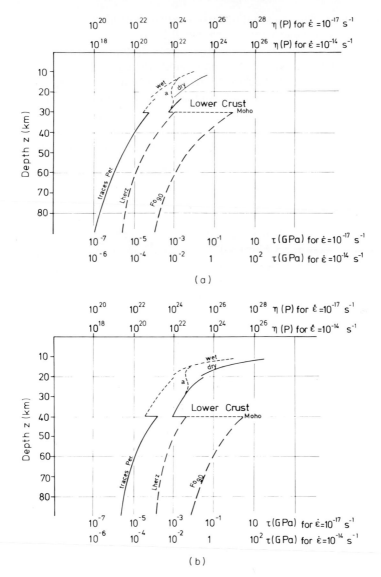

Fig. 4.32 Viscosity η and maximum stress τ as a function of depth for lower crust and uppermost mantle. (a) Young continent and (b) shield area. Traces represents traces of H_2O, lherz represents lherzolite, Per represents peridotite, Fo_{90} represents forsterite 90%, and a is a possible η (or τ) curve in the lower crust.

flexural rigidity. Such calculations are used successfully to obtain the elastic and anelastic parameters for an oceanic lithosphere bent by the load of an oceanic island or by a complicated stress system near trenches. Such calculations, which might give some reliable average values for the oceanic plate, must fail when applied to the thicker and highly variable continental lithosphere. Even the many data from the postglacial uplift in various regions of the continents can only provide rough viscosity values for the astheno-sphere (or lower crust) from extended, previously loaded areas and for the average crust from small areas of loading (Walcott, 1973; Meissner, 1980).

4.5.3   THE SPECIAL ROLE OF FAULT ZONES

Crustal fault zones, their shape, and types of rupture processes have been mentioned in Section 3.3.2 and will be discussed again in Section 6.6.3. Here, the rheological aspects of fault zones will be addressed.

Starting with *in situ* observations near the surface along fault zones, it is well known that slippage along them may occur violently, causing earth-quakes, or gradually by "fault creep" (Steinbrugge and Zacher, 1960). Creepmeter recordings have shown many episodic creep events with a creep propagation speed of 1 to 24 km/day. Such movements are too slow to generate normal earthquakes, although the equivalent "rupture length" of several kilometers is comparable to that of shallow earthquakes of magni-tudes 4–5 (King *et al.*, 1973). The offsets connected with creep events are generally an order of magnitude smaller than those of shallow earthquakes with comparable rupture length. Apparently, near-surface faults can be brittle *and* ductile. These observations imply that special conditions, e.g., the presence of special minerals or mineral assemblages such as montmorillonite and other clay minerals, must be responsible for the episodic creep events. A few features of prominent fault zones beginning with macroscopic peculiari-ties, are listed below:

(1)   Mylonitization, an intense fragmentation of host rocks into small pieces with some preferred orientation.

(2)   The variable width of fault zones, from less than 1 m to some tens of meters at and near the surface. Mylonite zones from greater depth may today be observed at the surface having several branches and reaching widths of 4 km (Sibson, 1974, 1982).

(3)   For the San Andreas fault system stresses of 10 MPa have been obtained for 15–20 km depth (Zoback and Roller, 1979); stress drops of earthquakes are generally smaller than 2 MPa.

(4)   Microscopic investigations of samples from fault zones (fault gouge)

show deformed, elongated grains, correlatable to the stresses during deformation.

(5)   Reduction of grain size, caused by recrystallization (see also Section 4.4.2).

(6)   Unique orientation of quartz c axes.

(7)   General increase of diffusivity.

(8)   General hydrolytic weakening and often a high pore pressure.

(9)   Thermal weakening of various degrees, depending on the intensity of the rupture process.

Some more characteristics have been obtained through geologic observations of faulting and folding of many strata now exposed at the surface. Seismic observations, especially from reflection seismics, have shown that some fault zones really cut the whole crust and even reach the mantle.

In other areas, fault zones seem to end in the middle crust and appear again in the upper mantle. Although the detection of fault zones of high dip angles is rather problematic in seismic observations, there are a few observations which show that some "fault zones" in the lower crust really act as reflectors. Petrological evidence obtained by comparing the microstructure (points 4 to 6 on the list) of gouge samples with artificially induced microstructures in previously intact samples of host rocks under high p–high τ–high T conditions shows that mylonite zones created at depths of 20–40 km suffered a stress of up to 10 MPa at $T = 500°C$ with a strain rate of 10^{-8}–10^{-11} s^{-1}.

The special kind of strong mylotinization found in the previously mentioned samples was not created by a sudden rupture process but by a strong dislocation glide (Etheridge and Wilkie, 1979). It is interesting to decipher these kinds of fault zones in the lower crust, which are apparently aseismic, but sometimes appear as seismic reflectors of appreciable strength. Apparently, these fault zones of the lower crust are zones of limited width and high strain rates in which the elastic and inelastic properties differ remarkably from those of the surroundings. The observed (elastic) reflection coefficient, caused by a different acoustic impedance ρV, can be explained by about 10% lower ρ and lower V values, as will be discussed in Section 6.6.3.

Whereas in the brittle regime of the upper crust seismic and aseismic slip concentrate in fault zones, the lower ductile zone seems to concentrate events only as aseismic creep. The observed high strain rates and the weakening processes listed earlier make these deep fault zones also zones of strongly reduced effective viscosity. Assuming equal stresses for these zones and their environment, we may conclude that

$$\tau = 2\dot{\varepsilon}_F \eta_F = 2\dot{\varepsilon}\eta, \tag{4.32}$$

where the index F stands for the fault zone values. From the deduced high creep rates of 10^{-8}–$10^{-11}\,s^{-1}$ for the fault zone area compared to the estimated creep rates of 10^{-14}–$10^{17}\,s^{-1}$ in the unfaulted surrounding, it follows that

$$\eta_F = \eta\dot\varepsilon/\varepsilon_F \qquad\qquad (4.33)$$

must be at least three orders of magnitude lower than that of the fault zone's vicinity. Such fault zones or zones of weakness in the ductile regime must play an important role for the stress loading of the transition zone and the brittle upper crust. Creep movements concentrate on these relatively narrow zones. It is the sharp peak of the STRESSMAX curve in seismic areas which strongly argues for a continuous reloading of the high-stress zone from below, as illustrated in Fig. 4.31. The viscosity isolines, as assessed from Eq. (4.33), show a strong updoming in the fault zone area (Fig. 4.33b) while the isovelocity lines, depicted in Fig. 4.33a (Feng and McEvilly, 1983) for the San Andreas fault, show an equivalent downwarping.

Crustal fault zones may also serve as a favorable path for the ascent of magmas, which are known to prefer zones of weakness. Ascending magmas, on the other hand, may create new fault zones themselves. The zone of maximum strength in the upper crust acts as a barrier for ascending magmas. Once the barrier is broken, magmas should find easy access to the surface. Such processes will be discussed again when we deal with rifting processes in Section 6.4.4.

4.5.4   RELATIONS BETWEEN VISCOSITY η AND ATTENUATION Q^{-1}

Both high-temperature creep and the attenuation of seismic waves seem to be activated processes; both are possibly defect-controlled and related to the glide of dislocations across subgrains (Anderson and Minster, 1980). Hence, it seems reasonable to try to establish η and Q relations both from the theoretical point of view and from *in situ* measurements. As mentioned in Section 4.3, Q^{-1} as a function of frequency is characterized by a Debye peak or by a number of closely spaced Debye peaks with the two flanks separated by a plateau-like behavior of Q^{-1}. From a number of observational data it seems that the frequency dependence of Q^{-1} is weak over most of the seismic frequency bands, but Q^{-1} seems to decrease for frequencies greater than about 1 Hz (Kanamori and Anderson, 1975; Sipkin and Jordan, 1979). This results in

$$Q \sim f \qquad \text{or} \qquad Q^{-1} \sim f^{-1} \qquad\qquad (4.34)$$

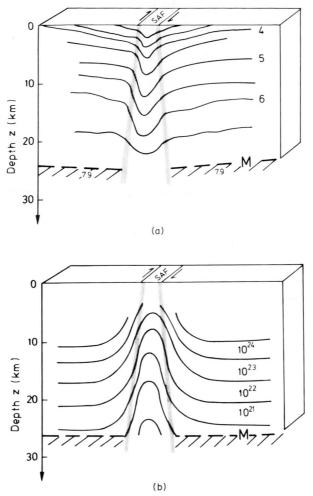

Fig. 4.33 Velocity and viscosity behavior in a fault zone area. (a) Velocity isolines around the San Andreas fault, based on seismic reflection and refraction data from Feng and McEvilly (1983) and (b) viscosity isolines as assessed from Eq. (4.33) and previous figures.

for $f > 1$ Hz, with $f = \omega/2\pi$, in accordance with the right-hand side of the relaxation band. The same result is achieved by solving Eq. (4.11) in Section 4.2.1 for high frequencies, i.e., for $\omega\tau \gg 1$. We obtain

$$Q^{-1} = \tilde{M}\omega\tau \qquad (4.35)$$

with $\tilde{M}$ as in Eq. (4.11). If Q^{-1} is intrinsically similar to steady-state creep, i.e., an activated process, it should follow the same formulas.

Recalling Eq. (4.20) in Section 4.4.3,

$$\dot{\varepsilon} = C_n \tau_S^n \exp(-gT_m/T), \tag{4.36}$$

where τ_S is the shear stress, we may also formulate:

$$Q^{-1} = \tilde{M}\omega\tau C_Q \exp(-g_Q T_m/T), \tag{4.37}$$

where τ is the relaxation time and g_Q a constant.

Recalling Eq. (4.21)

$$\ln \eta = (1/n)[(g^* T_m/T) + (1 - n)\ln \dot{\varepsilon} - \ln C_n], \tag{4.38}$$

we may also write

$$\ln Q^{-1} = -g_Q T_m/T + \ln \tilde{M} + \ln \omega\pi + \ln C_Q. \tag{4.39}$$

These two last equations describe the strong relationship between η and T and between Q^{-1} and T. In order to arrive at a quantitative approach we may look for the derivations

$$\partial(\ln \eta)/\partial(T_m/T) = (1/n)g^* \tag{4.40}$$

with $n = 1$ or 3 and

$$\partial(\ln Q)/\partial(T_m/T) = g_Q. \tag{4.41}$$

By looking for different temperature provinces where Q has been measured, g_Q can be obtained and the relation of Q with T_m/T can be established (while g^* is supposed to be known from the creep experiments, Section 4.4.3). From a collection cf calculated Q and η values, e.g., from those of Fig. 4.13, an empirical relationship has been established for the subcrustal lithosphere (Meissner and Vetter, 1979)

$$\ln \eta = 4.4 \ln Q + 22 \tag{4.42}$$

with η in poise.

This relationship is shown in Fig. 4.34. It is interesting to note that Berckhemer and Hsue (1982) arrived at an equation identical to Eq. (4.42) from their relaxation experiments. Transforming g^* and g_Q back to the original equation for activation processes, i.e., Eq. (4.19), the activation energies are

$$U + pV = E^* = 125 \quad \text{kcal/mol} = 523 \quad \text{kJ/mol} \qquad \text{for} \quad g^* = 29,$$

$$E_Q = 28 \quad \text{kcal/mol} = 117 \quad \text{kJ/mol} \qquad \text{for} \quad g_Q = 7.$$

Anderson and Minster (1980), by comparing measured Q values in the oceanic and in the continental lithosphere (supposedly 200° cooler), arrived

Fig. 4.34   Effective viscosity η versus quality factor Q; ln η = 4.4 ln Q + 22.

at an E_Q value of 210 kJ/mol. It is certainly doubtful whether Eq. (4.42) can be applied for the continental crust. If so, one would get a picture similar to that of Fig. 4.34 with a low-Q lower crust. Here, where η = 10^{20} poise, Q would be 2–300, and would increase to Q = 1000 for η = 10^{23} poise in the area below the stress peak in the upper crust and in the upper mantle. Only very warm lower crusts with η between 10^{18} and 10^{19} poise would exhibit Q values around 100. Until more data are available, Eq. (4.42) might be used as a very first estimate for the conversion of Q into η and vice versa. Certainly, both processes (creep and attenuation) depend on tectonic stress and are exponentially dependent on temperature, although the controlling activation energies are different.

4.6   ABSOLUTE AGE DATING AND ITS IMPLICATIONS

Age determinations have a long history. It was known in Greece about 2500 years ago that fossils in some sedimentary rocks were witnesses of former life and that these rocks were on the sea bottom "a long time ago."

Lyell (1833), comparing the evolution of shells from the Lower Pliocene to the Quaternary and from the Quaternary to the Present, determined an age for the Tertiary of 80 Myr and for the Paleozoic of 240 Myr. The first inaccurate U/Pb (Boltwood, 1907) and U/He (Rayleigh, 1887) determinations proved the hypothesis of geologists, that rocks older than 2000 Myr do exist.

Although radioactive processes and their significance with regard to the decay of certain "unstable" atoms and their time of origin were detected in 1905 by Rutherford, it was nearly 50 more years before radiometric age dating could be considered as an accepted and reliable tool in geoscience. Since then geochronology has been developed into one of the most spectacular branches of science. By determining the ages of minerals and rocks, of the Earth, Moon, and meteorites, and with additional knowledge about geochemical properties and isotopic variations, the nature of planetary differentiation processes and the associated time scales can be inferred. For the Earth's crust, they provide the basis for all models of crustal evolution, crustal recycling, and stability.

4.6.1   BRIEF TREATMENT OF THE PHYSICAL BACKGROUND

It is convenient to describe the composition of atoms by the number of protons and neutrons in the nucleus. The mass number A is the sum of the number of protons Z (atomic number), which is also the average number of electrons, and the number of neutrons N, i.e.,

$$A = Z + N. \tag{4.43}$$

Only nuclei in which Z and N are nearly equal are stable. Moreover, most of the stable nuclides have even numbers of Z and N. The nuclei of unstable atoms undergo spontaneous transformations by the emission of particles and energy, resulting in changes of Z and N, hence transforming the original (parent or mother) element into another (daughter) element. The daughter itself may not be stable and may decay into another daughter, a process which may continue until a stable daughter is finally reached.

There are basically three different types of rays which are emitted during radioactive decay: α-, β-, and γ-rays. A large group of radionuclides with $Z \geq 58$ decay by the spontaneous emission of α-particles from their nuclei. *Alpha-particles* are made of two protons plus two neutrons, i.e., they have a charge of 2, reducing both Z and N by 2 and hence A by 4 (see also Fig. 4.35a). Another large group of unstable atoms decay by emitting β-particles from the nucleus. There are β-rays that are streams of particles identical to electrons. These β-particles from the nucleus are negative (β^-) (see Fig. 4.35b). The β^- process is related to a transition from a neutron to a proton in the nucleus, including the emission of a neutrino. There are other decay processes which are of some importance, such as electron capture and nuclear fission, as described in detail by Faure (1977). *Gamma-rays* are high-energy electromagnetic waves from the nucleus. They are similar to x rays. They do not change Z or N. Some decay schemes of useful elements in geoscience are presented in Table 4.2.

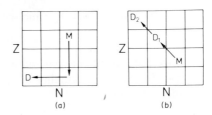

Fig. 4.35 Two radioactive decay processes in a proton (Z)-neutron (N) diagram; M, mother; D, daughters. (a) α-decay and (b) β^--decay.

The decay from a parent (or mother) element M is expressed by dM/dt and is proportional to the number of M remaining, i.e.,

$$-dM/dt = \lambda M. \tag{4.44}$$

The proportionality constant λ is by now well known for a large number of decay processes. It represents the probability that an atom will decay within a certain time t. Integrating Eq. (4.44), we obtain

$$-\int \frac{dM}{M} = \lambda \int dt \tag{4.45}$$

or

$$-\ln M = \lambda t + \ln M_0, \tag{4.46}$$

where M_0 is the initial concentration of the mother element, i.e., M at $t = 0$:

$$M = M_0 e^{-\lambda t}. \tag{4.47}$$

Table 4.2

Decay Scheme of Useful Isotopes and Their Half-Lives[a]

Nuclide	Isotopic abundance (%)	Decay type	Decay product	Half-life (yr)
Primordial Nuclides				
^{87}Rb	27.83	β^-	^{87}Sr	4.99×10^{10}
				4.88×10^{10}
^{232}Th	100	a	^{208}Pb	1.39×10^{10}
^{238}U	99.28	a	^{206}Pb	4.50×10^9
				4.47×10^9
		sp. fission	$^{131-136}$Xe^x, tracks	8.2×10^{15}
^{235}U	0.72	a	^{207}Pb	7.13×10^8
				7.04×10^8
^{40}K	0.0119	(β^+) K-capture	$(^{40}$Ca$)$ ^{40}K	1.31×10^9
^{147}Sm	15.0	a	^{143}Nd	1.06×10^{11}
^{187}Re	62.6	β^-	^{187}Os	$\sim 5 \times 10^{10}$

[a] From Kirsten (1976).

Solving for t, we obtain the time when the radioactive process started:

$$t = (1/\lambda) \ln(M_0/M). \tag{4.48}$$

The constant λ is often expressed by the half-life t_H, the time at which half of M_0 has decayed. Introducing t_H for t and $M_0/2$ for M in Eq. (4.47), the half-life t_H becomes

$$t_H = (\ln 2)/\lambda. \tag{4.49}$$

Only M in Eq. (4.47) can be determined by laboratory measurements and expressed as an atomic concentration (e.g., moles per cubic centimeter) or, better, a relative concentration, related to a stable isotope, while M_0 remains unknown (except for the ^{14}C method). But fortunately, the radiogenic stable daughter element can be determined by introducing

$$D = M_0 - M \tag{4.50}$$

or

$$D/M = M_0/M - 1. \tag{4.51}$$

Combined with Eq. (4.47), this gives

$$D/M = e^{+\lambda t} - 1 \quad \text{or} \quad D = M(e^{\lambda t} - 1), \tag{4.52}$$

and finally

$$t = (1/\lambda) \ln(1 + D/M). \tag{4.53}$$

Figure 4.36 illustrates the behavior of mother M and stable daughter D according to Eqs. (4.47) and (4.50). For this example, two uranium decays are depicted. Although in practice there are other complications, such as the existence of more than one stable daughter or nonradiogenic daughters in the mineral (rock) or the migration of daughter and mother elements during the tectonic history, Eq. (4.53) remains one of the basic formulas for age

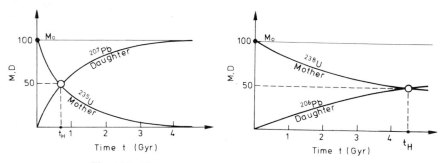

Fig. 4.36   The two uranium (U)–lead (Pb) decay schemes.

determination. One of the main problems of radioactive dating is that the daughter isotope determined in the rock (mineral) originates only partly from the parent present in the rock; another part, D_0, is preexisting (primordial). This modifies Eq. (4.52) to

$$D = D_0 + M(e^{\lambda t} - 1). \tag{4.54}$$

As mentioned previously, for radioactive dating it is preferable to use relative concentrations, because a mass spectrometer measures such relative concentrations and because of geochemical considerations. Thus, D, D_0, and M are related to a stable and nonradiogenic isotope of the element D_n. In the U–Pb method this stable daughter element is ^{204}Pb; in the Rb–Sr method it is ^{86}Sr. When introducing such reference values it is tacitly assumed that any diffusion or migration of daughters D and D_n takes place in equal proportions because of their chemical similarity. Introducing the nonradiogenic daughter D_n changes Eq. (4.54) to

$$(D/D_n) = (D_0/D_n) + (M/D_n)(e^{\lambda t} - 1). \tag{4.55}$$

From this equation t can be determined in two ways. If D/D_n is $\gg 1$, then D_0/D_n can be neglected or assumed to be known; for instance, the value ^{87}Sr/^{86}Sr $= 0.704$ for recent mantle-derived basaltic rocks. In the second and more general case, t can be calculated from Eq. (4.55) if *several* minerals are present in the rock all of which have different M/D_n ratios. In this case it is assumed that the initial values of D_0/D_n are all the same. As $e^{\lambda t} - 1$ is approximately λt for all methods with $\lambda < 10^{-10}$ yr^{-1} (or half-lives larger than 10^{10} yr), Eq. (4.55) represents a straight line in a D/D_n versus M/D_n diagram. Different minerals determine the slope of the straight line λt in a D/D_n versus M/D_n diagram:

$$(D/D_n) = (D_0/D_n) + (M/D_n)\lambda t. \tag{4.56}$$

(See also Fig. 4.37a.) Not only can t be determined from the slope of the straight line, the initial value $I = D_0/D_n$ can also be obtained from the intercept point.

Equation (4.56) can be used for a number of minerals or rocks. To show the development of the D/D_n ratio a second diagram is used in which the time is plotted on the abscissa. In such a diagram (see Fig. 4.37b), Eq. (4.56) splits into several straight lines, e.g., into four lines if four minerals (or rocks) are used for the age determination (as shown in Fig. 4.37a). In Fig. 4.37b the different slopes of the four straight lines resemble the different values of M/D_n ratios in the minerals (or rocks). This diagram shows the growth of the daughter elements with time. It is especially useful for explaining the effect of resetting the clocks of a mineral by thermal events. Figure 4.38a shows an example of two isochrones with D/D_n in the form of ^{87}Sr/^{86}Sr in the Rb–Sr

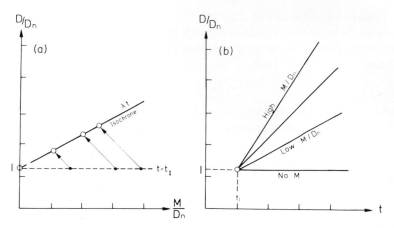

Fig. 4.37 Different growth of radiogenic daughters as shown in the two basic diagrams; abscissas are (a) M/D_n and (b) t, where M is the mother, D the radiogenic daughter, D_n the stable daughter, and I the initial D/D_n value. The λt curve is the isochrone; its gradient is λt and permits easy calculation of t.

method. One is the "whole-rock isochrone" and the other the "mineral isochrone". It is generally found that during high-grade metamorphism the Rb–Sr isotopes mix with each other; the metamorphic event resets the Rb/Sr clock. Figure 4.38b shows such a case history. The granite formed originally at t_i. At t_{Me} a metamorphic event mixed the Sr isotopes. From the present ^{87}Rb and ^{87}Sr content in the minerals only t_{Me} can be determined; t_I, the absolute age of the rock, can be determined only if several rocks are

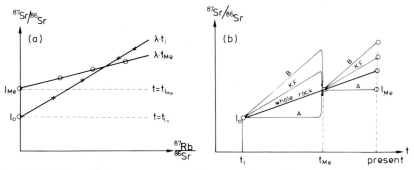

Fig. 4.38 Rb–Sr method at one metamorphic event (Me). (a) λt_i is the whole-rock isochrone showing correct I and t_0 values, and λt_{Me} is the mineral isochrone, showing concentration and time of metamorphic event. (b) Different growth of $^{87}Sr/^{86}Sr$ in three minerals and in the whole rock; A, apatite; B, biotite; KF,K-feldspar resetting of clocks at $t = t_{Me}$. [From Faure (1977).]

investigated. The metamorphic event has reset the clocks of various minerals while the whole rock is still a closed system which keeps all mother and daughter elements in it. Consequently, in the whole rock the growth of the daughter elements is independent of the metamorphosis, while a new initial value I_{Me} is created for the minerals at t_{Me}. This implies that different laboratory measurements of mother and daughter elements for *several* minerals *and* for several rocks lead to the determination of the time when the rock was formed, I_0, and the time of the metamorphic event, I_{Me}.

4.6.2 THE MEANING OF "TIME" AND "CLOSED SYSTEMS" IN RADIOMETRIC AGE DATING

The results of the previous section, especially the different ages obtained by careful analyses of rocks and minerals, afford a better understanding of the time t which is to be calculated. Simply speaking, t_{Me} is the time when the radiogenic clocks started to tick. Before that time, of course, the radioactive processes also worked. However, because of high temperatures, e.g., in a melt or/and in a chaotic or widely dispersed state of matter, any daughter could move anywhere. When mothers and daughters were incorporated in a "closed system" the whole family had to stay together, and this makes age dating possible. The smallest closed system is a mineral; the largest might be the solar nebula. Examples of closed systems, roughly in the order of increasing size, are

a mineral	the continental crust (and parts of it)
a rock	the oceanic crust (and parts of it)
a meteorite	the mantle (and parts of it)
a rock unit	
a class of meteorites	the whole Earth (or Moon)
a geologic province	the solar system

The time of origin of all these systems can be determined, provided representative samples can be collected. Whether samples are really representative and whether a system is really closed are the basic questions of geochronology. Many of them can be solved with the help of geochemical constraints and increased accuracy of the physical laboratory measurements.

4.6.3 THE BASIC PRINCIPLES OF LABORATORY MEASUREMENTS

After the detection of radioactivity in the beginning of this century instruments were first developed to perform a chemical analysis of the ratio of mothers to daughters. Using ordinary methods of chemical analysis, the uranium and thorium decays were obtained by measuring the amounts of U, Th, and Pb in samples. However, these chemical methods could not distinguish between different Pb isotopes, e.g., those from the two uraniums, those from Th, and nonradiogenic Pb. Age dating greatly increased in accuracy with the development of the mass spectrometer. This instrument can separate charged atoms and molecules of different masses. A beam of artificially charged ions of a gaseous sample passes through electric and magnetic fields in a vacuum system. In the electric field with a potential difference U, the ions of mass m and electric charge z are accelerated and acquire the energy

$$E = zU = \tfrac{1}{2}mv^2, \tag{4.57}$$

where v is the velocity of the ions. Solving for v, we get

$$v = (2zU/m)^{1/2}. \tag{4.58}$$

In the magnetic field with strength H the ions are forced into circular paths because here

$$HzV = mv^2/r. \tag{4.59}$$

Eliminating v from Eqs. (4.57) and (4.59), we finally arrive at

$$m/z = (H^2/2U)r^2. \tag{4.60}$$

The magnetic field H, therefore, deflects the ions into circular paths whose radii r are proportional to the square root of the isotope mass m. The beam, hence, becomes divergent; the separated ions fly through the analyzer tube to the collector.

The ion collector behind a split plate collects only those ions that have the right radius r, i.e., the right mass. The electric and magnetic fields are systematically adjusted to the ions to be measured. A voltage difference is measured across the terminals of a resistor. Therefore, a mass spectrum can be obtained by continuous variation of the magnetic or electric field. The heights of the spectral peaks are proportional to the abundances of the respective isotopes. Figure 4.39 shows the principles of a mass spectrometer; the circular path here is realized by a 60° deviation.

The precision and the sensitivity of mass spectrometers have steadily improved. Since about 1955 it has also been possible to investigate solid (and fluid) samples, and this initiated the Rb–Sr and many other new methods. Recent developments include the incorporation of digital data-collecting systems and computer evaluation and isotope dilution analyses. In the latter method an exactly known quantity of a known second isotope is added to the sample. The resulting two peaks are analyzed, and their ratio is directly related to the quantity of the unknown isotope in the sample (Faure, 1977).

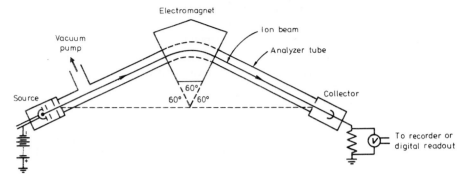

Fig. 4.39   Schematic diagram of a 60° sector mass spectrometer.

Neutron activation has also improved the accuracy of isotopic measurements.

Another exciting development was the extension of mass spectrometry to some of the rare earth elements (REEs). The importance of this step stems from the fact that the REEs occur in virtually all types of rocks (and meteorites). Because their ionic radii are similar, all REEs show strong geochemical coherence. During most fractionation processes they act differently from other mother–daughter pairs. This will be a topic of Section 4.6.4. To summarize, mass spectrometry with all its recent developments is the most important tool for age determination and studies of evolution of all closed systems. Only the carbon dating and tritium hydrological methods do not make use of mass spectrometry. The carbon dating method has been extended to measure the past 100,000 years by analyzing radioactive ^{14}C in materials containing dead plants, counting the decay particles of the ^{14}C mothers.

4.6.4   THREE IMPORTANT METHODS OF AGE DATING AND SOME IMPLICATIONS

It is beyond the scope of this book to give a systematic treatment of all radiometric decay systems and methods. Instead, three examples are presented, two of them classical and the third rather new, which are widely used in studies of crustal evolution; these are the uranium–lead (U–Pb) method, the rubidium–strontium (Rb–Sr) method, and the new samarium-neodymium (Sm–Nd) method. All three have contributed greatly to geochronology, all three make use of Eqs. (4.55) and (4.56), and each of them has certain special merits. The development of new isotopic methods such as the lutetium–hafnium (Lu–Hf) method will certainly lead to new tools for additional investigations.

The U—Pb methods provide two independent decay schemes. Uranium-238 decays by eight α and six β mechanisms to the stable isotope ^{206}Pb. Uranium-235, after seven α and four β decay processes, finally ends up at ^{207}Pb. As described in Section 4.6.1, the ratio of the radiogenic to a stable, nonradiogenic isotope, in this case ^{204}Pb, is used to describe the decay schemes, as outlined in Eq. (4.55):

$$\frac{^{206}\text{Pb}}{^{204}\text{Pb}} = \left(\frac{^{206}\text{Pb}}{^{204}\text{Pb}}\right)_0 + \frac{^{238}\text{U}}{^{204}\text{Pb}}(e^{\lambda_{238}t} - 1), \tag{4.61}$$

$$\frac{^{207}\text{Pb}}{^{204}\text{Pb}} = \left(\frac{^{207}\text{Pb}}{^{204}\text{Pb}}\right)_0 + \frac{^{235}\text{U}}{^{204}\text{Pb}}(e^{\lambda_{235}t} - 1). \tag{4.62}$$

Both decays were already shown in Fig. 4.36. Both are independent of each other, but an additional constraint is provided by the ratio $^{235}\text{U}/^{238}\text{U}$. At present this ratio is 0.007257 in all natural rocks. Dividing Eqs. (4.61) by (4.62) and introducing $^{235}\text{U}/^{238}\text{U} = C$, one obtains the basic equation for the so-called lead–lead (Pb–Pb) method:

$$\frac{^{207}\text{Pb}}{^{204}\text{Pb}} = \left(C\frac{e^{\lambda_{235}t} - 1}{e^{\lambda_{238}t} - 1}\right)\frac{^{206}\text{Pb}}{^{204}\text{Pb}} + \left(\frac{^{207}\text{Pb}}{^{204}\text{Pb}}\right)_0 - \left[\frac{e^{\lambda_{235}t} - 1}{e^{\lambda_{238}t} - 1}\left(\frac{^{206}\text{Pb}}{^{204}\text{Pb}}\right)_0\right]. \tag{4.63}$$

No measurements of U are required in this method. Plotting $^{207}\text{Pb}/^{204}\text{Pb}$ as a function of $^{206}\text{Pb}/^{204}\text{Pb}$ again gives a straight line because of the small λ values. Figure 4.40a gives an example of this method and also shows the small I values provided by the last two terms of Eq. (4.63).

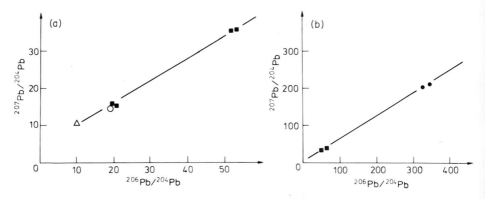

Fig. 4.40   Two examples of the Pb–Pb method. (a) and (b) Isochrone gives $t = 4.66$ Gyr for achondrites, lunar regolith, and marine sediments; ■, stony meteorites; ○, marine sediments; △, iron meteorites; and ●, lunar regolith. [From Faure (1977). Copyright 1977 by John Wiley and Sons, Inc.]

The Pb–Pb method provides an important contribution to the calculation of the age of the Earth. Isotope ratios of "average" terrestrial lead as found in marine sediments are on the same isochrone as those of meteorites and lunar regolith, as shown in Fig. 4.40b. As mentioned before, the chemical similarities of the daughter isotopes, e.g., lead in this case, keep their ratios constant through all fractionation and differentiation processes. Therefore, representative average samples of lead in well-mixed marine sediments, not influenced by variations of the mother element, may really represent average terrestrial values. For more details see Faure and Powell (1972) and Faure (1977).

The Rb–Sr method has been applied to determine the age of minerals, rocks, and meteorites. It was especially useful for deriving times of metamorphic events of crustal rocks, as mentioned in Section 4.5.1 and shown in Fig. 4.38. The different methods of obtaining "mineral ages" and "whole-rock ages" and their relation to different degrees of thermal effects during metamorphic events were developed when working with the Rb–Sr methods. After each metamorphic event, the $^{87}Sr/^{86}Sr$ ratio versus time is the same as before and is proportional to the $^{87}Rb/^{86}Sr$ ratio, i.e.,

$$d(^{87}Sr/^{86}Sr)/dt = \lambda^{87}Rb/^{86}Sr. \qquad (4.64)$$

The slope depends on the Rb content of the closed system and may be extremely different depending on the geochemical constraints of the individual minerals, as indicated in Fig. 4.38. The age of diagenesis of sedimentary layers can be determined in a similar way to that of a metamorphosis in a metamorphic rock.

Like the U–Pb methods, Rb–Sr age dating has contributed greatly to the understanding of early differentiation processes in crustal material. In particular, the initial value $I = (^{87}Sr/^{86}Sr)_0$ was interpreted as being representative of different reservoirs or (semi-) closed large systems, like that of the solar nebula before accretion, like various classes of meteorites, or like the crust–mantle mix which produced the very first crustal rocks, dated by the Rb–Sr method. The evidence that crustal material can recycle into the mantle and that older rocks are often involved in later metamorphic processes is based on I values from the Rb–Sr method. The I-values from the oldest, mantle-derived granitic rocks are about 0.700 to 0.702. They are only slightly larger than the "Basaltic achondrite best initial" (BABI) value of 0.69899 which describes the $^{87}Sr/^{86}Sr$ ratio at the time of formation of the solar system $4.6 \pm 0.1 \times 10^9$ years ago. The I-values from mantle-derived crustal rocks steadily increase with time. If the mantle could be considered a closed system, the increase of $I = (^{87}Sr/^{86}Sr)_0$ in mantle-derived samples should be linear as the $^{87}Sr/^{86}Sr$ values of minerals increase linearly as a function of time. Observations of I values, however, show a nonlinear increase with time. These observations clearly indicate a varying degree of extraction of crustal material from the mantle. In the beginning much more crustal material was created by extraction from the mantle than was the case later.

The situation of mantle evolution in view of the $^{87}Sr/^{86}Sr$ values is shown in Fig. 4.41. Present-day $^{87}Sr/^{86}Sr$ values, which determine the I values for all mantle-derived crustal rocks, are about 0.704. Older mantle-derived rocks, e.g., basaltic or granitic units, have I values smaller than 0.704. The mantle itself does not seem to be homogeneous. There are parts which are strongly depleted in Rb because of strong differentiation and extraction of magmas for the formation of crustal units. Partial melting of the lower or middle crust, on the other hand, must produce much higher I values because magmas were derived from a point along the steep slopes of the "crustal" lines, which have higher $^{87}Sr/^{86}Sr$ values and a slope proportional to $^{87}Rb/^{86}Sr$ [see also Eq. (4.64)]. The steeper slopes of these lines are explained by the fact that much more Rb than Sr enters the crust during differentiation processes in the mantle. All mantle-derived magmas take a much larger amount of Rb than Sr with them to the crust.

Returning to the $^{87}Sr/^{86}Sr$ versus $^{87}Rb/^{86}Sr$ diagram, it is possible to distinguish between similar material extracted from the mantle and from the crust, although both rock units may show the same age. Figure 4.42 shows the parallel slopes, indicating the same age, but different I values for two granitic rock complexes, one derived from the crust and one from the mantle.

Following these lines of evidence, the evolution of different crustal complexes can be studied by the Rb–Sr method. Both the U/Pb and Rb/Sr ratios

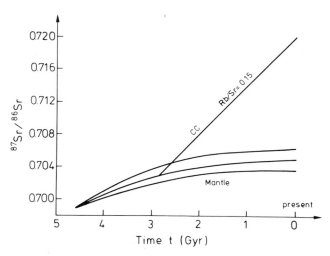

Fig. 4.41 Isotopic evolution of terrestrial strontium. [From Faure (1977). Copyright 1977 by John Wiley and Sons, Inc.]

changed remarkably during accretion of the solar nebula. Rubidium is much more volatile than strontium, and a high Rb/Sr ratio is still found in the sun's photosphere, while the Earth became depleted in the mother element Rb. For U–Pb an opposite kind of effect took place during accretion of solid bodies. The mother element U was enriched because of its earlier condensation as a consequence of its higher condensation temperature, while the daughter Pb remained stronger in the solar nebula. Core formation might have had a similar effect in enrichment of U in the mantle, because Pb may have entered the Earth's core in form of sulfides. Differentiation of crustal material from the mantle, however, started with isotope ratios very different from those of the solar nebula, chondrites, and achondrites and from the whole-Earth composition. The main reason for these differences is the difference in chemical properties between mothers and daughters.

The samarium–neodymium (^{147}Sm/^{143}Nd) method differs from the two other methods in several ways. First, both Sm and Nd are rare earth elements, and they are found as trace elements in all rocks and minerals. They are nonvolatile and nonmetallic. Moreover, they are chemically so similar that both mother and daughter must have condensed from the solar nebula at the same time and with the same ratio. They could not have entered the Earth's core in different proportions. Only fractionation in magmatic processes seems to play a role in producing different mother–daughter ratios. During these processes the Sm–Nd ratio also behaves differently from the ratios in the two other methods, since the daughter element, in this case Nd, is

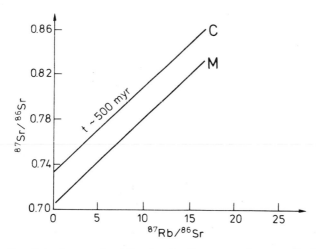

Fig. 4.42   Two Sr isochrones of granites showing the same age (same gradient) but different initial ratios. Here C is the crustal origin and M the mantle origin.

the enriched isotope in the crust. In other words, while the continental crust is enriched in the mothers Rb and U it is depleted in the mother Sm. This is no handicap for the age dating methods, which are similar for all three decay schemes, including the relation of the radiogenic daughter ^{143}Nd to the nonradiogenic ^{144}Nd.

Figure 4.43 gives an overview of the growth of the isotopic ratios as a function of time (De Paolo, 1981a). Note that the slopes of the curves for the daughters' ratio are the same and are not affected by accretion and core formation for the Sm–Nd method. The slope represents the same mother/daughter ratio as that of the curve for a chondritic uniform reservoir (CHUR). Only the magmatic differentiation of the mantle, i.e., the segregation of the crust, modifies the slope of the CHUR curve. Deviations from it may be interpreted directly in terms of depletion of the mantle caused by crustal segregation. In this context it is especially noteworthy that the initial

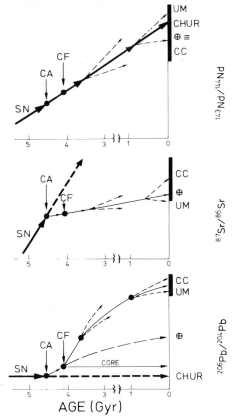

Fig. 4.43   Growth of isotopic ratios of Pb, Sr, and Nd as a function of time; UM, upper mantle, average; CC, continental crust, average; CHUR, chondritic uniform reservoir; SN, solar nebula; CA, cosmic accretion, ⊕, whole Earth average, and CF, core formation. [From De Paolo (1981a).]

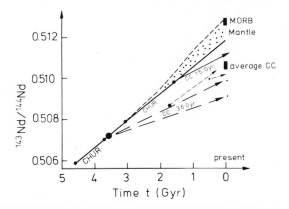

Fig. 4.44 Isotopic evolution of Nd ratios for samples from mantle and crust showing deviations from the CHUR curve since about 3.7 Gyr; MORB, mid-oceanic ridge basalts. Other symbols are defined in Fig. 4.43.

ratios $I = {}^{143}\mathrm{Nd}/{}^{144}\mathrm{Nd}$ of old crustal rocks from Eqs. (4.54) and (4.55) and the plot of ${}^{143}\mathrm{Nd}/{}^{144}\mathrm{Nd}$ versus ${}^{147}\mathrm{Sm}/{}^{144}\mathrm{Nd}$ all lie on the CHUR curve and no whole-rock ages have been found greater than about 3.8 Gyr. Hence, the Sm–Nd method provides important confirmation of the age gap between accretion at 4.6 Gyr and the first appearance of crustal rocks that were not recycled any more. Figure 4.44 shows the general development of the crust/mantle system in the light of the Sm–Nd method. It should be compared with Fig. 4.41, the analogous description by means of the Rb–Sr method.

4.6.5   DEVIATIONS FROM THE AVERAGE INITIAL ISOTOPIC RATIOS IN THE MANTLE AND THEIR IMPLICATION FOR THE SEGREGATION OF CRUSTAL COMPLEXES

In the previous section the enrichment of the crust in the daughter Nd and the mothers Rb and U and the corresponding depletion of these isotopes in the mantle were mentioned. For the Sm–Nd method deviations in the initial ${}^{143}\mathrm{Nd}/{}^{144}\mathrm{Nd}$ from the CHUR standard, and for the Rb–Sr method deviations from the mantle's ${}^{87}\mathrm{Sr}/{}^{86}$ average curve, have been carefully studied as a function of time for various mantle-derived crustal rock complexes. Deviations from the standard curves are generally symbolized by the Greek letter ε, e.g., $\varepsilon_{\mathrm{Nd}}$ or $\varepsilon_{\mathrm{Sr}}$. In general:

$$\varepsilon = \frac{(D_0/D_n)_{\mathrm{obs}} - (D_0/D_n)_{\mathrm{st}}}{(D_0/D_n)_{\mathrm{st}}} \times 10^3, \qquad (4.65)$$

which gives

$$\varepsilon_{Nd} = \frac{(^{143}Nd/^{144}Nd)_{obs} - (^{143}Nd/^{144}Nd)_{CHUR}}{(^{143}Nd/^{144}Nd)_{CHUR}} \times 10^3 \qquad (4.66)$$

for the Nd–Sm decay, the indices obs, st, and CHUR indicating observed, standard, and CHUR values. As already shown by the scatter of the I values of the Nd ratios from the CHUR standard for the past 2 Gyr, the corresponding ε_{Nd} values increase toward the present. However, the deviation ε_{Nd} is different for different crustal rocks. Investigations so far have concentrated on basaltic rocks in different environments. The upper mantle today seems to have an average ε_{Nd} of $+12$ and an ε_{Sr} of -30. Young mantle-derived basalts can be divided into different groups based on histograms of their ε values, as seen in Fig. 4.45. The mantle below the source of the mid-oceanic ridge basalts along present-day rifts is much more depleted in Nd (and enriched in Sr) than that below continental flood basalts. The positive ε_{Nd} and negative ε_{Sr} values represent the integrated effect of making low-Sm/Nd and high-Rb/Sr crust over the past 3.8 Gyr. The good correlation of the ε-values from these methods (and also those from the new Lu–Hf method) is a reliable indication that the assessment of the enrichment and depletion processes of magma generation from the mantle is reasonable. The data from the U–Pb method are less well understood. Some leakage from crust to mantle and even from mantle to core has been invoked for Pb isotopes (Dupré and Allègre, 1980). For the other methods mentioned earlier the mantle appears as a rather well-mixed system with only small deviations from the standard curves, although prominent deviations were recently observed (Waenke, 1984). Basalts derived from the Moon's mantle show very much greater scattering. Hence, the lunar mantle must have been much more heterogeneous in the time of mare formation than was the early Earth.

The ε_{Nd} and ε_{Sr} values generally show a clear (negative) correlation for most rocks, as shown in Fig. 4.46. The $\varepsilon = 0$ values, i.e., the values for undifferentiated (standard) mantle, correspond to each other. The island arcs,

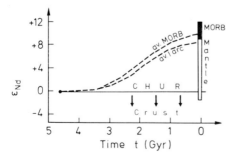

Fig. 4.45   Deviations of Nd ratios from the CHUR curve (ε values) for samples from different mantle reservoirs; I arc, island arcs ε values (fractional differences) are 100 times that of CHUR.

which are thought to be associated with the generation of new continental crust, are slightly outside the general trend of ε_{Nd}-ε_{Sr} values. This suggests an involvement of ocean water in the formation of these rocks, because there are larger amounts of Sr than of Nd in ocean waters. Ocean floor basalts also have such anomalous Sr ratios. Hence, it might be assumed that island arcs are generated from the melting of basalts and gabbros in the ocean floor (De Paolo, 1981a,b; Allègre *et al.*, 1982). The difference between the ε values for island arcs, which are supposed to represent major parts of continental crust, and for the mid-oceanic ridge basalts (MORBs), which are supposed to represent oceanic crust derived from "average" mantle, has been found to exist for at least the past 2 Gyr. According to De Paolo and Johnson (1979), it is nearly 20% over this time period and indicative of mantle evolution and crustal recycling. If no crustal recycling had taken place the difference between the MORB and island arc curves would be much greater than calculated from the decay schemes and the mass of the continental crust (De Paolo, 1983). However, estimates of recycling in the order of 35–40% of the mass of the continental crust per billion years, or 3.5–4% per 100 million years, are definitely overestimates when compared to the rates of accretion on a global scale (see also Chapter 7). Although the relation between the chemical composition of the crust and the complementary changes of the underlying mantle seems so straightforward and simple, the models for the depth of the different magma sources are highly controversial. Continental flood basalts with their ε_{Nd} and ε_{Sr} values near zero may come from a nearly undifferentiated, i.e., CHUR-like mantle, possibly from a lower reservoir. Alternatively, they might originate from an Nd-depleted average upper mantle with $\varepsilon_{Nd} = +12$ together with some complicated mixing with the continental crust, where ε_{Nd} is about -15 (McCullogh and Wasserburg, 1978; Allègre *et al.*, 1982). The large masses of MORB with ε_{Nd} between 8 and 12 may originate by direct derivation from the upper mantle with $\varepsilon_{Nd} = 12$. A more complicated model has been suggested by Anderson (1981, 1982) in which the different magmas come from a stratified mantle and the MORB

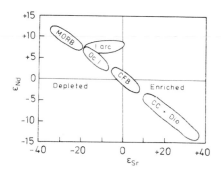

Fig. 4.46 Different samples in an ε_{Nd}-ε_{Sr} diagram; MORB, mid-oceanic ridge basalts; I arc, island arc basalts; Oc I, oceanic island basalts; CFB, continental flood basalts; CC, continental crust, average; and Dio, mantle-derived diopsides.

masses come from a huge eclogite cumulate layer resulting from crystallization of an early picritic melt fraction. Another reservoir, more primitive and less depleted, is considered responsible for island arc kimberlites, continental xenoliths, and alkali basalts. All magmas are mixtures from these two (or more) reservoirs. Ongoing geochemical and isotopic research should reduce the range of possible models in the near future. Some of the implications of age dating and geochemistry for the evolution of the continental crust will be discussed again in Chapter 7.

Chapter 5 | The Composition of the Continental Crust

5.0 INTRODUCTION

It is evident that continental crust has been extracted from the mantle, either directly or indirectly, by a long history of melting events. Magmatic and metamorphic processes change the primordial crusts episodically. Geological, petrological, geophysical, and geochemical studies of extrusives, intrusives, xenoliths, and sites of exposures of ancient, deep crustal complexes together with age determinations of the formation of rocks or metamorphoses are the main tools for studies of crustal composition. The old idea of a "granitic" layer on top of a "basaltic" layer overlying an ultrabasic mantle was the fundamental concept until the 1940s and 1950s. It was mainly based on correlations of seismic velocities, measured in the laboratory on granitic and basaltic rocks, with those found in refraction surveys. In the 1960s more detailed seismic information together with a growing knowledge of petrological details led to a more complicated picture of continental structures. Figure 5.1 shows some of the models which could well explain petrological and seismic observations. A discontinuity between 12 and 22 km depth, called the Conrad discontinuity in Europe and the Reel discontinuity in Canada, could be traced under most continental regions. Formerly, it was thought to mark the boundary between granitic and basaltic material. Today it is still considered to mark the boundary between the upper and the lower crust, but it was found to be rather elusive in many areas. Seismically, it may be defined as the boundary where velocities change from values of 5.7 to 6.3 km/s in the

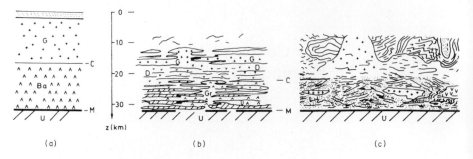

Fig. 5.1   Structural–compositional models of the continental crust. (a) Classical view, (b) model of Meissner (1967), and (c) model of Smithson (1978); G, granitic; D, dioritic; Ba, basaltic; Gr, gabbroic–granulitic; U, ultramatic; C, Conrad; and M, Moho.

upper crust to those of 6.5 km/s or more in the middle and lower crust. In reflection seismics in Caledonian and Variscan areas the Conrad seems to coincide with the beginning of a high reflector density, as will be described in Sections 6.2.3 and 6.2.4. Petrologically, it might be related to a change from gneissic–granitic, i.e., sialic, material of the upper crust to an amphibolitic or granulitic facies below. All these rock types are within the range of the observed seismic velocities. It might be added that the old idea of a gabbro–eclogite transition from lower crust to mantle is not in agreement with the observation of a rather sharp and narrow transition zone between crust and mantle or with the observed velocities in the lower crust.

The gneisses of the upper crust are generally coarse-grained metamorphic rocks, rich in feldspar with streaks of mica or hornblende. Many of them are metamorphosed sediments (so-called metasediments); others contain major contributions of volcanics. Within these gneisses granitic or granodioritic diapirs are found, often teardrop-shaped with a lower density and a slightly lower seismic velocity. It was, of course, the lower density of quartz-, plagioclase-, and feldspar-containing granites which made them enter the upper crust in the first place, not only as low-density granitic melt but also by diapirism via plastic flow. As mentioned in Section 4.6.4, granitic melts may be created either in the mantle or in the lower crust under appropriate conditions.

This chapter should provide a rough introduction to the mineralogical and petrological background written by a geophysicist. For a thorough and detailed study of crustal composition the reader is referred to textbooks such as those by Press and Siever (1982), Verhogen *et al.*, (1970), Gass *et al.* (1971), and Ringwood (1975), from which much information has been extracted and incorporated in the following sections.

5.1   ROCK-FORMING MINERALS AND THEIR ELEMENTS

Minerals are naturally occurring, mainly crystalline, chemical compounds. In their aggregates they form three classes of rocks:

(1)   the igneous rocks, solidified from a molten stage,
(2)   the metamorphic rocks, formed under high-pressure, high-temperature (p–T) conditions,
(3)   the sedimentary rocks, formed by erosion and redeposition.

The most important minerals for the continental crust are the silicates, the aluminosilicates, and some oxides. All silicates exhibit great structural variety, and there is a close correlation between structure, symmetry, and physical properties (Verhoogen *et al.*, 1970).

In general, silicates are built of SiO_4 tetrahedra with a central (small) Si-atom in four-coordination with respect to the large O atoms. Quartz (SiO_2) is an extreme case in which every O atom is shared by two tetrahedra in a three-dimensional framework. Tetrahedra are arranged in different ways, with such cations as sodium (Na^+), potassium (K^+), calcium (Ca^{2+}), magnesium (Mg^{2+}), ferrous iron (Fe^{2+}), and ferric iron (Fe^{3+}) in spaces between tetrahedra. Many silicate minerals contain aluminum, which is the third abundant element in the Earth's crust (after oxygen and silicon). Only some silicates that seem to be important for various crustal studies will be mentioned in the following short summary.

The feldspars belong to the so-called frame silicates. In general, there is sufficient variation in the way the three elements Si, Al, and O can combine to produce a large number of compounds with a wide range of melting temperatures and pressures. Thus, the feldspars are found in many igneous, metamorphic, and sedimentary rocks. They seem to be the most abundant minerals in the crust. They belong to the aluminosilicates of K, Na, and Ca, the O atom also being shared by two tetrahedral groups. There are the alkali feldspars orthoclase ($KAlSi_3O_8$) and albite ($NaAlSi_3O_8$). Quartz and feldspars have a three-dimensional framework structure. In contrast, the pyroxenes form chains. Pyroxenes are silicates of Ca, Mg, and Fe. Closely linked to them are the amphiboles, which form a rather complicated double chain. Pyroxenes have high melting points and are usually found in igneous and high-temperature metamorphic rocks that are rich in Fe and Mg. The great stability of amphiboles over a wide range of p–T conditions is related to the compactness of their structure and to the bond strengths between various cations. The micas belong to the sheet silicates with very weak bonds between sheets. They are low-density minerals and light in color, except for the biotites, the iron species. Micas are apparently formed—like the amphiboles—over a large range of p–T conditions (Press and Siever, 1982). The phyllosilicates show tetrahedra linked into sheets of large lateral extent (Cox, 1971).

Cations with similar ionic radii and coordination numbers tend to substitute for each other (Press and Siever, 1982). They form mixed compounds, called solid solutions. Some of these compounds which are important in relation to igneous rocks and the study of melting processes are the minerals olivine and plagioclase feldspar.

Olivines are mixed Fe/Mg orthosilicates and the main constituents of the peridotitic rocks of the uppermost mantle. Their composition varies between fixed limits, marked by Mg_2SiO_4 (forsterite) and Fe_2SiO_4 (fayalite). The ions Fe^{2+} and Mg^{2+} have similar radii and the same charge and can be easily substituted. As will be discussed in more detail in the next section, the composition of olivine depends on the circumstances of crystallization from a silicate melt. The replacement of Mg^{2+} by Fe^{2+} and vice versa is a general phenomenon and is also found in pyroxenes and amphiboles.

Plagioclase feldspars are also products of melting processes. However, since they are less dense than the olivines and even less dense than a basaltic melt under the p-T conditions of the lower crust, their abundance generally increases toward the upper part of the crust. Like the olivines they form certain end members in which the ion Na^+ is replaced by Ca^{2+}, which has a similar ionic radius. These end members are the albite $NaAlSi_3O_8$ and the anorthite $CaAl_2Si_2O$, which have already been mentioned. Figure 5.2 shows the various feldspars occurring at high and low temperatures. Table 5.1 gives an overview of major silicate structures.

Among the nonsilicates the carbonates and the sulfides play a major role. The carbonates of Ca, Mg, and Fe are the main constituents of limestones and marbles. The sulfides are represented in all kinds of rocks in minor amounts, mainly as FeS_2 in the form of pyrite. In a certain concentration the sulfides are the sources of many important metals, e.g., ZnS, PbS, CnS, or $CaFeS_2$.

5.1.1   SOME PHYSICAL PROPERTIES OF MINERALS

Physical properties of minerals are especially important for geophysical investigations. The *density* of minerals varies from 2.68 for quartz to 4.37 for olivine. Density provides the key to many differentiation processes. Other important parameters are *seismic velocity* and *seismic anisotropy*. As crystals cannot be spherically symmetric, their specific structure provides the basis for anisotropic behavior in a rock unit which is formed predominantly of a certain single class of minerals. The anisotropy of the olivine minerals, for instance, results in the observed seismic anisotropy of dunites and peridotites,

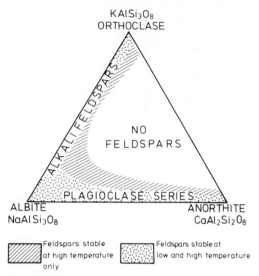

Fig. 5.2   Approximate limits of solid solutions of feldspars in an orthoclase–albite–anorthite diagram. [From Cox (1971). Copyright 1972 by MIT Press.]

which is found in the laboratory as well as in the uppermost mantle (Kern, 1982a; Morris *et al.*, 1969). The structure of the minerals is important for thermal conductivity and hence for heat flow. Their compressibility and elasticity in general provide the basis for the seismic velocity in the appropriate rocks. The relation of these values to p and T, together with the thermal expansion, forms the basis for many zones of lower (or higher) crustal velocities. In general, density ρ is proportional to the bulk and the shear modulus, as seen in Section 3.1.4. An important exception to this is the inverse relation of ρ and V_P for the transition from fayalite, (Fe_2SiO_4) to forsterite (Mg_2SiO_4). The higher V_P values of forsterite are related to lower ρ values, and the opposite relationship holds for fayalite. Plasticity of minerals and rocks provides the basis for the different kinds of flows in response to stresses. These phenomena, which are important for the rheological behavior of crustal material, were discussed in Section 4.4. Finally, susceptibility determines the magnetic properties of rocks, mainly as a function of percentage of magnetite, Fe_3O_4, as discussed in Section 3.7.1.

5.1.2   THE AVERAGE ELEMENT COMPOSITION OF THE CONTINENTAL CRUST

Weaver and Tarney (1984) attempted to develop a model for the element composition of the continental crust which takes into account boundary

Table 5.1

Major Silicate Structures[a]

Geometry of linkage of SiO_4 tetrahedra		Si/O ratio	Example mineral	Formula
Isolated tetrahedra: linked by bonds sharing oxygens only through cation		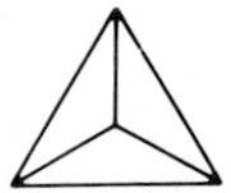 1:4	Olivine	$(MgFe)_2SiO_4$
Rings of tetrahedra: joined by shared oxygens in 3-, 4-, or 6-membered rings		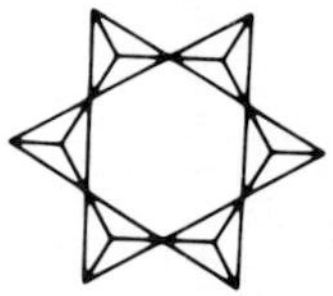 1:3	Beryl	$BeAl_2(Si_6O_{18})$
Single chains: each tetrahedron linked to two others by shared oxygens. Chains bonded by cations		 1:3	Pyroxene	$(MgFe)SiO_8$

Structure	Si:O ratio	Mineral	Formula
Double chains: two chains joined by shared oxygens as well as cations	4:11	Amphibole	$(Ca_2Mg_5)Si_8O_{22}(OH)_2$
Sheets: each tetrahedron linked to three others by shared oxygens. Sheets bonded by cations or alumina sheets	2:5	Kaolinite	$Al_2Si_2O_5(OH)_4$
Frameworks: each tetrahedron shares all its oxygens with other SiO_4 tetrahedra (in quartz) or AlO_4 tetrahedra	3:8	Feldspar (albite)	$NaAlSi_3O_8$
	1:2	Quartz	SiO_2

conditions from (1) geophysics, (2) isotopic evidence for the rate of crustal growth with time, and (3) the density of known components of the continental crust. From geophysical reasoning, a quartz diorite or granodiorite is postulated on the average from the seismic velocities. Heat flow studies suggest that the lower continental crust is depleted in the radiogenic heat-producing elements, suggestive of refractory granulite-facies material in the lowest part of the crust. The model of Weaver and Tarney (1984) is further based on the suggestion that 75% of the present continental crust was generated by the end of the Archean and 25% by the formation and incorporation of andesitic–tonalitic magmatic arcs in post-Archean times. See also Sections 6.5.1, 6.5.2, 7.2, and 7.3. Table 5.2 shows the composition of crustal components for the major elements. The proportions of upper, middle, and lower crust are assumed to be 2:1:3. These proportions have been found to be extremely variable for continental crusts of different age and history, as will be discussed in Chapter 6.

Table 5.2

Composition of Major Elements in the Continental Crust

Compound	Archean			Post-Archean middle and lower crust	Average continental crust	
	Upper crust	Middle crust	Lower crust			
SiO_2	66.0	66.7	61.2	59.2	63.2	58.0
TiO_2	0.6	0.3	0.5	0.9	0.6	0.8
Al_2O_3	16.0	16.0	15.6	17.2	16.1	18.0
FeO	4.5	3.2	5.3	6.1	4.9	7.5
MnO	0.08	0.04	0.08	0.12	0.08	0.14
MgO	2.3	1.4	3.4	3.4	2.8	3.5
CaO	3.5	3.2	5.6	5.9	4.7	7.5
Na_2O	3.8	4.9	4.4	4.0	4.2	3.5
K_2O	3.3	2.1	1.0	2.4	2.1	1.5
P_2O_5	0.17	0.14	0.18	0.27	0.19	—

[a] Extracted from estimates by Weaver and Tarney (1984). Last column contains estimates by Taylor and McLennan (1981).

5.2   IGNEOUS ROCKS

Igneous rocks provide the primary material of the crust. The other two types of rocks, metamorphic and sedimentary, were created from igneous rocks by a large variety of processes. Igneous rocks may first be subdivided into intrusives, generally coarse-grained, and extrusives, generally fine-grained or even glassy. Another important subdivision is that into felsic (or sialic) and mafic rocks, according to the relative amount of light (usually less dense and rich Si) and dark (usually denser and poorer Si) minerals. The words felsic (from feldspar plus silicon) and mafic (from magnesium and ferrous iron) provide only a rough description of the variety of igneous rocks. There is generally a continuous transition from the lightest rocks, the granites, via the diorites to the mafic gabbros and to the ultramafic rocks. The different kinds of felspar vary systematically from felsic to mafic rocks. Granite, for instance, is rich in alkali feldspar, whereas plagioclase feldspar dominates the diorites and gabbros. The gabbros consist mainly of pyroxenes, amphiboles, and olivines; i.e., they are rich in Mg and Fe. Peridotite has a still lower content of plagioclase and consists mainly of olivine and pyroxene, while dunite contains olivine only.

Some types of igneous rocks are shown in Fig. 5.3 in a so-called alkali–silica diagram, where their $Na_2O + K_2O$ content is plotted against their SiO_2 content. The names of the generally coarse-grained intrusive rocks are in capital letters, while those of their related (generally fine-grained) extrusive counterparts are in parentheses. A certain rock type is characterized by its alkali–silica weight percentages, with the Mg (or Fe) contribution marked as a parameter. Nearly all igneous rocks fall into the $Na_2O + K_2O$

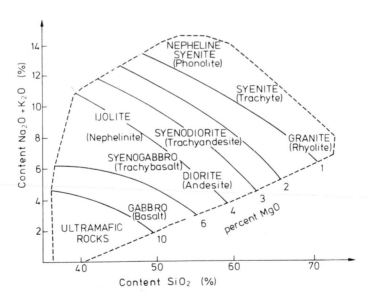

Fig. 5.3   Important igneous rocks in an alkali–silica diagram with MgO as parameter (extrusiva in brackets). [From Cox (1971). Copyright 1972 by MIT Press.]

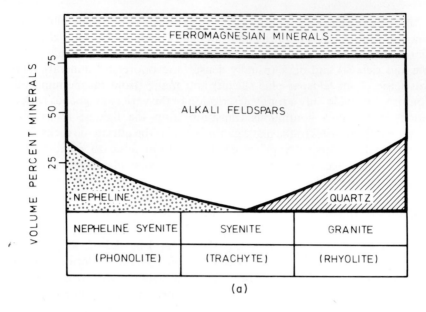

(a)

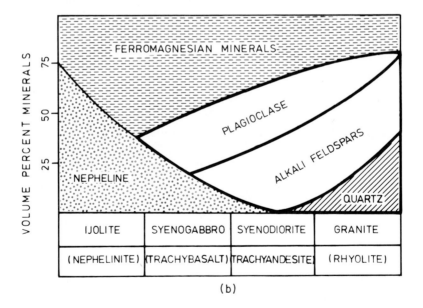

(b)

Fig. 5.4 Generalized mineralogical composition of igneous rocks (extrusiva in brackets). (a) Volume percent minerals of dominant crustal rocks, (b), and (c) volume percent minerals of other igneous rocks. [From Cox (1971). Copyright 1972 by MIT Press.]

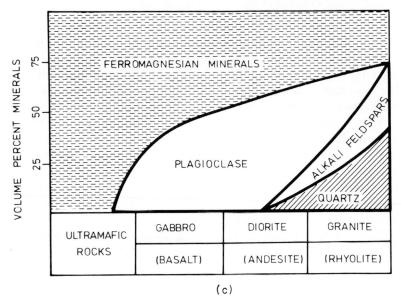

Fig. 5.4   (*Continued*)

versus SiO_2 area which is surrounded by the dashed line in Fig. 5.3. Contributions other than those shown in this figure could also be contoured. They generally behave in a smooth and reasonable way. Except for the ultramafic rocks, Al_2O_3 makes an important contribution to most igneous rocks, as seen already from the composition of the various feldspars.

An example of a possible series of igneous rocks in the Earth's crust is shown in Fig. 5.4a. The method of rock formation, by crystallization of liquids or by the formation of cumulates, which will be discussed in the next section, may enrich or deplete a rock unit considerably in certain elements or minerals. Figure 5.4 shows examples of mineralogical compositions of igneous rocks in certain geological provinces which might have experienced complicated or repeated metamorphic or melting events. In general, there is no sharp division between volcanic and plutonic rocks, which differ mainly in their cooling history.

5.2.1   FORMATION OF IGNEOUS ROCKS

There are basically two different mechanisms of formation of igneous rocks:

(1)   crystallization of liquids, and
(2)   formation of cumulates.

Unfortunately, a wide range of mixtures between these two extreme processes is possible. Liquids may be enriched by the addition of crystals from elsewhere. Repeated thermal events under slightly different $p-T$ conditions may contribute to the variety of igneous rocks; different degrees of partial melting and H_2O and CO_2 contributions may lead to quite different types.

The most important process in the formation of igneous rocks is fractional crystallization. Crystals, generally denser than magma, tend to sink to the bottom of a magma chamber. Evidence for this natural crystal settling, sometimes also called crystal sedimentation, has greatly increased in the last several years. The products of crystal settling are called cumulates or residues.

The layered structure of such rocks reminds the observer of sediments. It may create seismic reflections very similar to those obtained from layered sedimentary rocks. Multiple intrusives with multiple processes of fractional crystallization may produce thick bands of seismic reflections in the continental lower crust, as found in Caledonian and Variscan areas (see Section 6.2 and 7.3). Figure 5.5 gives an impression of the large Muskox intrusion (Irvine, 1980). This thick and highly stratified sequence of some silicic but generally mafic and ultramafic products provides a complicated seismic reflection pattern with a strong interference. It is also an example of the diversity of basaltic magmas and their rock types.

Only three types of basaltic magmas will be mentioned here. They are:

(1)   komatiitic basalts or magnesium-rich basalts, with the highest density among the basaltic lavas, found predominantly in Archean (greenstone) areas;

(2)   tholeiitic basalts, strongly saturated or oversatured in silica, found in the oceanic ridges, the ocean floors, and certain continental provinces; and

(3)   alkaline basalts (or olivine basalts), more alkaline than komatiitic basalts and strongly undersaturated in silica, found in oceanic islands and continental provinces.

The boundary between these three magma types has become rather soft. Additional magma types and subdivisions have been found, such as the high-aluminum magmas of andesitic provinces, the nephelinite types of alkaline terrains, and the oceanic tholeiitic magmas with their high Na/K ratio, representing the most abundant magma types on Earth. Basaltic provinces are also found on the continents, where basaltic lavas have formed the most voluminous volcanic rocks and have also intruded the continental crust in huge amounts. In addition, there are large areas of andesitic and rhyolitic lavas and tuffs with only a minor association of basalt. Plutonic bodies seem to be dominated by granodiorite and granite series. Rocks and magmas in the

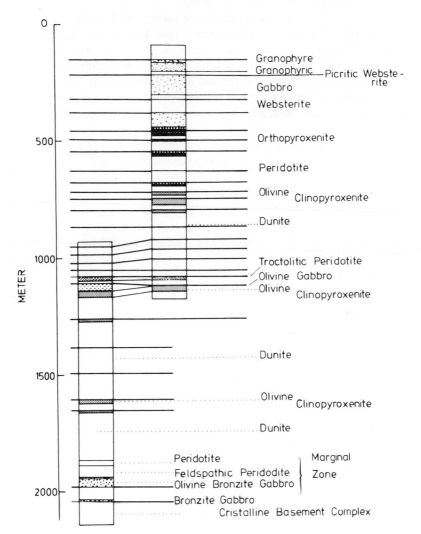

Fig. 5.5 The Muskox intrusion, simplified. An example of mafic–ultramafic cumulates with a strong lamellation. [From Irvine (1980).]

andesitic–rhyolite and diorite–granite range of compositions are called calc–alkaline.

5.2.2 REACTION SERIES AND MAGMATIC ACTIVITY

According to Bowen's theory of chemical differentiation and crystallization, which was already established in 1915, magmas of granitic composition can be obtained from a high-temperature magma of basaltic composition. There are two Bowen's reaction series, (Bowen, 1928), the continuous plagioclase–feldspar series and the discontinuous mafic mineral series. As seen in Fig. 5.6, the end product of both series is a low-temperature melt of granitic composition. At each stage the magma loses some of its mafic and heavy components through crystal settling, thus leaving the remaining melt lighter and more sialic. Recently, Kushiro (1980) investigated the density and viscosity of basaltic melts and found a strong increase of the density with confining pressure. The density of the melt is higher than that of plagioclase for pressures higher than about 0.5 GPa (equivalent to about 15 km crustal depth). Only for smaller values of p and z, i.e., in the upper crust, is it lower. Figure 5.7a shows the densities of olivine tholeiitic melt compared to that of plagioclase with 60 and 90% anorthite. The consequences of the density differences are remarkable. When basaltic magma intrudes the lower crust the lighter plagioclase floats while the heavier olivine and pyroxene minerals sink. While lighter anorthosite is formed on top of the magma chamber, ultramafic cumulates are formed at the bottom. The bottom part with its

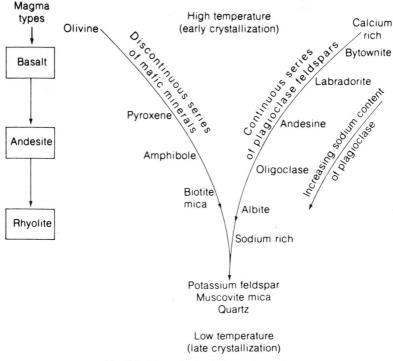

Fig. 5.6   Bowen's classical reaction series.

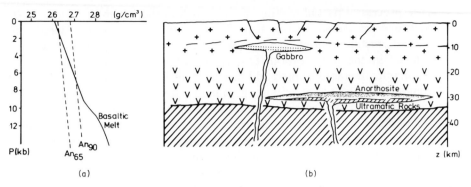

Fig. 5.7 Two different products of differentiation. (a) Density–pressure relationship for an olivine tholeiitic melt and plagioclase with different percentage of anorthite and (b) intrusions into lower crust with formation of plagioclase-rich rocks with ultramafic cumulates. Upper crust contains gabbroic cumulates. [After Kushiro (1980).]

ultramafic composition will show a high seismic velocity and may be considered as part of a—newly created—mantle. In this way, basaltic intrusions into the lower crust may form new and shallower Mohos and may be responsible for the layered structures and the strong overall velocity gradients of the lower continental crust in many areas. Anorthosite, as the lighter material on top of the ultramafic layers, if not melted again, may be responsible for some of the low-velocity zones of the middle and lower crust and also for the strong reflection coefficients.

Basaltic magmas intruding the upper crust, on the other hand, will act in a very different way. Because plagioclase is heavier than the melt in this low-pressure regime, it sinks together with the ferromagnesian silicates and forms gabbro. Figure 5.7b is a sketch of the two contrasting cumulates at two selected depths. The higher the content of ferromagnesian components in the magma, the lower will be the pressure at which melt and plagioclase show the same density.

Anorthosite formation is of major importance for the lunar crust and for some terrestrial Precambrian shield areas, where magmatic activity was great and nearly complete melting took place.

Returning to the generally more complicated reaction series of Bowen, there is no doubt that plagioclase-rich sialic magmas and finally granitic magmas can be created by one of Bowen's reaction series. However, granitic liquids and granitic rocks existing under low-temperature conditions are often found in equilibrium with quartz and alkali feldspars. If crustal rocks containing these two minerals are heated, they begin to melt and give rise to granitic magmas as partial melting products. The heat for such processes may come from zones of (possibly basaltic) igneous activity from below.

The process of partial melting of rock units of the upper or middle crust and its upward rise is called anatexis. The melting can also be considered as a fusion of crustal rocks. It may involve a stage of metamorphism, as will be discussed in the next section. Deeply buried geosynclinal sediments are certainly first metamorphosed and then melted. The isotopic composition of strontium in many young granites is so high ($^{87}Sr/^{86}Sr > 0.70$) that they are certainly not derived from the mantle, as discussed in Section 4.6.5.

The appearance of many large volcanoes with acidic lavas and massive eruptions of dacitic–rhyolitic material, independent of any andesitic or basaltic material in both space and time, also provides a strong argument for the independence of such acid–magma-forming processes.

The andesitic magmas are considered to be generated as primary magmas and not necessarily as a member of Bowen's reaction series. They dominate in many orogenic provinces. Compared to tholeiitic basalts, they contain higher percentages of SiO_2, Na_2O, and Al_2O_3 and lower amounts of K_2O. Andesitic magmas, which are ubiquitous around the Pacific margin, may originate from a melting of oceanic crust, sediments, and mantles in the Benioff zones by rather complicated dehydration and phase transition processes (Ringwood, 1975), where even parts of the continental mantle might be involved. Depending on the depth of the Benioff zone a wide range of liquids from basalts to rhyolites is formed, with only the mean (but most common) composition being that of (calc-alkaline) andesite.

In general, the continental crust below a sedimentary layer is made of igneous and metamorphic rocks and has undergone igneous and metamorphic processes. Processes such as partial melting or partial fusion leading to magmatites have contributed to the wide variety of rock types. The more often these processes took place, the more differentiated the crust became. An extreme differentiation has apparently taken place in large parts of the Variscan crusts, where "granitic" material with V_P velocities below and around 6 km/s makes up nearly the whole crust. The shallow mantle in these areas may be made of ultramafic cumulates of a strong or repeated fractional crystallization. See Sections 6.2 and 7.3.

5.3 METAMORPHISM AND METAMORPHIC ROCKS

Under high-temperature–high-pressure conditions (high p–T) a new assemblage of minerals appears and new rocks are created. The process by which igneous or sedimentary rocks are transformed into the new assemblage is called "metamorphism." The effects of metamorphism are widespread. A considerable proportion of rocks at or near the surface is metamorphic, and most probably the bulk of the continental crusts consists of metamorphic rocks.

In addition to temperature and pressure, nonhydrostatic stresses contribute considerably to many metamorphic processes, resulting in a mineralogical and structural anisotropy. As mentioned before, the lower part of the crust and the uppermost mantle are supposed to be highly anisotropic.

5.3.1 CONTACT METAMORPHISM AND DYNAMIC METAMORPHISM

The effect of temperature alone can be observed around intrusions of igneous rocks into near-surface layers. A rock in the contact zone loses its original texture, becoming coarse-grained and crystalline. Such contact metamorphism or thermal metamorphism causes the disappearance of low-temperature minerals like clay and the appearance of new, generally larger, minerals like pyroxenes, characteristic of high temperatures (Turner, 1968; Press and Siever, 1982). The zone of metamorphosed rocks around an intrusion is termed the contact aureole. Aureoles may be 2–3 km wide around granodioritic plutons 10–15 km in diameter, while those of basic plutons seem to be generally smaller. Within the aureole the grade of metamorphism decreases from the zone of contact toward the periphery. The term "grade" is often used to refer to the temperature attained during metamorphism, as indicated by the mineral assemblage. High-grade assemblages are found near the contact zone, while the low-grade rocks are restricted to the outer zones of the aureole, where a smooth transition to the unaltered surrounding rock units may be observed.

The effect of pressure and shear stress may often be observed along fault planes. The rocks in such areas may be extremely sheared, pulverized, and mylonitized, although their mineralogical composition may be relatively less affected. Such mechanically fragmented metamorphic products are called cataclastic rocks or simply fault gouge material. They are products of dynamic metamorphism, related to strong faulting and folding. The effects and structures of fault zones will be the subject of Section 6.6.3.

5.3.2 REGIONAL METAMORPHISM

Compared to thermally dominated contact metamorphism and stress-related dynamic metamorphism, it is regional metamorphism which is the most widespread type of metamorphic process. It is related to a forced descent of rocks in the collisional regime of subduction and sutures and is accompanied by intense deformation. Regionally metamorphosed rocks are found in deeply eroded, folded mountain belts, often together with granitic batholiths. They generally exhibit distinct structural and textural features. The texture is a result of recrystallization or conversion from one mineral to another.

Schistosity is one of the most common metamorphic structures. It refers to a parallel or subparallel orientation of mineral grains, often enlarged and visible with the naked eye. Rocks with such a coarse foliation are called schists. Some rocks exhibit linear schistosity, produced by the parallel orientation of elongated minerals such as amphibole (Cox, 1971). Schistosity is usually parallel to the original bedding in sedimentary rocks. Often a segregation of minerals into lighter and darker bands is observed. If there are sets of parallel planes that cut the rocks at an angle to the original bedding the term foliation is used. This denotes a structure in which certain minerals are aggregated into layers or streaks during the metamorphic processes under stresses different from those of the formation of the original rock. Rocks showing a coarse foliation are the gneisses, usually belonging to the high metamorphic grades. Gneisses are the most widespread types of rocks in the upper crust.

5.3.3   METAMORPHIC FACIES OF THE CONTINENTAL CRUST

One of the most spectacular achievements of petrological and mineralogical investigations of metamorphic rocks was the development of pressure–temperature schemes for metamorphic facies. Rapidly subsiding crustal regions come under a high p–low T environment (because heat conduction is a very slow process). In high-T areas, on the other hand, such as zones of contact with plutons, metamorphism may take place under relatively low-p conditions. The situation is schematically described in Fig. 5.8. A rapid and deep subsidence of sediments in oceanic trench subduction zones during continental collision will create blueschists (or glaucophane schists). They are considered to be the typical high-grade high-pressure metamorphic rock. In contrast, in active zones of basaltic upwelling at oceanic or continental rifts, in volcanic zones, and in island arcs, localized heating transforms the newly extruded basalts into greenstone, a low-grade metamorphic rock which is also found in many very old, i.e., Archaen and Proterozoic, belts incorporated later into shield areas (Windley, 1983). Very often, belts of both types of metamorphism, i.e., high-grade and low-grade, are found adjacent to each other. The existence of such belts is interpreted as one of the major signs of ancient collision zones. The high-grade metamorphic belt is supposed to be a transformed mixture of marine sediments and oceanic crust, a so-called mélange, which was carried rapidly to depths of several tens of kilometers in subduction zones, later to be scraped off and shuffled back to the surface by mountain-building processes and/or buoyancy. The low-grade belt is interpreted as being a weakly metamorphosed zone of the igneous rocks of magmatic arcs, where the upcoming basaltic, andesitic, or even rhyolitic lavas are recrystallized under high-T, low-p conditions. The occurrence of such paired belts of different grades of metamorphism is a strong indication of subduction processes. The concept of these paired belts leads to an extension of modern plate tectonic concepts far into the past, as will be shown again in Chapter 7. Regional metamorphism in general determines the facies of all rocks of the continental crust. During the evolution of the crust several phases of thermal and/or tectonic events generally provide the conditions for various grades of metamorphism: a general heating and/or a deep burial. Erosion, especially strong in areas of uplift, may carry parts of the upper crust away, leaving layers of the middle and lower crust exposed at the surface.

This great variability of crustal rocks is also reflected in the large number and variety of metamorphic facies. Different processes take place under the same p-T conditions for rocks of only slightly different composition. To

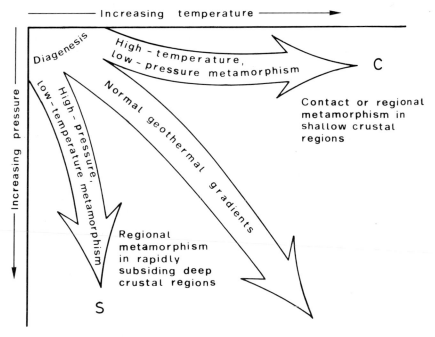

Fig. 5.8 Schematic view of development of metamorphic facies. High p–low T branch S, e.g., in subduction and sutures, leading to blueschist facies. Low T–high p branch C, e.g., at hot spots, plumes, and rapid uplifts, leading to greenstone facies. [After Press and Siever (1982).]

avoid unnecessary complication, a reduction to about seven or eight facies is used. Their names are derived from metamorphosed basic igneous rocks. There are the zeolites and the glaucophane schists (or blueschists) on the low-temperature side, followed by the greenschists, amphibolites, and granulites with increasing temperature of metamorphism. Eclogite is formed under very high pressure. Figure 5.9 shows a simplified p–T scheme for metamorphic facies. Facies boundaries and the beginning of melting, i.e., the "wet solidus" for granitic material, are also shown (compare with Fig. 5.8). Because amphibolites and granulites are thought to be the main rock types of the middle and lower continental crust, they will be described below.

Amphibolites are generally rather hydrous rocks, consisting mainly of the amphibole hornblende, but also of plagioclase feldspar and garnet. Their composition, hence, is within the intermediate to mafic range with 45–60% SiO_2. According to Fig. 5.9 amphibolites may be present at lower temperatures than the granulites. Seismic P-wave velocities of the middle (and sometimes the lower) crust in the range of 6.5 to 7.2 km/s are compatible with those measured on amphibolite samples (Brown and Musset, 1981).

A widely accepted candidate for a dry lower crust is granulite. It forms at even higher temperatures than amphibolite. It is among the highest metamorphic grades of mafic volcanics, containing pyroxenes and anorthitic plagioclase. The metamorphic mineral assemblage is similar to gabbro. Pyroxene granulites are the most common products of a high T–moderate p

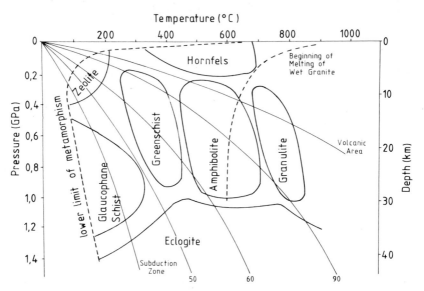

Fig. 5.9 Different metamorphic facies in a pressure–temperature diagram and some geotherms; surface heat flow densities in mW/m².

metamorphosis. Their seismic velocities are around 7 km/s and only slightly higher than those of amphibolites on the average.

Both granulite and amphibolite facies are found in deeply eroded crustal regions of the Precambrian. Granulite is more abundant than amphibolite, and there are several reasons for the assumption that it plays a major role in the lower continental crust. Natural granulites are depleted of many volatile and heat-producing elements, an observation compatible with model calculations of heat flow values which require a depleted lower crust. This fits rather well with the observed geophysical, especially seismic, observations. The relationship between granulites and amphibolites, on the other hand, may be a transient one. At lower confining pressures and lower temperatures the addition of water might result in rehydration, leading to a transition to amphibolite, a process which belongs to the known phenomenon of remetamorphism. The transition of (wet) amphibolites to (dry) granulites may indeed be gradual and transitional in some crusts, while in others prominent seismic reflectors might develop. In some areas, such as the Mesozoic Variscides, the gneissic–granodioritic upper crust makes up nearly the whole crust. It was mentioned in Section 4.6.5 that the strongly lithopile elements concentrate in the upper crust and that they are much more sensitive to melting and fractionation processes than the major elements.

A final remark concerns the eclogites (see Fig. 5.9). They are formed at very high pressure over a wide range of temperatures and consist mainly of pyroxene and garnet. It was mentioned already that the transformation of gabbro to eclogite was postulated to mark the transition from crust to mantle. However, neither the observed jump in velocities at the Moho nor the velocity or density values agree with this assumption for average crustal depths. It cannot be ruled out that in very thick crusts at depths ≥ 40 km a gradual transition to eclogites may play a certain role, but in general, velocities and elastic constants of granulite facies rocks, possibly below an amphibolite facies, give a better fit to the observed data from the lower continental crust. It will be shown, however, in Chapter 6 that seismic investigations indicate a large variation in the composition and thickness of the lower crust.

5.4 SEDIMENTATION AND SEDIMENTARY ROCKS

Sedimentary basins of various types may reach thicknesses of more than 12 km. Sedimentary rocks are formed from erosional products of igneous or metamorphic rocks with contributions of living organisms. Erosion, transportation, and deposition result in sedimentation. Unconsolidated sediments are referred to as sedimentary deposits. They may become consolidated and lithified and transformed into sedimentary rocks, mostly by diagenetic processes. Diagenesis is a process by which sediments change physically and chemically as a result of increasing p–T conditions during burial. The boundary between diagenesis and weak metamorphism is transitional. Metamorphosed sediments are called metasediments. They play a major role in the creation of the metamorphic rocks of the continental crust. It is known that metasediments have been transported to great crustal depth by tectonic processes, as in case of the Ivrea body. Here they were uplifted again and still show a layered appearance with V_P velocities greater than 7 km/s (!). They have a granulitic facies (Christensen and Fountain, 1975). Tectonic processes control the degree of erosion through the rate and the amount of uplifted area. Subsidence in the depositional area is also controlled by geodynamic (or tectonic) forces from below. The pulling of sediments into subduction zones and their metamorphism, fusion, and contribution to rising magmas in the magmatic areas constitute the "sedimentary circle." See also Section 7.3.3.

Sedimentary rocks are known from all ages. The oldest rocks found in the nuclei of old shield areas are metasediments, showing that the process of erosion and sedimentation (possibly in shallow seas) has been going on for at least 3.8 Gyr. The materials that make up sediments and sedimentary rocks may be divided into the two categories: clastic and chemical. The clastic component consists of all kinds of transported debris, e.g., sand, gravel, and shell fragments. The chemical component includes inorganic precipitates from water, e.g., calcite, gypsum, and silica, as well as biogenic material extracted from the water or air, partly in the form of skeletons of organisms.

5.4.1 CLASTIC SEDIMENTS

Clastic sediments, including the shales, sandstones, and conglomerates, are far more abundant than the chemical precipitates, possibly constituting about three-fourths of the sediment mass. Of these, the shales are the most abundant rock types. The strength of water (or wind) currents determines sedimentation, especially the size of the particles. Mud, or its lithified equivalent (shale), is related to quiet waters, where even the smallest particles are deposited. Sand and sandstones imply moderate currents like those in rivers or strong winds, creating dunes. Gravel, or its lithified equivalent conglomerate, is related to strong currents such as those of fast rivers or high waves. Cumulative grain-size distributions are used for a more sophisticated subdivision of the three basic categories: fine, medium, and coarse.

The fine-grained sediments, mud and shale, the most abundant and widespread species, very often contain pieces of organic matter, and many are mixed with chemical sediments, creating calcareous shale or shaly limestone. Cherts and evaporites are also often mixed with mud. From their mineral composition clay is the most abundant constituent and provides some basis for studying weathering conditions.

Medium-grained sediments are sand and sandstones. Sandstones often resist erosion and are observed as outcrops. Subgroups of sandstones are the arenites, made of rounded and rather uniform quartz grains (quartz arenites) or of fine-grained fragments of shales, schists, or slates (lithic arenites); the arkoses, containing much feldspar with a rather inhomogeneous distribution of grain sizes; and the graywackes, made of quartz, feldspar, and fragments surrounded by a fine-grained clay matrix.

The coarse-grained sediments consist of gravel and conglomerate. Pebbles may be up to 25 cm in diameter and generally are very rounded. Because pebbles are not so frequently observed their area of origin can often be inferred (Cox, 1971).

5.4.2   CHEMICAL SEDIMENTS

Limestone, $CaCO_3$, and dolomite, $CaMg(CO_3)_2$, the carbonate rocks, are the most abundant chemical sediments. Under surface conditions, calcite is the stable form of $CaCO_3$; the polymorph aragonite requires higher pressures. Marine organisms precipitate both calcite and aragonite. The resulting carbonate sediments consist of the shells of organisms rather than of inorganic precipitates. The organisms extracted $CaCO_3$ from the water to build their shells, a process which is strongly related to water temperature (Press and Siever, 1982). Oolithes containing nuclei of shell fragments are special carbonate sands formed in an environment of strong tidal currents. Stromatolites are a layered mixture of algae and carbonate sediment, found in many tidal-flat limestones, even in Precambrian rocks. Nannofossil oozes are made up of very small organisms; if deeply buried they are transformed into chalk.

Besides the calcium carbonates there are some other chemical sediments that are even more controlled by biological processes. There is a large variety of silica, SiO_2, of biological origin, secreted by single-celled organisms. There are the peat layers, created in an environment where oxygen is scarce, and their lithified equivalents coal, with its most variable appearance. The evaporites are created in arid climates in basins with restricted circulation, by the evaporation of seawater. Evaporites are sediments composed of the mineral halite or rock salt. Repeated formation of barred basins may result in huge salt sequences, which become unstable under an increased sediment load. Tectonically spectacular uprisings of salt deposits in the form of salt domes, diapirs, and walls determine the tectonic pattern of the Permian evaporites in northern Germany and the North Sea, as well as in western Texas and New Mexico (Press and Siever, 1982).

5.4.3   SEDIMENTARY ENVIRONMENTS

Sedimentary environments are generally described in terms of geomorphology. A prominent clastic sedimentary environment is related to sediment transport in rivers; it is the so-called alluvial environment, including meander belts, floodplains, and alluvial fans, in which the sorting-out of material takes place. Others are the desert environment, containing windblown, well-sorted fine sand, often formed into dunes. The glacial environment is responsible for a heterogeneous, unsorted character of sedimentation in front of, or below, the ice, often being alluvial, modified by the ice. The formation of eskers or moraines or drumlins is a consequence of the ice movements. Delta environments include alluvial delta plains with a characteristic transition from freshwater, coarse-grained sedimentation upstream to carbonate sediments or fine-grained silts and muds. Shallow marine environments favor medium- to fine-grained sands with ripple marks and muds, all of them showing signs of marine life. Of particular importance for revealing paleogeographical information are the turbidity environments, related to turbidity currents at continental slopes or trenches. A thick assemblage of interbedded coarse sands and shale is created, coarse and structureless at the lower end transforming to medium- and fine-grained material at the upper end, and often topped by silts and mud. These thick deposits of turbidity currents are also called flysch, first observed in the Alpine–Himalayan chain, but very common in all other mobile belts. Pelagic environments are deduced from nonturbidite deposits of fine-grained colored clays in the deep sea, often interbedded with carbonate oozes (Verhoogen *et al.*, 1970).

Several environments can be observed in the range of chemical sedimentation. There are the coral reefs or islands (atolls) or the large carbonate platforms, which often also produce the algal reefs. There are huge shallow-water platforms attached to the continents, which are produced by slow subsidence of a shelf area and rich deposition of carbonate sediments keeping pace with the subsidence.

5.4.4   SUBSIDENCE OF MARGINS

Subsidence is a consequence of tectonic forces from below, sometimes also related to the cooling and contraction of a formerly heated crustal section. The increasing sediment load may reinforce the subsidence and support the accumulation of huge sediment thicknesses. Further subsidence might be generated by increased metamorphism and even by phase transitions in the sinking crust below a huge sediment column. Such a process with its increasing density might accelerate the subsidence and might be the trigger for a conversion from a passive to an active margin.

Continental margins have been analyzed and resolved by extended application of marine seismic methods, especially reflection seismology. A special branch of geoscience, seismic stratigraphy, reveals many details of sedimentation. A hiatus in the sedimentation, possibly caused by a regression during which the sea withdrew, is seen in very strong bands of reflections. Transgressions of the sea can also be clearly defined. The whole sedimentary deposition with its eustatic sea-level changes and its differential subsidence is revealed by multichannel, high-coverage seismic reflection surveys. Often, the same geologic layer can be followed laterally to great depths. Boreholes provide points of tying in the various layers. The final results of seismo-stratigraphic studies are relationships between the time and speed of subsidence (or an episodic uplift) and the depth of specific layers. Such a correlation is of utmost importance for the formation of hydrocarbons, which require a small window in temperature and pressure for their maturation. Figure 5.10 is a sedimentation diagram showing the depth for three different layers as a function of time. For corrections due to compaction see Section 4.4.3.

It is interesting to note that many subsiding shelves seem to be isostatically compensated. The sediments, with their weight deficit compared to an adjacent shield area, are compensated by dense mantle material at depth, an "antiroof." Rather often no signs of high-velocity, lower crustal material can be detected. The reason for the origin of this peculiar crustal structure is controversial and will be discussed later.

5.4.5 SEDIMENT ACCUMULATION: PLATFORMS AND GEOSYNCLINES

Major accumulations of sediments take place in three distinctly different categories: platform, basin, and geosynclinal. Horizontal, essentially nontectonized, sedimentary accumulations on top of very old areas are called platforms. They are generally flat-lying and relatively thin. Deep-water sediments are absent and shallow carbonate sediments are dominant. The shape of the platforms is sometimes nearly circular.

Subsidence and sedimentation are generally slow, and platform thickness seldom reaches 4 km. Such a mild subsidence apparently often takes place far away from active plate boundaries and must have its origin rather deep in the mantle.

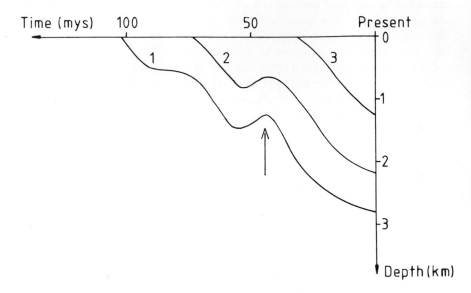

Fig. 5.10 Time–depth diagram of sedimentation showing the speed of vertical movements for three different layers with an uplift at ∼45 Myr, affecting layers 1 and 2.

In contrast to platforms, geosynclinal accumulations form thick and elongated bodies with abrupt lateral variations. A geosyncline is a sediment-filled trough. Closer to the continental platform the shallower miogeosyncline is divided from an outer belt closer to an ocean, i.e., the thick eugeosyncline. While limestones, shales, and sandstones are found in the shallow miogeosynclines, flysch, pelagic sandstones, and sometimes mixtures with volcanic rocks, basalts, and ash are observed in the eugeosyncline. See Fig. 5.11 for a simplified overview (Press and Siever, 1982). Sedimentary basins, not specifically shown in this figure, might appear as molasses or intermontane areas of subsidence. Their magnitudes are generally between those of platform and geosynclinal features. All sediment basins and especially the deep ones will develop a thermal blanketing effect. The low thermal conductivity of the sediments results in a delayed heat release and an increasing temperature. The additional influx of magma from below might occasionally contribute to a high heat flow, a general weakening of the sediment pile, and even periods of inversions. Such inverted basins, caused by intrusions or a limited period of stress reversal, may form interesting structures for the accumulation of hydrocarbons.

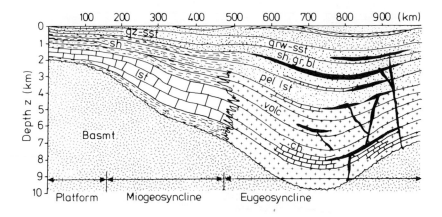

Fig. 5.11 Schematic cross-section through a geosyncline; grw–sst, graywacke–sandstone; pel–lst, pelagic limestone; volc, volcanic rocks; sh, gr, bl, shales (gray and black); ch, cherts; lst, limestone; sh, shales; qz-sst, quartrose sandstone; and Basmt, basement. [After Press and Siever (1982).]

The origin of geosynclines is intimately related to processes at active and passive continental margins. A miogeosyncline may be considered as representing the young stage of a passive margin, where subsidence is affected by cooling and contraction of the crust. When a formerly passive margin transforms into an active one, turbidities, volcanic activity, and the so-called fore-arc sediments are mixed with scraped-off pelagic sediments on top of subduction zones. These processes result in the generally thick geosynclinal sediment piles along the trenches and in strongly negative gravity anomalies. According to the Wilson cycle (next section), geosynclines can be considered as the forerunners of mountain-building processes during the collision of continents (see also Fig. 5.11). In a strongly compressional system, all sediments are deformed, metamorphosed, folded, and faulted, in many places injected by magmas and plutons. Huge sedimentary nappes are stacked together, and major overthrust and strike–slip faults develop. Young mountain systems like the Alpine–Himalayan chain and older ones like the Appalachians, the Caledonides, or the Western Cordillera are examples of these developments. Erosion strips off the highest and youngest peaks of the mountains and sometimes uncovers the igneous and metamorphic rocks below. The sediments are transported to the continental margins and help to restore and keep up the sedimentary cycle.

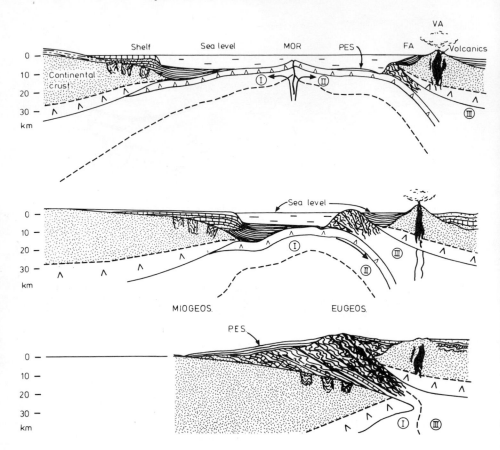

Fig. 5.12   Schematic development from active rift and eugeosynclines to continent–continent collision with orogenesis and miogeosynclinal and eugeosynclinal masses; I, II, and III, different plates or parts of plates; PES, pelagic sediments; FA, fore arc; VA, Volcanic arc; and MOR, mid-ocean ridge. [After Press and Siever (1982).]

5.4.6   THE ROLE OF SEDIMENTS IN THE PLATE TECTONIC CYCLE

The sedimentary cycle was described in the previous section; here a short summary on the suggested plate tectonic cycle seems appropriate. The closing of an ocean with the associated continent–continent collision and formation of intercontinental mountain belts has been called the Wilson cycle (Wilson, 1966) (Fig. 5.12), who first suggested that the Atlantic Ocean closed and then opened again. This interpretation describes only a semicycle, although its concept, as outlined in the previous section, is on solid ground and compatible with the classical description of geosynclines and their modifications by compressional forces, metamorphism, and different kinds of volcanism (e.g., Stille, 1944). Trying to close the Wilson cycle into a full plate tectonic cycle, we may start at any plate tectonic feature and develop a sequence such as:

(1)   rifting of a continent along old zones of weakness, often oceanization of lower crust;

(2)   development of one of the rifts into a spreading ocean, formation of passive margins, with increasing sediment thickness, passive drifting of continents;

(3)   conversion of passive to active margins, subduction of oceanic lithosphere; formation of Benioff zones, island arcs, volcanic chains, eugeosynclines, and deep trenches;

(4)   retreat of subduction zones, closing oceans, and approaching continental fragments or microplates;

(5)   collision of microplates, island arcs and ridges with continents; formation of ophiolites;

(6)   continent–continent collision; orogenesis, interstacking of segments of continental crust, formation of an intracontinental belt (e.g. the Urals or the Caledonides); and

(7)   Erosion of the mountain belt, disappearance of mountain roots, and remaining zones of weakness.

Such a full plate tectonic cycle will be called a Wilson cycle throughout this book. The physical reasons for the various processes (1)–(7) may be found in convection currents, plumes, and contraction by cooling. Many suggestions and model calculations for convection patterns have been developed, but they will not be considered here. Only some aspects of convection related to rift/margin formation and to lithosphere thickness will be discussed in Sections 6.4 and 7.2.

Chapter 6 | Crustal Structure in Various Geologic Provinces

6.0 INTRODUCTION

Compared to the oceanic crust with its simple broad-scale geophysical coherence, the continental crust looks more complicated in several ways. It contains a record of time that is 20 times as long as that of the oceans. Most of the continental rocks have been subjected to a large variety of tectonic processes, such as folding, faulting, metamorphism, repeated partial melting episodes, intrusions, fusion, and fractional differentiation, as indicated in Chapter 5. Information on the composition of the crust is generally obtained from deep drill holes, xenoliths, outcrops of various low- and high-grade metamorphic rocks, and laboratory studies that compare experimentally determined physical parameters with those obtained from indirect geophysical studies in the field (Chapter 4).

At the margins and inside the continents there are relatively narrow mountain chains, young ones like the Alps, the Himalayas, and the Andes and older ones like the Caledonides, the Appalachians, and the Urals. These belts are considered orogenic and their origin was (or is) in collisional zones, generally related to active plate boundaries. Other kinds of specific features in the continents are the basins, depressions, rifts, and grabens. The larger of these structures are created by slow, so-called epeirogenic movements, which are often quasi-continuous over hundreds of millions of years. There are different tectonic reasons for the small epeirogenic modifications of continental units which will be discussed in the appropriate section.

There is no doubt that the various seismological methods provide the densest information, the best resolution in depth, and the key parameters for tying in other geophysical data sets like density, resistivity, and heat flow. Seismic explosion studies on land probe the whole lithosphere along refraction profiles (Fuchs and Vinnik, 1982). Velocities can be calculated from refracted, diving, and wide-angle reflected waves and compiled as a function of depth. Characteristic velocity patterns in different geologic provinces can be derived and major boundaries can be delineated. Deep continental reflection studies, on the other hand, provide a resolution of crustal fine structure. As mentioned in Sections 3.4.2 and 3.4.4, Vibroseis-type sources or explosions may be used for fast exploration. The limited energy of vibrators and limited aperture of the recording arrays—generally less than 10 km—also limit the penetration depth and the calculation of velocities from the moveout times. In principle, reflection techniques with a spread length of about 20 km can provide better velocity resolution than do refraction studies (Bartelsen *et al.*, 1982; Meissner and Lueschen, 1983). Reflection studies for continental structure are most efficiently performed in shelf areas or large lakes by marine multichannel surveys, using air gun arrays and a towed hydrophone streamer several kilometers in length. In any case, seismological studies form the basis for delineating crustal structure and velocity and will also form the background of this chapter.

The old shield areas without a thick sediment cover will be reviewed first. Although they are by no means homogeneous, they have been tectonically stable over a long period of time and have undergone little warping, in contrast to the strong folding of geosynclinal belts. They have generally thick crusts. A "platform" is a submerged shield, i.e., a shield area covered with sediments of varying, but also limited, thickness.

6.1   PRECAMBRIAN SHIELD AREAS AND PLATFORMS

Several nuclei of the continental crust—the Archean cratons—began their existence between 3.8 and 3.0 Gyr ago. Figure 6.1 shows the position of these cratons. Their origin was the consequence of mantle differentiation on a great scale. Such differentiation must have competed with recycling of crustal material due to drag forces of a vigorously convecting mantle. Cratonization, however, occurred at a higher rate, especially between 3 and 2 Gyr ago.

One of the few reliable physical parameters of the Precambrian period is the heat production by the radionuclides. Figure 6.2 shows the temporal decrease of heat production for different types of rocks. The consequences of the higher heat production in the past, however, are much less clear. Certainly the temperature gradient was higher and the temperature T was

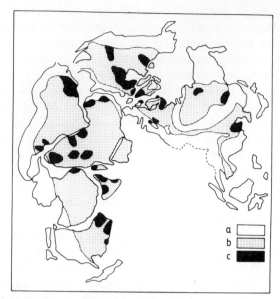

Fig. 6.1 Old cratons in a pre-Permian reconstruction of continents; (a) Phanerozoic, (b) Proterozoic, and (c) Archean rocks. [From Windley (1977). Copyright 1977. Reprinted by permission of John Wiley and Sons, Ltd.]

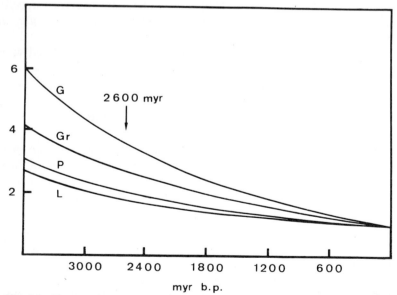

Fig. 6.2 Heat production as a function of time for several types of rocks normalized to today's value (= 1): G, granitic; Gr, granulitic; P, peridotitic; and L, lherzolitic.

also higher, but apparently only *on the average*. Convective forces might have led to an accumulation of crustal material on top of downgoing convection cells, thereby creating zones with low T gradients and low T values. In fact, thermal gradients must have been very different. A low p–high T metamorphism is reported for the Australian protoshield (Binns *et al.*, 1976) and a high p–low T metamorphism with $dT/dz > 35°C/km$ for African granulites (Clifford, 1974).

Real blueschists, however, as frequently seen today at active margins, have not yet been found in rock units older than 800 Myr (Windley, 1977). Positive signs of steep-angle subduction of the Benioff zone type seem to be missing. Greenstone belts, on the other hand, indicating low p–high T environments, are abundant in the Archean areas, but are dying out with the early Proterozoic. The contribution from the early and very transient oceanic regime also remains unclear. Probably, thin oceanic plates moved much faster and underplated the growing continents in a shallow-angle subduction. This would result in an additional thickening of the continental nuclei and possible lead to a back-arc structure where ensialic belts might develop similar to those postulated by Kröner (1977) (compare also Fig. 7.3).

Precambrian events have produced such diverse phenomena as linear mountain belts, similar to today's orogenic belts, and high-grade domains with rather meager deformation but high-grade metamorphism and some magmatism. The magmatic processes in the Precambrian especially provide a great deal of important information. They offer a second explanation for the seismically observed thick crusts of old shield and platform areas. Because of the high temperature gradient and the high temperature at shallow depth a larger percentage of melt must have been extracted from the underlying mantle. This extraction must have resulted in a thicker and more basaltic-gabbroic crust than would be possible in the present p–T environment (Vetter and Meissner, 1979). Considering erosion of several kilometers, the original crustal thickness (and petrology) might not have been too different from the crusts of the Moon and Mars (compare Fig. 1.2 again). Subsequent thermal and tectonic events have modified and enlarged the old nuclei on Earth. Figure 6.3 shows the different age provinces in the North American continent with the shield proper in the north and the sediment-covered shields—the platforms—around it. The growth of the oldest nucleus by episodic accretion of younger terrains—most probably magmatic arcs or real continents—is clearly defined. These events were apparently episodic in nature and could be related to a change in convection patterns in the mantle. The old shields show massive granitoid emplacements which were particularly strong around 2.5 Gyr ago, i.e., at the Archean–Proterozoic boundary. However, the younger shields are also pierced with syn- and postorogenic granitic plutons.

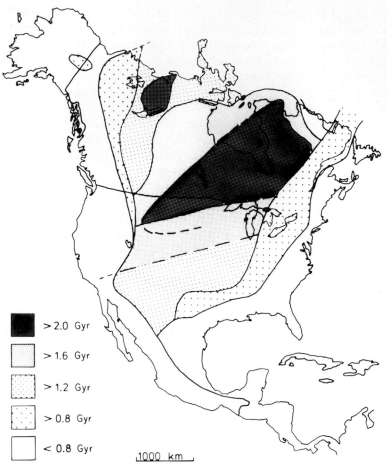

Fig. 6.3   Age provinces in the North American continent. [After Dott and Batten (1976) and Press and Siever (1982).]

6.1.1   REFRACTION MEASUREMENTS—DEEP SEISMIC SOUNDING IN SHIELDS

As mentioned in Section 3.4.1, large explosions from quarry blasts or other sources provide the seismic energy for refraction observations up to several hundred kilometers in length. The P_g wave generally provides velocity information about the uppermost 8–12 km. Reflected phases from the Moho and the Conrad or other intracrustal discontinuities are used to calculate the interval velocities for the middle and lower crust, respectively. Diving waves

from strong–gradient zones, often associated with wide-angle Moho reflections, provide additional data points. Sometimes, shear waves are observed
and interpreted in addition to the P-wave seismogram sections, thus providing Poisson's ratio as a function of depth. The structure of shields in the
Siberian and Russian area has been reviewed by Kosminskaya (1971) and
Kosminskaya and Pawlenkowa (1979), the Canadian shield by Smith *et al.*
(1966) and Berry and Mair (1980), the Ukrainian shield by Pawlenkowa
(1973), Sollogub *et al.* (1973), and Sollogub and Chekunov (1983), and the
Indian shield by Kaila *et al.* (1981a,b). Differences between the Eurasian
shields and the part of Europe that was accumulated later have been analyzed
by Vetter and Meissner (1976) and Theilen and Meissner (1979). The
Eurasian shields all have rather thick crusts, i.e., about 40–50 km on the
average. Velocities may reach 7–7.2 km/s in the lower crust, indicating a
mafic or granulitic composition. The parts accumulated later in western
Europe all have thinner crusts and are much more sialic than the shield areas.
Even where crustal thicknesses are comparable to some shield areas, e.g.,
those of the West Siberian platform, the average crustal velocities are much
smaller and reveal a different composition. The shields themselves show
a general crustal thinning from their central parts to their periphery
(Kosminskaya and Pawlenkowa, 1979). This thinning is mainly a result of the
thinning of the (mafic) lower crust with its high-velocity part, an observation
which will be related to the dynamics of the lower viscosity of deep crustal
layers. Often the heat flow increases and the P_n velocity (i.e., that of the
uppermost mantle) decreases from the center to the periphery. All these
observations are compatible with the concept of an episodic accretion of
magmatic arcs to the old nuclei, as was shown in Fig. 6.3. The younger these
units are, the more sialic they appear, i.e., the larger is the percentage of low-
velocity material. This is a sign of the increasing differentiation of the
continental crust. As mentioned in Section 4.6.5 and shown in Figs. 4.43 and
4.45, the increase of crustal material reflects the extraction of an increasing
amount of melt from the mantle and its growing depletion in the strongly
lithophile elements.

In the following the velocities are presented in constant steps as shown in
Table 6.1. These steps have been made uniform and equidistant in order to
lead to velocity intervals of the same length if only the (linear) pressure
dependence of velocities (not a material change) is involved. Changes in
composition, on the other hand, are indicated by a missing (or a very small)
interval of that specific velocity range, having a high potential as a reflecting
boundary. Major problems for the compilation of reliable velocity values are
the differences in quality of the field data collected and the different ways of
interpreting them. Both factors are responsible for the simple and steplike
velocity–depth (V-z) models of early researchers and for the more detailed

Table 6.1

Standardized V_p Values Used for the Velocity Columns in Various Figures of Chapter 6 and Their Possible Petrological Meaning

V_p (km/s)	Possible rock types
< 5.7	Sediments, uppermost (or stretched) part of crystalline basement
5.7– < 6.4	Upper crystalline basement = upper continental crust, made of granites and low-grade gneisses
6.4– < 7.1	Intermediate rocks such as diorites, light gabbros, norites; metamorphic, medium to high-grade amphibolites and granulites; serpentinites
7.1– < 7.8	Dense gabbros, eclogites, highest-grade granulites; "crust–mantle mix", serpentinites
> 7.8	Ultramafic rocks such as peridotites, pyroxenitic peridotites, dunites, harzburgites, lherzolites; eclogites

and often more transitional V–z curves of modern analyses of data that have been gathered with a station spacing of less than 3 km. Sometimes old, but good, data are just underinterpreted and have been reinterpreted. Sometimes, however, modern data seem to be overinterpreted, giving details in the V–z structure that are not justified by the travel time–distance (t–x) data or could be explained by much simpler V–z models. Many thin low-velocity zones (LVZs) are seen in the interpretations of many authors or groups of authors. Such zones not only provide another problem for the compilation of V–z data, they also do not seem to be convincing. They are generally obtained by an inversion of t–x into V–z data sets, where delays of travel time branches are interpreted as LVZs. There are at least two reasons to mistrust such zones. First, arrivals of such layers are energetically not realistic for refraction or diving waves if the layer's thickness is in the range of the dominant wavelength, i.e., 1 km or smaller. Second, in the inversion to the V–z field it is tacitly assumed that the velocity is only a function of depth; every delay in travel time curves then is automatically transformed into an LVZ. Knowing the heterogeneous nature of the upper crust, i.e., the upper 8–12 km with its gneissic and diapiric granitic structure, a small delay in the P_g-wave travel time and the onset of a new branch may very well be related to a lateral inhomogeneity and to a corresponding lateral bending of ray paths creating a travel time delay. Even a small *vertical* layering may result in an energy transfer to later phases of the seismic arrival, also resulting in an apparent small LVZ. There is no doubt that significant LVZs do exist under certain conditions, i.e., in zones of high pore pressure, in special granitic laccoliths and plutons, and generally in warm crusts with a high temperature gradient where the temperature influence exceeds the pressure influence; but they are generally not proved, and certainly not expected, in the old and cold shield

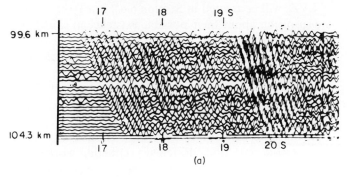

(a)

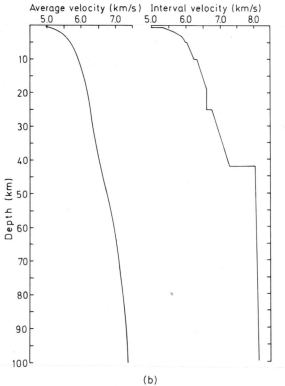

(b)

Fig. 6.4   (a) Field record and (b) V–z profiles from a DSS survey in the Indian shield. The field record covers the distance between 99.6 and 104.3 km; note the strong P_mP arrivals between 19 and 20 s travel time. [From Kaila *et al.* (1981a).]

areas. Since LVZs will be treated in Section 6.6.2 it may be sufficient here to invoke some suspicion about their possible existence in shield areas.

The following examples from various shields will show that there is certainly a great variety of V-z structures, even after a critical filter has been applied regarding specialities of interpretation groups.

Dense coverage combined with modern interpretation techniques along two profiles in the Indian shield resulted in the V-z section of Fig. 6.4b, while an example of the field records is shown in Fig. 6.4a. The first-order boundaries in the otherwise rather smooth V-z curve are based on some observed near-vertical reflections. Extensive refraction work in the Russian, Siberian, and Ukrainian shields (and platforms) has been going on for the last 30 years, and a tremendous amount of data has been accumulated. As an example, profiles across the Ukrainian shield are presented in Fig. 6.5. These early measurements (Sollogub et al., 1969) were performed with a dense geophone spacing. The interpretation was first done by assuming several layers with a constant boundary and/or interval velocity. Later, more refined analyses of t-x data and a ray inversion to smooth V-z curves were applied (Pawlenkowa, 1980, 1984; Kosminskaya and Pawlenkowa, 1979; Koshubin et al., 1984). These data will be compared when discussing the platform areas in Section 6.1.2.

Rather dense data acquisition and a detailed interpretation were recently carried out for the Voronesh shield in the central part of the east European platform (Tarkov and Basula, 1983). Although the interpretation of the x-t data was made with small stepwise V-z intervals, the V-z curves could equally well have smooth transitions. Two out of six V-z profiles show small LVZs in the upper crust. A velocity around 7 km/s is obtained at about 30–35 km depth (see Fig. 6.6). As the main result of these and many other investigations the relatively high velocities in the middle and lower part of the shield areas are an important discovery. Material with velocities around 7 km/s, which is

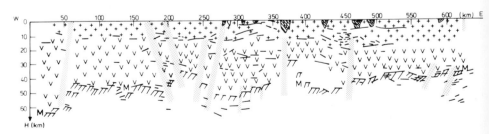

Fig. 6.5 Cross-section through the Ukrainian shield. The crust is seen to consist of a granitic–gneissic upper crust ($+$) and a basaltic–gabbroic lower crust ($\vee$) with a rough Moho boundary. Dotted regions represent suggested fault zones. [After Sollogub et al. (1969) and Pawlenkowa (1973, 1979).]

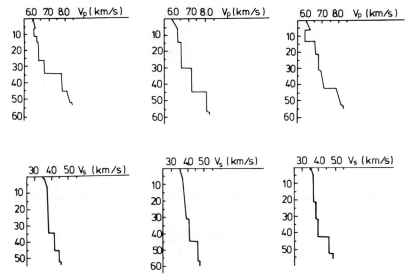

Fig. 6.6 Corresponding V_P and V_S functions for three DSS profiles in the Voronesh shield. [From Tarkov and Basula (1983).]

supposed to correspond to gabbroic–granulitic layers, is generally found below 25 or 30 km depth, comparable to the V–z curves of the Indian shield (Fig. 6.4). The indicated fault zones marked in the sections do not always seem to be substantiated by the original data. The inferred block structure for the shield must still be considered hypothetical.

Figure 6.7 shows the different age provinces in the Baltic shield. Note the decreasing ages of crustal units from north to southwest, showing the episodic accretion of crust. Extensive deep seismic sounding (DSS) investigations have also been carried out in the Baltic shield, and the situation of only two profiles is shown in Fig. 6.7. The crust is between 40 and 58 km deep; the V–z relationship seems to be rather smooth except for a slightly stronger gradient on top of the Moho, which gives rise to relatively low-energy $P_m P$ waves. Figure 6.8 shows three V–z functions based on the work of Hirschleber et al. (1975), Lund (1979), and Avedik et al. (1984). A velocity of 6.5 km/s is already reached at 15 km depth, while 7 km/s is found at a depth of 35 km. The strong variation of the crustal thickness along the 2000-km-long Fennolora profile is remarkable (Bittner, 1983). Some of the rather abrupt changes in depth may be related to a change in the age provinces, i.e., to a very old plate boundary, but the mechanism of accretion, i.e., suturing or shallow subduction, remains unclear. Recently, the Arabian shield (Fig. 6.9a) was also investigated by a long DSS profile. The interpretation was performed by an international group of researchers. The V–z curves along the

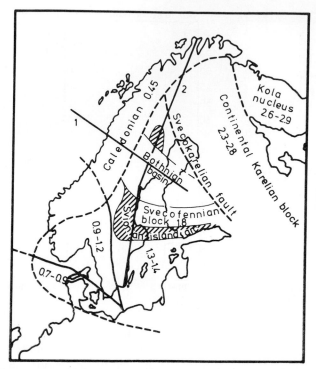

Fig. 6.7   Age provinces in the Fennoscandian–Baltic shield; 1, the Blue Norma–Blue Road geotraverse; 2, the Fennolora profile.

profile did not vary too much from the results for the previously mentioned shield areas. Figure 6.9b shows some examples of the V–z data obtained. The Moho generally does not produce strong $P_m P$ arrivals, in accordance with the more transitional crust/mantle boundary in shields (Fig. 6.9c).

All DSS measurements in shields mentioned so far did not show any convincing LVZ. Only sporadically some rather small LVZs are reported in the Russian shield as well as in the Baltic shield.

A rather smooth V–z relation is also reported from the North Australian craton (Finlayson, 1982). Figure 6.10 presents some of these data. The upper parts of the curves show some small—but not very convincing—LVZs. A velocity around 7 km/s is again reached at about 30–35 km depth, and crustal depth is around 50 km.

Extensive DSS measurements were also made in the Canadian shield. The early data collected by Berry (1973) show a very diverse structure. This may be partly an effect of the different crustal structure in the different age provinces of the huge Canadian shield. The Superior Province is 2.3–2.6 Gyr

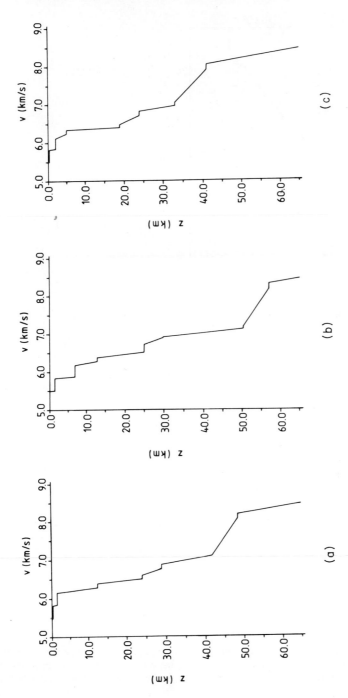

Fig. 6.8 Three V–z profiles from the Fennoscandian shield. [From Hirschleber *et al.* (1975), Lund *et al.* (1975), and Bittner (1983).]

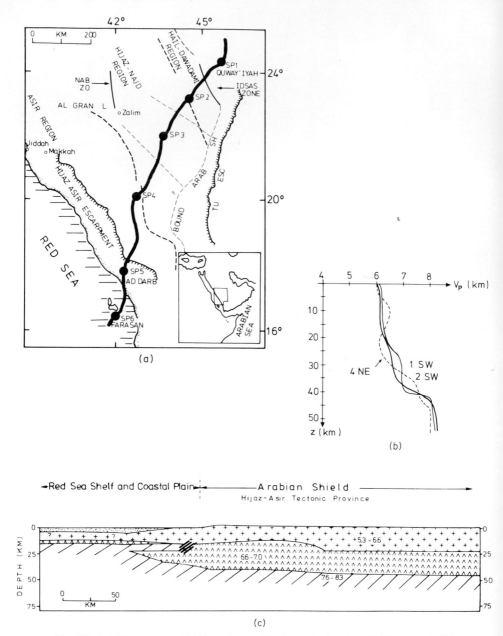

Fig. 6.9   (a) Situation map. (b) Three V–z profiles from the Arabian shield. Curves 1 SW and 2 SW are from the shield proper, and curve 4 NE is approaching the margin of the Red Sea. [After Mooney (1981).] (c) Crustal cross-section between SP6 (Red Sea) and SP4. Note transitional structure with double Moho at the margin. [After Milkereit and Flüh (1984).]

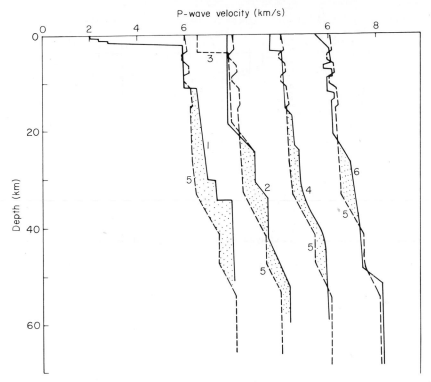

Fig. 6.10   Six different velocity–depth functions from the northern Australian craton; curve 5 is used as a reference function. [From Finlayson (1982).]

old, while the surrounding Churchill Province in the northwest (1.6–1.9 Gyr) and the Grenville Province in the southeast (0.8–1 Gyr) are considerably younger. In Fig. 6.11 some of the early data with stepwise V–z intervals are compared with data from a more modern interpretation (Berry and Fuchs, 1973). It should be stressed that this seems to be the only case in which, in addition to some small-scale LVZs in the upper crust, the data from a real shield area—although only those from the youngest structure, the Grenville Province—are interpreted as showing (small) LVZs. The transition to a 7 km/s velocity is rather deep, i.e., around 40 km, in the young Grenville shield. This young shield seems to be a bit more "sialic" than the older ones.

As a result of the compilation of the DSS data from the old shields we may summarize that a rather continuous increase in velocity is observed in the whole crust without prominent LVZs and first-order discontinuities. The crust in shield and platform areas, therefore, is differentiated into a more silicic (or low-grade) upper crust and a more mafic (or high-grade) lower

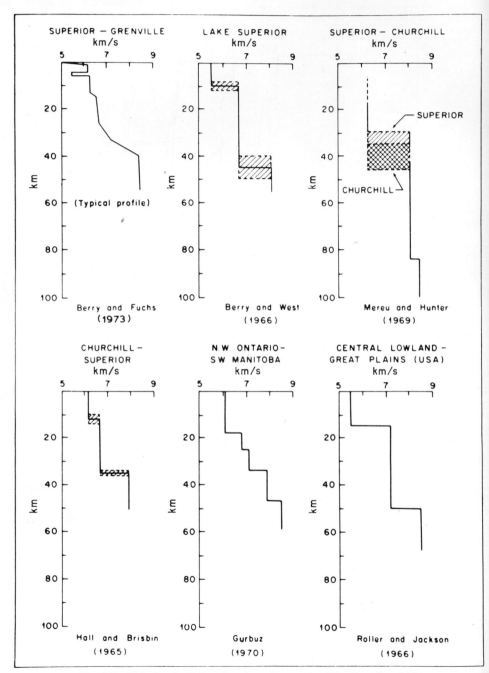

Fig. 6.11 Velocity–depth functions in various parts of the Canadian shield.

crust. However, this transition is gradational and certainly not as strong and not as extreme as in younger crusts. Several age-dependent processes, such as a long-time pressure-dependent prograde or retrograde metamorphism, have homogenized the old cratons. Only the transition from crust to mantle shows some steeper velocity gradients. A compilation of typical V-z curves of various shields and platforms in the form of block diagrams will be given later.

6.1.2   DEEP SEISMIC SOUNDINGS IN PLATFORMS

A change in the structure of a shield often appears if the shield is submerged. Cooling or some convective movements in the asthenosphere create some areas in the shields where subsidence starts. Certainly the reasons for such large-scale and slow subsidence lie in the mantle below. In contrast to continental margins, where sediments replace water and really may contribute to subsidence, sediments in platform areas replace basement rocks. They create an area of buoyancy and therefore are a consequence of, not a reason for, subsidence. Sometimes, subsidence areas in old shields are nearly circular; sometimes they form deep and narrow, rather asymmetric areas, so-called aulacogens. An example of a slow epeirogenic downward movement of a large area without much folding or faulting is the Michigan basin, a bowl-shaped, nearly circular depression between the Appalachian and the Rocky Mountains which subsided throughout Paleozoic time and accumulated a sediment thickness of up to 3 km. The V-z structure does not seem too different from that of the surrounding shield area. Only a rather rough interpretation of the U.S. American platform area was given by Healy and Warren (1969), showing the V-z structure by means of three or four crustal layers with constant velocities. A number of authors from the Soviet Union report a remarkable correlation between the thickness of the upper crust and that of the whole crust in platform areas. A negative correlation seems to exist between the thickness of the sediments and the thickness of the whole crust, indicating a kind of compensation in form of an "antiroot" below subsiding areas (see also Section 6.4).

An example of an aulacogen is the Dnjepr–Donetz depression, as shown in Fig. 6.12 (Alekseev *et al.*, 1973; Pawlenkova, 1973). It divides the Ukrainian and Voronezh shields and may be considered as a once-mobile transition zone. Sediments reach the extreme thickness of 15 km. A considerable thinning of the crust is observed, especially as shown in Fig. 6.12b, the Moho being shallower below the depression. Apparently, extensive stretching of the shield has taken place, the lower crust–upper mantle transition being modified either by a primordial mantle plume or by a residue of large magma

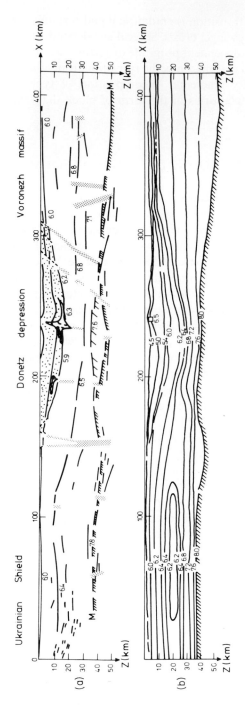

Fig. 6.12 The Dnjepr–Donetz depression. (a) From Alekseev *et al.* (1973), and (b) Pawlenkova (1973). The figure shows parallel profiles with slightly different methods of interpretation. Numbers are V_P in kilometers per second.

bodies which intruded the lower and middle crust during the process of stretching. Both interpretations (Fig. 6.12) show signs of strong modifications of the velocity field in the depression.

Such an aulacogen in shield areas apparently has a long history and developed slowly, thus showing a tendency for isostatic equilibrium. Its structure is similar to that of rifts and passive continental margins, and there may be a genetic relationship. As will be mentioned when dealing with the passive margins (specifically in Sections 6.4.3 and 6.4.4), there might be some difference in the velocity structure, both units generally having lost their high-velocity lower-crustal material.

In both interpretations of Fig. 6.12 the high-velocity layers of the lower crust are very thin. Magmas have also intruded the sediment basin. Whatever the exact V-z structure in such aulacogens might be, they may be considered a "failed attempt" of the underlying mantle to split an old continent apart. The V-z data from shields and platforms are compiled in Fig. 6.13.

We may summarize:

(1)   Shields and platforms have generally thick crusts (>40 km) with a relatively thick lower crust of mafic and/or high-grade composition.

(2)   They show only a moderate differentiation into low-velocity and high-velocity material.

(3)   Boundaries inside the crust are generally smooth; the Moho is a good refractor (strong P_n!).

(4)   Wide-angle reflections like P_mP or P_cP are generally weak, probably partly because of (2) and (3). Therefore, strong near-vertical reflections are not to be expected either.

6.1.3   CRUSTAL THICKNESS IN SHIELDS, PLATFORMS, AND THEIR SURROUNDINGS

Although so far we have discussed only shield and platform structures as revealed by DSS studies, two maps of crustal thickness will be shown which also contain data from younger areas that were later attached to the old nuclei. As was shown in Fig. 6.3, the percentage of younger areas along the western, southern, and eastern margins of the North American continent is rather small. Figure 6.14 shows a map of crustal thickness that was recently compiled by Allenby and Schnetzler (1983) from DSS data in the United States (see also Table 6.2). Note that only the younger areas along the margins have crustal thicknesses less than 35 km. Shields and platforms have average thicknesses of 40 to 50 km (with only small undulations), while the young mountain ranges in the west show elongated and narrow crustal roots.

In order to compare this Moho map with that of Europe, a compilation of

260   6 Crustal Structure

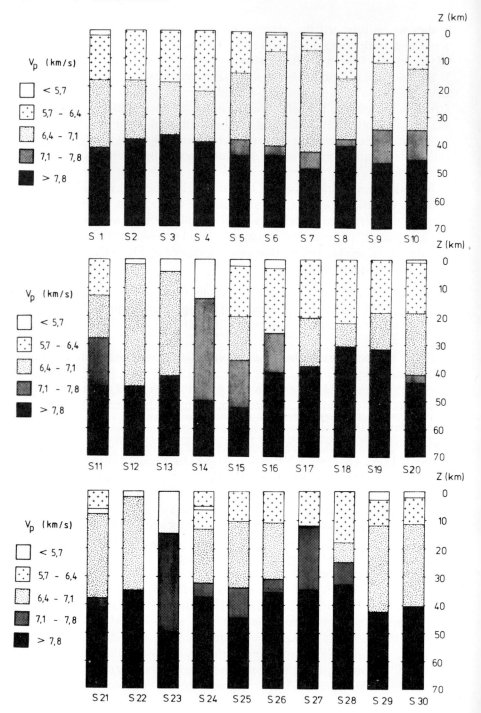

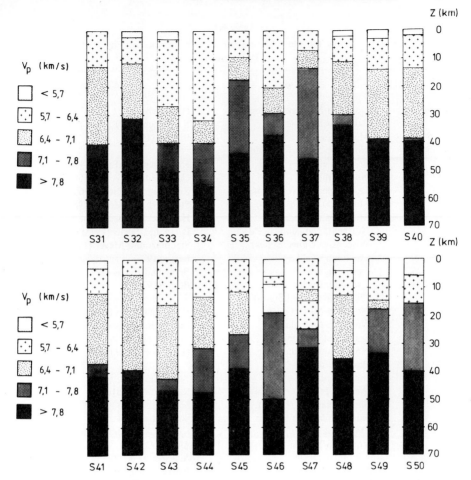

Fig. 6.13 Compilation of standardized velocity–depth functions from shields and platforms. For locations of the columns see Table 6.2.

DSS data from eastern and western Europe was performed. In Fig. 6.15 a simplified tectonic map is shown with the shield and platform areas in the north and east. The areas accumulated later are in the south and west. Their percentage is certainly larger than it is in the United States, but they are generally covered by a very dense network of DSS lines. The major line of separation between the Baltic shield/Russian platform and the area accumulated later in the southwest is the Tesseyre–Thornquist lineament (TTL), which may also be observed in the following map of crustal thicknesses and will be discussed again later. Such a map of crustal thickness in Europe, based on compilations of Giese (1983), Stegena (1984), and Meissner et al. (1983a), is presented in Fig. 6.16. In this map the separation between the stable shields

Table 6.2

Locations and References for the Standardized $V(z)$ Functions of Fig. 6.13 (Shields and Platforms)

Function	Location	Reference
S1–S4	Northern Norway/Sweden	Hirschleber *et al.* (1975)
S5	Baltic shield	Korhonen and Parkka (1981)
S6–S7	Middle Sweden	Bittner (1983)
S8–S10	Northern Norway/Sweden	
S11	Baltic shield	Korhonen and Parkka (1981)
S12–S13	North America	Cohen and Meyer (1966)
S14	Lake Superior (North America)	Roller and Jackson (1966)
S15–S16	Montana (United States)	Steinhart and Meyer (1961)
S17–S19	Manitoba (Canada)	Hall and Hajnal (1973)
S20–S21	Northeast Canadian shield	Berry and Fuchs (1973)
S22	Wisconsin (United States)	Steinhart and Meyer (1961)
S23	Canadian shield	Berry (1973)
S24–S25	Canadian shield	Berry and Fuchs (1973)
S26–S28	Canadian shield	Berry (1973)
S29	Cliff Lake (United States)	Smith *et al.* (1966)
S30–S32	Northern Australia	Hales and Rynn (1978)
S33–S34	Northern Australia	Finlayson (1982)
S35–S37	Southwest Australia	Mathur (1974)
S38	Northern Australia	Hales and Rynn (1978)
S39	Brazilian shield	—
S40	Indian shield	—
S41	Western India	Kaila *et al.* (1981a)
S42	Skifian platform (USSR)	Sollogub (1969)
S43	Russian platform	Sollogub (1969)
S44–S45	Donetz basin	Kosminskaya and Pawlenkova (1979)
S46	Voronesh Massif	Alekseev *et al.* (1973)
S47	Ukrainian shield	Alekseev *et al.* (1973)
S48	Ukrainian shield	Jentsch (1978/79)
S49–S50	India	Kaila *et al.* (1981b)

and platforms in the northeast with only smooth undulations in Moho depths and the Paleozoic areas in the west or the Cenozoic features in the southwest and the south can be clearly observed. The younger the orogeny, the steeper are the gradients in Moho depth. The Pyrenees, Alps, Carpathians, Dinarides, Hellenides, and the Caucasus are all marked by elongated and steep crustal roots. The Variscan area in central and western Europe, with its last great tectonic activity between 380 and 300 Myr ago, has already lost its crustal roots and shows a rather uniform crustal thickness of about 30 km. This is about 15 km less than the crustal thickness of the old cratons, which, however, are attacked and modified by rather young tectonic pro-

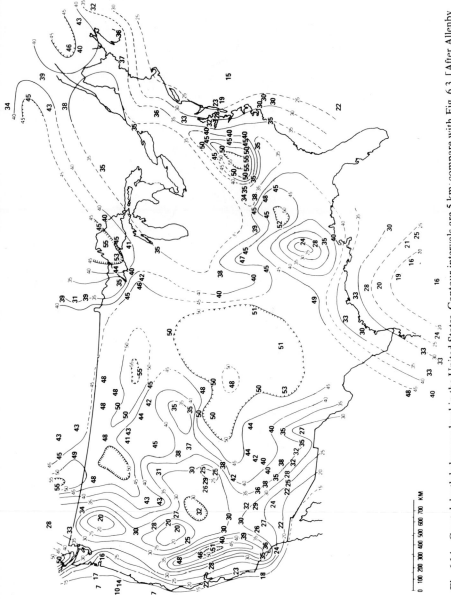

Fig. 6.14 Crustal depth below sea level in the United States. Contour intervals are 5 km; compare with Fig. 6.3. [After Allenby and Schnetzler (1983).]

0 100 200 300 400 500 600 700 KM

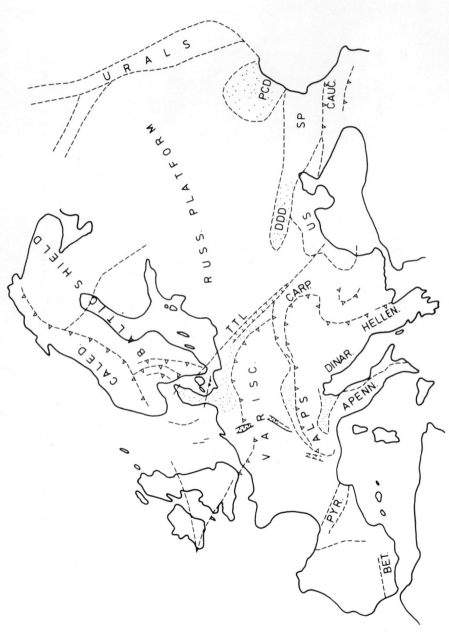

Fig. 6.15 Simplified tectonic map of Europe modeled after Berthelsen (1983) and Marçais *et al.* (1981); Caled, Caledonides; Varisc, Variscides; Pyr, Pyrenees; Bet, Betics; Apenn, Apennines; Dinar, Dinarides; Helen, Helenides; Carp, Carpathians; Cauc, Caucasus; TTL, Tesseyre–Thornquist line; US, Ukrainian shield; DDD, Dnjepr–Donetz depression; PCD, pre-Caspian depression; and SP, Scythian platform.

Fig. 6.16 Crustal depth in Europe, compiled from data by Stegena (1984, private communication), Meissner *et al.* (1983a), Mostaanpour (1984). and Panza *et al.* (1984).

265

cesses at their southern margin. Here not only compressional structures but also areas of extension, subsidence, and elevated Mohos, e.g., the Pannonian basin and the Black Sea, are clearly observed in the map. Comparing the Moho maps of Fig. 6.14 and 6.16 there are many common features, such as the smooth depth pattern of shields and platforms with average Moho depths z_M of 40–50 km and that of Paleozoic areas with $z_M = 30$–35 km, versus the rugged appearance of young orogenies with maximum $z_M > 55$ km and young basins with minimum $z_M < 25$ km.

6.1.4 NEAR-VERTICAL REFLECTION DATA FROM SHIELDS AND PLATFORMS

A number of near-vertical reflection profiles have been obtained by the COCORP group in the North American platform area. In the USSR some near-vertical reflection data from platform areas have been gathered, but these were collected more sporadically during DSS surveys and not by a routine, industry type of continuous profiling. Sollogub *et al.* (1975) obtained some weak bands of reflections along an 8-km-long reflection profile in the Ukrainian shield. In the following discussion the COCORP data from platform areas will be selected because they provide a uniform data set from quite a number of localities. In general, near-vertical reflection profiling with its high resolving power is not applied to a structureless part of a shield or platform area, but to some outstanding geological structures which have modified the platform area by additional tectonic activity. This might generate some problems in the interpretation and must be carefully considered.

Figure 6.17 shows a situation map of the COCORP profiles in the United States (Brown *et al.*, 1981) combined with the map of the age provinces of Fig. 6.3. There are four profile networks in the 1.6–1.9-Gyr-old Penokean platform, but the two western ones are across major and younger compressional fronts, i.e., the Wind River thrust fault and the Rocky Mountains thrust, and the eastern one is in the Michigan basin near the Grenville Province boundary. Also, in the 1.2–1.5-Gyr-old central province all profile networks are across special mountain ranges, uplifts, rifts, or basins.

The northernmost profile, and probably the one least disturbed by later tectonic activity, is the Minnesota profile near the boundary of the 2.3–3.3-Gyr-old Superior Province and the 1.6–1.9-Gyr-old Penokean age province (Fig. 6.18a). The location is slightly west of the large ENE–WSW-striking midcontinent gravity high (see Section 3.1.6) and should have a crustal thickness of 46–50 km according to the map of Fig. 6.14. According to our experience from DSS investigations the crust should be rather mafic or

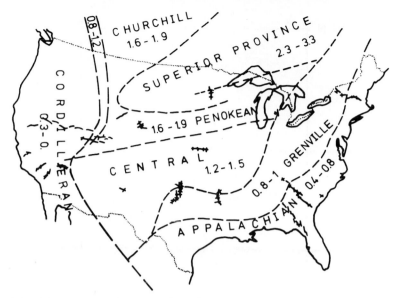

Fig. 6.17   COCORP reflection lines in the United States related to age provinces (as of 1984).

strongly metamorphic and possibly not too much differentiated. The COCORP reflection work was undertaken to study a specific boundary between the northern granite–greenstone belt of the Superior Province and an older Archean gneissic–granulitic area in the south (Gibbs *et al.*, 1984). The survey revealed many continuous and discontinuous, generally north-dipping reflectors in the upper crust, i.e., down to about 4–5 s two-way travel time (TTT), equivalent to 12–15 km depth. Farther down, reflections become more and more discontinuous and scarce. Only weak indications of the Moho or other crustal boundaries are seen. A careful statistical analysis by Wever (1984) provides histograms for the occurrence of reflections per 1-s TTT interval, a distribution which seems typical for old shield areas with a small amount of crustal differentiation (see Fig. 6.18b). Although some influence of the source (Vibroseis) may be present, this kind of anaysis will be used for a comparison with reflection densities from other locations.

The next location to be discussed is the one in the Michigan basin (Brown *et al.*, 1982). It is a nearly circular depression in the old shield area of about 500 km diameter and filled with Phanerozoic sediments (Fig. 6.19a). The Moho should be at about 35 km only, according to Fig. 6.14. As seen in the example of Fig. 6.19b there are strong subhorizontal reflections from the sediments in the basin. As described by Brown *et al.* (1982) and Brewer and Oliver (1980) there is a 60-km-wide, 3-km-deep trough-shaped structure

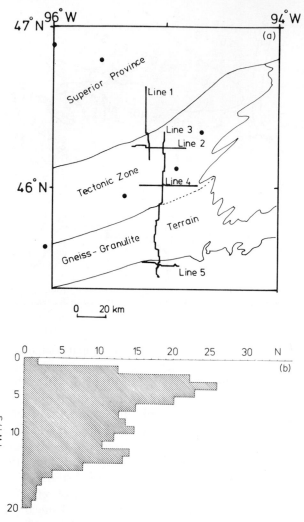

Fig. 6.18   (a) Situation map [after Gibbs *et al.* (1984)] and (b) histogram of the occurrence of reflections per 1-s travel time interval for COCORP's Minnesota profiles. [After Wever (1984).] This kind of analysis is frequently used in the following for discriminating between more or less differentiated crusts. The number of reflections is normalized to the number of common depth points (CDP) per kilometer. The abscissa is $N(\Delta t) = (n/s) \sum_{\Delta t} l_{\Delta t}$, where n is the number of common midpoints (CMP) (trace intervals) in the section, s the length of the seismic section (in the same units as n), and $l_{\Delta t}$ the length of coherence of reflections in kilometers per two-way travel time Δt. In this case, $\Delta t = 1$ s. Solid dots represent earthquakes.

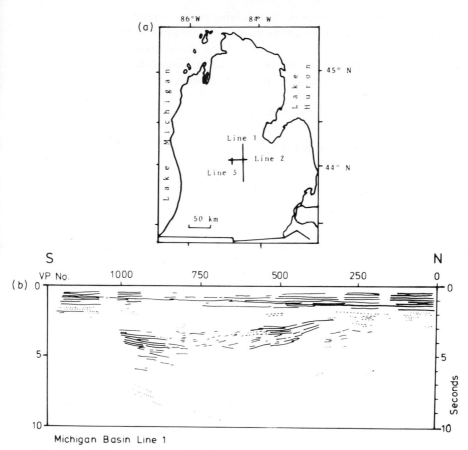

Fig. 6.19 (a) Situation map and (b) line drawings of COCORP's line 1 in the Michigan basin. Length of profile is approximately 80 km. [From Brown *et al.* (1982).]

about 6–9 km below the surface underlain by some discontinuous, curved reflections. As in the Minnesota profiles, reliable reflections from the deeper crust or the Moho are scarce or missing completely. As mentioned earlier, this would be expected from the rather smooth transition of *V–z* curves from DSS in other shield and platform areas (not necessarily from the old DSS interpretation in this area).

One of the most exciting COCORP profiles is that across the Wind River thrust fault. According to the Moho map of Fig. 6.14, crustal thickness should be only 35–37 km. The profile will be discussed when we deal with fault zones in Section 6.6.3. Although reflection quality is generally good, the Moho (or a Conrad) cannot be seen by continuous bands of reflections at the appropriate travel time of 11–14 s. While this lack of Moho reflections might be expected in old shield or platform areas, the resolving power of reflection seismology to delineate structural details like fault zones is certainly surprising.

There are four profile networks in younger shield areas, i.e., in the 1.2–1.5-Gyr-old Central Province. Two of them will be discussed here, while the Rio Grande (Socorro) lines will be included in Section 6.4.1 and the Quachita lines in Section 6.2.2.

Extremely strong reflectors at nearly 2.9 and 3.7 s two-way travel time corresponding to 9–12 km depth are seen in the record sections of Hardeman County, northern Texas (Oliver *et al.*, 1976; Schilt *et al.*, 1979) (Fig. 6.10a). These flat-lying reflectors were interpreted by Brewer *et al.* (1981) as representing the base of a late Precambrian basin. Sporadic and discontinuous but rather strong reflectors appear at 6, 9, 10, and 14 s (Fig. 6.20b). Some of these events, such as the one at 9 s, certainly are diffractions. According to the DSS surveys and Fig. 6.14, the Moho should be at about 50 km depth, i.e., at about 15–16 s two-way travel time.

A later extension of the network farther north into the Anadarko basin in Oklahoma across the Wichita Mountains revealed this boundary as a series of multiple northward-directed thrusts (Brewer *et al.*, 1983a). The whole region appears to have been tectonically active through the late Precambrian and Paleozoic and might not represent a true platform area. The two strong reflectors at 2.9 and 3.7 s represent continuous boundaries farther to the north up to the southern margin of the Wichita Mountains. Here they are lost and are probably "faulted out of the section" (Brewer and Oliver, 1980). Although the deeper parts of the new profiles show more continuous reflectors than the Hardeman County records, reflections from the Moho are absent. Again, the value of the survey lies in the identification and tracing of thrust faults, seen at the surface and in well data, down to at least 20 km depth under the crystalline basement of the Wichita Mountains. The Wichita uplift must have suffered a crustal shortening of about 15 km (Brewer, 1983). A large wedge of sedimentary rocks lies under the basement overthrust. The Quachita orogen in eastern Arkansas will be discussed when we deal with Paleozoic structures.

As a summary of the reflection work in shield and platform areas we may state:

(1)   Near-vertical reflections from the middle and lower crust are generally poor and discontinuous, and Moho reflections are often missing completely.

(2)   The reason for the absence of good information on the lower crust is thought to be partly the (only) moderate differentiation of the old crustal units, showing smooth velocity transitions as derived from appropriate DSS work (Section 6.1.2).

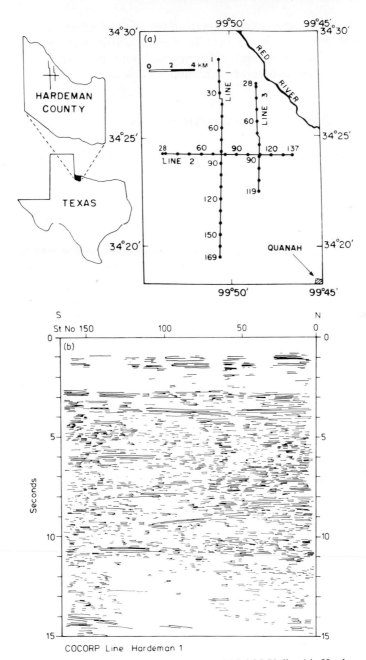

Fig. 6.20 (a) Situation map and (b) line drawing of COCORP's line 1 in Hardeman County, Texas. [From Oliver *et al.* (1976).]

(3)   The most useful information from this kind of reflection technique in shields and platforms is obtained in regions where later tectonic events created either an additional differentiation with horizontal layering or prominent fault zones by near-horizontal tectonics.

6.2   PALEOZOIC AND MESOZOIC AREAS

As discussed in several sections, especially in Section 6.1, the old cratons are surrounded by continental material accumulated later. The process by which this accretion took place is highly debated, but it seems that at least in the last 2000 Myr the formation of magmatic arcs with their basaltic–gabbroic, andesitic, and rhyolitic magmas played a dominant role. Compressed, deformed, and even distorted at the boundary of the old continental nucleus during an accretional tectonic phase of the Wilson cycle, the magmatic arcs are finally welded together with the old craton, sometimes along prominent fault zones. A convincing picture of a continent growing by episodic accretion of magmatic arcs (or some other continental or oceanic fragments) has evolved. It certainly seems apparent in those areas that were not split again by a later continental rifting.

The younger the accretional phase, the better it is observed by its tectonic consequences: mountain building, indentation, suturing, underplating, and thick- or thin-skinned nappe displacements. Some of these phenomena, e.g., the overthrust faults, some postorogenic normal faults, and the observable features of young mountain-building processes, are treated in separate sections.

At the beginning of the Paleozoic at about 570 Myr heat production was only a few percent higher than today (Fig. 6.2), and plate tectonics was well established in its present style. The Appalachian and Caledonian orogenies and later Hercynian events cover large intervals of the Paleozoic, while the Mesozoic from 225 to 65 Myr shows the beginning of the opening of the Atlantic Ocean and the first stages of the Alpine and Rocky Mountain orogeneses. Paleozoic areas have been well investigated both by refraction measurements, i.e., DSS, and by near-vertical reflection investigations in northern, central, and western Europe as well as in the North American continent.

6.2.1   REFRACTION MEASUREMENTS IN EARLY
          PALEOZOIC AREAS

During the Paleozoic the Caledonian was the first of five phases of orogeny in the North American–Eurasian continent (Windley, 1977). While the opening of the old Iapetus may have started as early as 1000 Myr, shrinking of the ocean started in the Late Ordovician, forming subduction zones at both sides of the Iapetus. The compressional regime of the final continent–continent collision during the Silurian to Middle Devonian created thrusting and huge nappe displacement, directed eastward against the Baltic shield and directed westward in Greenland and the northern Appalachians. For more details see Gee (1978) and GeoJournal, special edition 3.3 (1979).

The root zone of the European Caledonides seems to be missing everywhere. Several DSS profiles from the Baltic shield across Norway and into the Norwegian margin cross the thin nappe displacements which increase to the west, but cannot be identified by DSS measurements. Figure 6.21 shows the Blue Road–Blue Norma profile from the shield across the Caledonians, through a Tertiary rift (the Westfjord basin) across the old Lofoten horst and into the shelf area (see also Table 6.3). There is certainly no crustal root to be seen, and velocities below the Caledonides are similar to those of shield areas. Crustal depths decrease nearly continuously from the shield to the margin, apparently unaffected by the different tectonic structures in the crust's upper part. The Bouguer gravity minimum of the Caledonides does not reflect a crustal root zone but may contain some contribution of a low-density (oceanic) asthenosphere (Vetter and Meissner, 1979; Avedik et al., 1984).

The long lithospheric seismic profiles in Britain, also called LISPB profiles, i.e., north–south-directed DSS lines, also crossed the Iapetus suture zone and covered the English and Scottish Caledonides. Crustal depths here are about 30–35 km, and the average crustal velocities are only slightly lower than in shield areas (Bamford et al., 1976, 1978). Along the 1000-km-long LISPB lines generally strong and reliable $P_m P$ events are recorded in the Caledonian mountains of Scotland, i.e., in a high-grade metamorphic zone, as well as in the low-grade metamorphic zone in England. The $P_m P$ arrivals become

Table 6.3

Locations and References for the Standardized $V(z)$ Functions of Fig. 6.23 (Caledonian Structures)

Function	Location	Reference
K1–K2	Scotland	Assumpção and Bamford (1978)
K3–K5	Scotland	Bamford et al. (1978)
K6	Scotland	Bamford et al. (1978)
K7	Faeroe	Casten et al. (1975)
K8	Norway	Massé and Alexander (1974)
K9–K10	Southeast Australia	Finlayson et al. (1980)

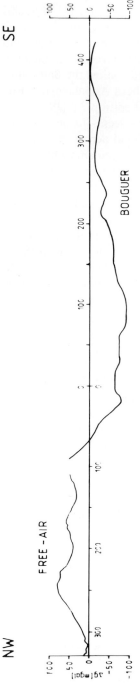

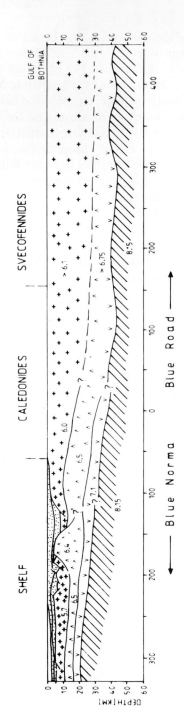

Fig. 6.21 Crustal cross-section along the **Blue Norma** profile from the **Baltic shield** (right) through the Caledonides and the Norwegian shelf. For situation of the profile see Fig. 6.7.

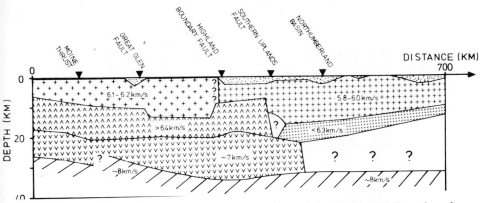

Fig. 6.22   Crustal cross-section of the northern part of the LISPB DSS lines through Caledonian Scotland and northern England. Compare with Figs. 6.31 and 6.32. [From Bamford *et al.* (1978).]

weaker and the crust thicker in the early Paleozoic to Precambrian basement in southern England, an observation which is comparable to the data from the DSS surveys in the United States. Figure 6.22 shows the northern part of the LISPB profiles in Britain crossing several major structures and faults of Caledonian age. Some of them can be recognized in the cross section. Compared to the Norwegian Caledonides, the crust is slightly shallower and less dense, possibly a consequence of post orogenic extension with heating and differentiation. A major contribution to the structure of the Scottish Caledonides was made by the British institutions reflection profiling syndicate (BIRPS) group, as reported by Brewer *et al.* (1983b) and discussed in Section 6.2.2. A situation map of the LISPB and BIRPS lines is shown in Figure 6.32.

At least three DSS profiles cross the northern Appalachians, i.e., the other side of the Iapetus suture. As seen in Fig. 6.12, crustal thickness in the Caledonian range is generally about 35 km between two lows of 46 km in the north, north of the US–Canadian border and one low of 50 km in the middle of the southern Appalachians. In general, however, no crustal root has survived. As will be discussed later, several collisions took place before the Iapetus was finally closed. Figure 6.23 shows a compilation of typical V–z curves of various Caledonian areas, similar to the compilation of Fig. 6.14 for the shields and platforms. We may summarize:

(1)   Crustal thickness is between 30 and 40 km on the average. Crustal roots are not observed.

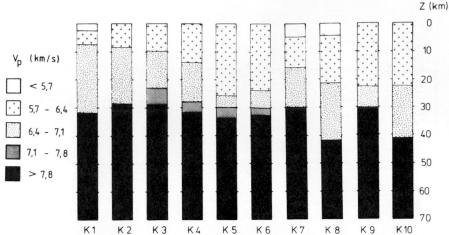

Fig. 6.23   Compilation of standardized velocity–depth functions from Caledonian structures.

(2)   Caledonian provinces show a certain differentiation with pronounced Mohos and sometimes also Conrads.

(3)   The percentage of high-velocity material in the lower crust is smaller than in the shield areas but greater than in younger areas. The velocity range from 7.1 to 7.8 km/s is mostly missing.

(4)   Wide-angle events like $P_m P$ or $P_c P$ are generally strong.

6.2.2   REFRACTION MEASUREMENTS IN MIDDLE TO LATE PALEOZOIC AREAS

In central and western Europe the next important orogenic event after the Caledonian was the *Variscan orogeny*. It took place in several phases, first in the south with the collision of the Moldanubian and the Saxothuringian; next at the Saxothuringian–Rhenoherzynian boundary. Finally, at about 300 Myr, the northern Variscan deformation front was created partly by a collision with the London-Brabant-Massif, producing a thin-skinned nappe displacement, as will be shown in Section 6.2.3. The development of the European Variscides is described in papers by Weber (1981), Ziegler (1978, 1982), Böger (1983), Murawski (1984), Murawski *et al.* (1983), and Lorenz and Nichols (1984). A dense network of DSS profiles has been established in the Variscan area of central Europe. Western Germany especially is covered by a great number of refraction lines. An extensive description of data collection, results, and interpretation of the data may be found in Giese *et al.* (1976). Most of the lines were observed with large quarry blasts and with a large set of homogeneous Mars 66 recording stations (Berckhemer, 1970).

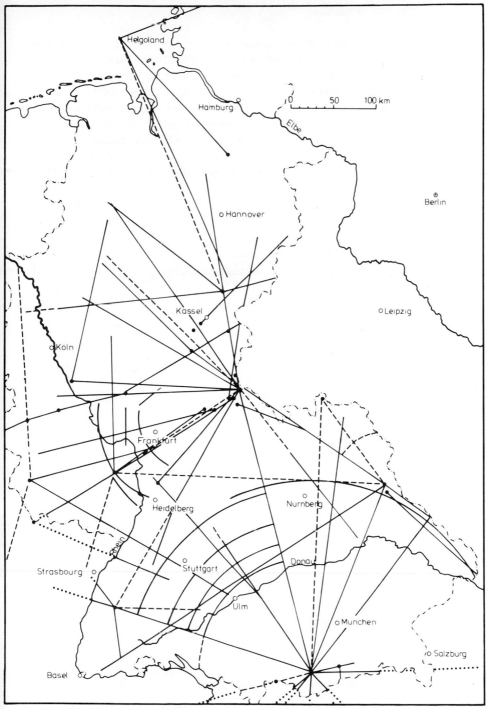

Fig. 6.24 Situation map of DSS lines in western Germany, showing the great density of the refraction lines. [From Giese (1976a).]

Occasionally, DSS lines with a common reflection element were observed together with near-vertical reflection work in 1964 in the Bavarian Molasse (Meissner, 1967) and in 1968 in the Hunsrück area. In general, very strong and reliable P_mP events are observed in the Variscan area of Europe. The LVZs seem to be limited to the Rhinegraben, the Alpine belt, and some geothermal areas, e.g., Urach (see Section 6.6.1). Figure 6.24 shows a situation map of major DSS lines in western Germany. Detailed contour maps of crustal thickness, P_n velocities, average crustal velocities, and other parameters emerged from these concentrated observations (Giese, 1976a,b). There are at least two important results which must be discussed: the rather uniform and shallow depth of the Moho in the Variscides, and the small average crustal velocity. Figure 6.25 shows a map of the crustal thickness.

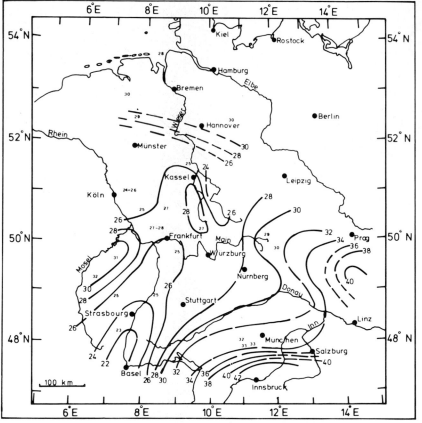

Fig. 6.25   Crustal depth in western Germany. Contour intervals are only 2 km. [From Giese (1976a).]

Only those areas approaching the Bohemian Massif in the southeast and the Alps in the south have crustal thicknesses in excess of 30 or 32 km; and only the area of the Rhinegraben systems shows smaller values. Later profiles have confirmed this result. The major orogenic events which are related to boundaries between the Moldanubian, the Saxothuringian, and the Rheno-herzynian at 400, 350, and 300 Myr, respectively, have not produced perman-ent variations in the crustal thickness. Only some minor variations in the crustal velocity $\bar{V}$ are seen in Fig. 6.26. With its value of 6 km/s, $\bar{V}$ is much

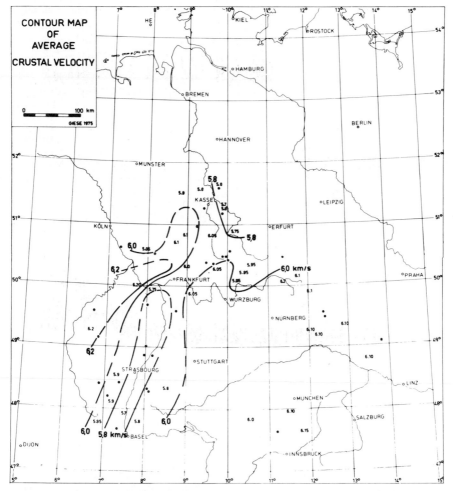

Fig. 6.26   Values of average crustal velocities in kilometres per second for parts of western Germany. [From Giese (1976a).]

smaller than in the shield or platform areas. Again, the homogeneity of the $\bar{V}$ values and the shallow crustal depth, which shows no sign of old crustal roots or variations from one age province to another, are puzzling. Although erosion might have contributed considerably to reducing the crustal thickness after the mountain-building process was over, the uniform Moho depths and the low $\bar{V}$ values indicate a thorough differentiation of the crust, the Moho probably being a new boundary, a "postorogenic Moho", formed by ultramafic cumulates as described in Section 7.3.3.

Voluminous post- and synorogenic granitic intrusions support this picture. As shown in Fig. 6.16, the crustal thickness is rather uniform in France, Germany, Britain, and the Iberian Peninsula. Age provinces and velocity values are also similar. An especially long DSS profile of 1200 km was observed across France from Brittany in the northwest to the Mediterranean coast near Toulon in the southeast (Sapin and Prodehl, 1973). The crust is generally about 30 km thick. Even in the Massif Central no considerable variation in crustal structure is observed. The gabbroic–granulitic component with $V > 6.4$ km/s is restricted to a smaller depth interval at the base of the crust than in the Caledonides.

The profile network in Spain (Banda et al., 1981), including some profiles into Portugal (Mendes Victor et al., 1980), does not show any dramatic variations from the observations made in France and Germany. In the Meseta the high-velocity part of the crust is larger, and toward the Atlantic coast in Portugal the thickness of the continental crust decreases to 26 and 23 km, undulating below a Mesozoic syncline of 5 km thickness. Even the Pyrenees, although a collisional zone of Variscan platelets, show on one side small crustal depths but a higher "mafic" part of the crust, possibly related to the specific orogenic mechanism, which might well have incorporated some material of oceanic provenance. In the American continent only the Atlantic coastal plains and narrow stripes along the Gulf and the Pacific coasts have comparable age, velocity values, and Moho depths. The Appalachians, although of the same age in the south as the European Variscides, have thicker crusts.

Neither the European Caledonides nor the Variscides show crustal roots (Figs. 6.15 and 6.16). The only place in which a kind of root zone of Variscan origin is observed is a narrow strip within the Tesseyre–Thornquist lineament in Poland. Here, a graben-like structure was found by a number of DSS studies, presented by Guterch (1977) and Guterch et al. (1976). Within the graben structure the crustal thickness increases to about 60 km. The graben along the lineament is so narrow that it could not be plotted in the Moho map of Fig. 6.16. However, the long lineament is certainly seen as an outstanding structure in the Moho map, dividing the younger areas in the southwest from the shield and platform area in the northeast. Extensional tectonics of the post-Variscan collapse might be responsible for the graben part along the Tesseyre–Thornquist lineament, while a wrench tectonic might have been dominant in the other parts.

Figure 6.27 shows a compilation of typical V–z curves (see also Table 6.4) of Variscan provinces in Europe in a way similar to Figs. 6.13 and 6.23. We may summarize for the middle to late Paleozoic areas:

(1) Crustal thickness is generally around 30 km. Crustal roots are not observed.

(2) Variscan provinces show a strong differentiation with very pronounced Mohos and sometimes Conrads.

(3) The percentage of high-velocity, i.e., mafic/granulitic, material in the lower crust is very small. In particular, the velocity range from 7.1 to 7.7 km/s is missing completely.

(4) Wide-angle events like P_mP or P_CP are generally very strong.

6.2.3 NEAR-VERTICAL REFLECTION SURVEYS IN EARLY PALEOZOIC AREAS

Several COCORP and some additional reflection profiles cover the Paleozoic Appalachian area and the Mesozoic of the coastal plains in the

Table 6.4

Locations and References for the Standardized $V(z)$ Functions of Fig. 6.27
(Variscan Structures)

Function	Location	Reference
V1	Spain	Payo (1972)
V2	Pyrenees	Dagnieres et al. (1982)
V3	Portugal	Mendes Victor et al. (1980)
V4	Spain	Banda et al. (1981)
V5–V7	Southern France	Sapin and Hirn (1974)
V8	Middle France	Sapin and Prodehl (1973)
V9–V10	Massif Central (France)	Sapin and Prodehl (1973)
V11	Brittany (France)	Sapin and Prodehl (1973)
V12	Bavarian Molasse (Germany)	Ansorge et al. (1970)
V13–V15	Northern Hessia (Germany)	Grubbe (1976)
V16	Saarland (Germany)	Edel et al. (1975)
V17–V20	Southern Germany	Edel et al. (1975)
V21	Black Forest (Germany)	Deichmann and Ansorge (1983)
V22–V23	Southern Germany	Angenheister and Pohl (1976)
V24	Hunsrück–Taunus (Germany)	Meissner et al. (1976a)
V25–V30	Frankfurt area (Germany)	Mooney and Prodehl (1978)

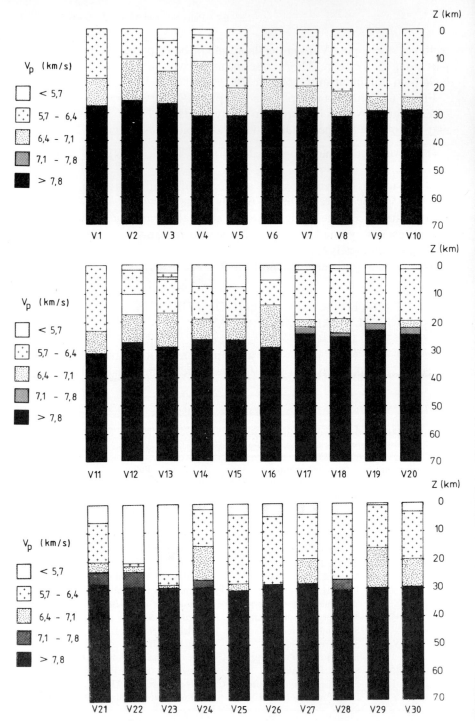

Fig. 6.27 Compilation of standardized velocity–depth functions from Variscan areas in Europe.

United States. In Britain the Caledonian shelf area has been extensively investigated by marine measurements, and in Germany several short profiles cross various remnants of Variscan mountain ranges. The COCORP profiles in Paleozoic terrains are in the southern Appalachians, in the northeast Appalachians, and in the Quachitas. In 1979 and 1980 COCORP acquired more than 650 km of profile on a line across the southern Appalachians, from Tennessee in the northwest through several collision zones through North Carolina and South Carolina down to the coastal plains in Georgia. The profile contains some of the most exciting and significant findings of COCORP (Cook *et al.*, 1980, 1981; Brewer and Oliver, 1980). Several thrust zones in the Appalachians compressional regime could be followed down along a 250-km-long "zone of detachment," which is about 6 km deep in Tennessee and about 12 km deep in Georgia. The discovery that a major portion of the crystalline basement in the orogen is allochthonous and that thin-skinned nappe displacement is virtually concentrated on a nearly horizontal "master fault" was certainly a surprise. This part of the survey will be discussed again in Section 6.6.3, which deals with fault zones.

Regarding the existence of reflections as an indication not only of such zones of detachment but also of zones of major differentiation, the southern Appalachian profile is also an interesting object. Figure 6.28 shows a situation map of the survey, and Fig. 6.29 presents a line drawing of profile 1.

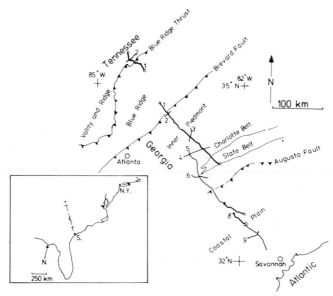

Fig. 6.28 Situation map of COCORP's southern Appalachian survey. [From Cook *et al.* (1981).]

The long zone of detachment is clearly defined across the whole profile. Deeper reflections are generally weak; the Moho seems to appear at the southeastern end of the line. On line 2, which crosses line 1 near the Brevard fault, some deep reflection can also be traced. Figure 6.30 shows the histograms, i.e., reflection density, along profile 1. The detachment zone provides the strongest peak of the histogram, but the Moho also comes out clearly at 10 to 12 s TWT while the lower crust remains an area void of reflections. Along profile 2 the lower crust shows many bands of continuous and discontinuous reflections. These results will be compared later with similar data sets from other Variscan (and Caledonian) age provinces.

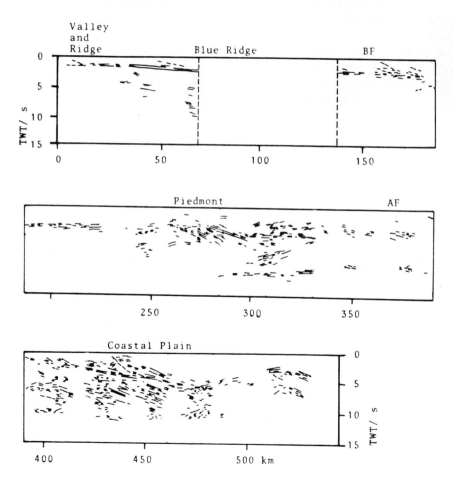

Fig. 6.29   Line drawing of profile 1 (Fig. 6.28). [From Brown, personal communication, 1983.]

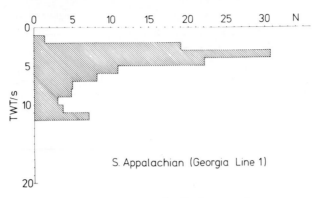

Fig. 6.30 Histogram of the occurrence of reflections per 1-s two-way travel time for COCORP's profile in the southern Appalachians. [From Wever (1984).]

In the northern Appalachians a network of lines was observed across the Adirondacks in the New York–Vermont area. Major orogenic phases are of Caledonian age. Although crustal shortening seems to have been similar to that of the southern Appalachians, a major zone of detachment is not observed with the same clarity as in the southern Appalachians. The six major profiles in the area all look rather different. Line 7 in the central Adirondacks shows extremely dense reflectors between about 15 and 23 km depth. It coincides roughly with a high-conductivity zone obtained by electromagnetic sounding (Nekut *et al.*, 1977; Connerney and Kuckes, 1980). Brown *et al.* (1983) suggested that a buried metasedimentary assemblage might be responsible for both kind of anomalies, the critical role of water in the lower crust being certainly debatable. As mentioned in Section 3.5, other explanations such as graphitic schists may also be employed for the anomalies.

In the eastern profiles through the Green Mountains the Precambrian complex is observed to have been thrust westwards along a subhorizontal zone of detachment, but the amount of basement overthrusting is smaller than in the southern Appalachians. Moreover, the many eastward-dipping reflectors in the eastern part of the profiles indicates several thrust plains that dip much more strongly and over a much greater depth interval than in the southern Appalachians; compare Figs. 6.29 and 6.31 (Ando *et al.*, 1985). As shown in Fig. 6.31, lower crustal material from the east seems to be thrust westward along a complicated system of faults along a ramp formed by the Grenville basement. Reflections from the crust–mantle boundary are observed only sporadically.

The Quachitas in Arkansas exhibit another Paleozoic compressional structure. The seismic sections show only some discontinuous reflections

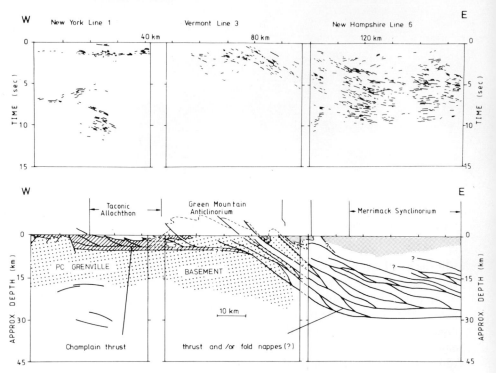

Fig. 6.31 Line drawing and crustal cross-section of COCORP's northern Appalachian survey along profiles 1, 3, and 6. [From Ando *et al.* (1985).]

from the lower crust. In contrast to the southern Appalachians, not a part of the crystalline basement but a large body of lower to middle Paleozoic deep-water sediments has been transported toward the continent and across the former continental margin, which now seems to form the core of the 8-km-thick antiform (Nelson *et al.*, 1982; Phinney and Odom, 1983).

Among the most spectacular crustal near-vertical reflection data are the MOIST and the WINCH lines, recorded by BIRPS. They were recorded by a multichannel marine technique using an extended air gun array 180 m long, towed at 8 m depth, and a 3-km-long 60-channel streamer, towed at 15 m depth. As mentioned by Brewer *et al.* (1983b), this array enhances deep reflections at the expense of resolution in the most shallow reflectors at less than 2 s TWT. The lines are definitely some of the best deep reflection profiles in the world and will be discussed in the following.

Figure 6.32 is a situation map of the MOIST line in the north and the WINCH lines in the northwest and west in the British Isles. The LISPB lines are also shown for comparison, and the most important geological structures

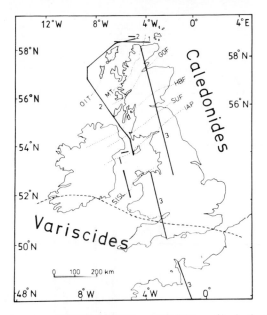

Fig. 6.32   Situation map of BIRPS MOIST and WINCH profiles in the Caledonian area of the shelf of the British Isles. The LISPB refraction lines on land are also shown; OIT, Outer Isles thrust; MT, Moine thrust; GGF, Great Glen fault; HBF, Highland Boundary fault; SUF, Southern Upland fault; IAP, Iapetus suture; SISL, southern Irish Sea lineament; 1, MOIST line; 2, WINCH lines; and 3, LISPB lines.

are indicated (Bamford *et al.*, 1976; Brewer *et al.*, 1983b; Blundell, 1981, 1983). Line drawings of unmigrated MOIST and WINCH profiles are shown in Figs. 6.33–6.35. It was a surprise to find the same strong density of reflectors in the lower crust of the Caledonides that had been seen in the Variscan area in Germany.

Profile section DC in Fig. 6.34 in the strike direction northwest of the last strong Caledonian fault system, the Outer Isles thrust, provides an un-

Fig. 6.33   Line drawing of BIRPS MOIST line. [From Smythe *et al.* (1982).]

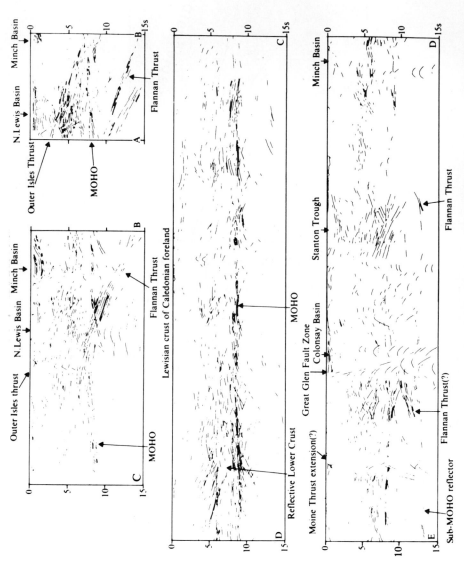

Fig. 6.34 Line drawing of BIRPS WINCH lines (northern part). [From Brewer *et al.* (1983b).]

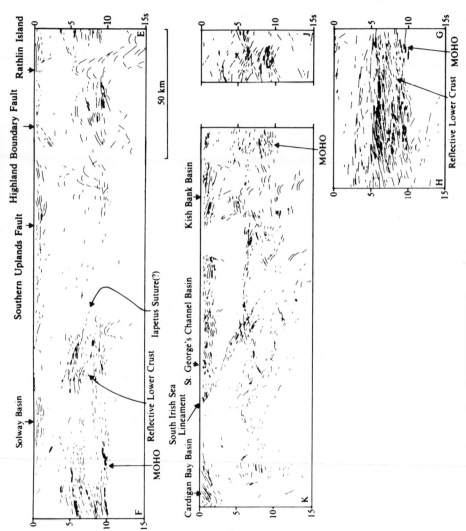

Fig. 6.35 Line drawing of BIRPS WINCH lines (southern part). [From Brewer *et al.* (1983b).]

disturbed and excellent example of the crustal reflectivity in the Caledonian foreland. Figure 6.36 shows histograms of the occurrence of reflectors for this line, while Fig. 6.36d shows a double-peaked histogram from the second profile of Fig. 6.35, thus illustrating the influence of sediments and possible faults in the upper part. Some sub-Moho reflections (TTT > 10 s) are also observed. Histograms like those of Fig. 6.36a–c are completely different from all COCORP lines and reminds us of similar peaks for the occurrence of reflections in the Variscan structures in Germany, especially the Urach geothermal anomaly (Bartelsen *et al.*, 1982; Meissner *et al.*, 1982). The observation that fault zones really can be traced down to the upper mantle was demonstrated by the BIRPS data with a clarity never seen before. For more details see Brewer *et al.* (1983b). Comparing the data with the northern section of the LISPB DSS profile, which was shown in Fig. 6.22, the different state of information from both techniques becomes apparent: the excellent structural resolution of the near-vertical reflection technique and the velocity information from the DSS line.

6.2.4   NEAR-VERTICAL REFLECTION SURVEYS IN THE LATE PALEOZOIC AREAS

Since 1964 several near-vertical reflection (NVR) lines between 30 and 80 km long have been observed in the Variscan (and Bavarian Molasse) area in the Federal Republic of Germany. Figure 6.37 shows a situation map of these lines together with the basic geologic provinces within the Variscan area. The technique of obtaining the data is somewhat different from those of the COCORP and BIRPS type of measurement. Explosions in shallow boreholes were used as the energy source for all measurements. The 1964 profile in the Bavarian Molasse trough and the 1968 profile in the Rhenoherzynian Hunsrück–Taunus mountains were a combination of NVR and wide-angle data with a common reflection element. The goal of this combination was an optimum gathering of velocity values (although only at the center of the profiles) and a correlation of reflections from the near-vertical with those from the wide-angle range. Both surveys showed the identity of major reflecting boundaries, as shown in Fig. 3.28. Reflectors appear at jumps (or within strong-gradient zones) of the V–z curve.

Another type of NVR measurement was carried out by continuous (near-vertical) profiling while refraction stations were emplaced along the line and in the wide-angle range, i.e., as fixed and mobile stations. This second type of operation was directed toward the investigation of special crustal structures such as the Ries impact crater (Angenheister and Pohl, 1976) and prominent fault zones like the Hunsrück–Taunus fault at the southern boundary of the

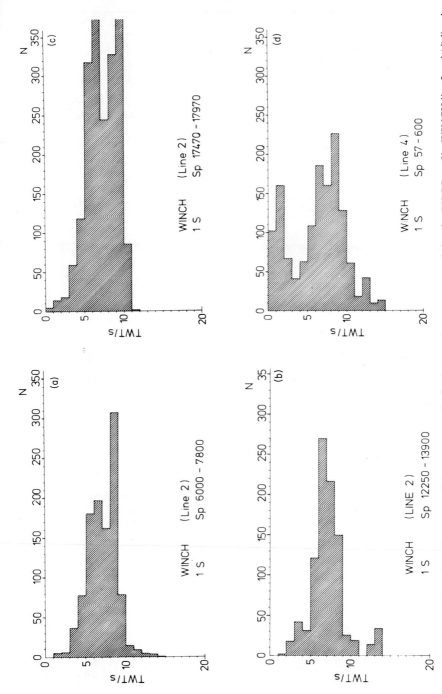

Fig. 6.36 (a), (b), and (c) Histograms of the occurrence of reflections per 1-s two-way travel time for **BIRPS** profile, WINCH line 2 and (d) line 4. Note the similarity of (a), (b), and (c) to Fig. 6.41. [From Wever (1984).]

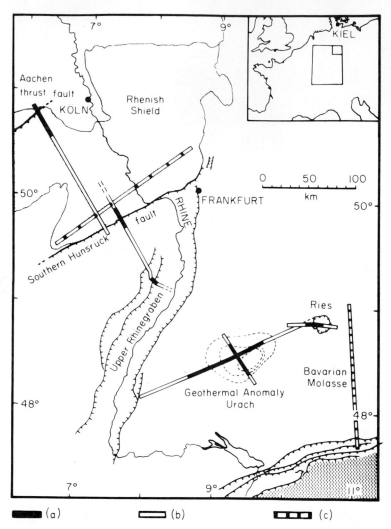

Fig. 6.37 Situation map of reflection profiles in the Variscan area in western Germany. (a) Near-vertical reflection surveys, (b) interconnected DSS surveys, and (c) mixed reflections—DSS measurements with a common depth point (CDP).

Rhenoherzynian or the Rhinegraben fault (Meissner *et al.*, 1980). Within this framework of operations some near-vertical measurements with extremely long spread lengths of 23 km were also carried out in 1978 in the geothermal Urach area and at the northern boundary of the Rhenoherzynian, i.e., across the North Variscan deformation front in the Hohes Venn area near Aachen. It turned out that the use of very long spread lengths provided an excellent tool for revealing velocities and velocity variations in areas with a predominantly horizontal layering. As in seismic prospecting, where velocities are much better determined by the moveout time of reflections than by refraction measurements, in seismic crustal studies the near-vertical observations with spread lengths of at least 15 km are superior in lateral resolution to the refraction–DSS observations. An example of such investigations is the Urach geothermal anomaly. Figure 6.38a shows a plot of a composite of all single-coverage records. This figure not only gives an impression of the amount of data and of the reflection density, it also shows that at least *some* reflections come from the upper mantle. They seem to concentrate below the center of the anomaly, where possibly new mantle is formed by ultramafic cumulates. Figure 6.38b shows a line drawing of profile Urach I (80 km). The low-velocity body (LVB) detected in the center of the anomaly is illustrated in Fig. 6.39. This body would not have been found (and actually was not found in a first attempt) with the lower resolution of refraction shooting. As explained by Meissner *et al.* (1983b), the interval velocities were calculated according to Krey (1976, 1978, 1980) by an iterative process from the stacking velocities obtained from the moveout times, which were still about 1 s for the strong Moho reflections. The velocity anomaly amounts to about 10% in the center of the LVB and could well be localized. The existence of the LVB is not only of paramount importance for explaining the geothermal anomaly, it is also a necessary boundary condition for depth calculations and migration: the apparent Moho dip was even reversed west of the LVB after applying the calculated crustal velocities. (Compare the Moho from Figs. 6.38b and 6.39.) Introducing this velocity structure into the travel time data obtained from the wide-angle measurements resulted in a surprisingly good fit. The wide-angle data actually helped considerably in mapping the lateral dimension of the LVB by means of laterally emplaced stations (Trappe, 1983). The high-quality data from the Urach area were also used for a check on the lamella concept of the high-reflectivity zone in the lower crust, which was explained in Sections 3.4.4 and 3.4.5. This lamellation seems to be especially strong in the Variscan area in general, but also along many sections of the WINCH lines through the Caledonian range (Figs. 6.34 and 6.35). When the polarity of some especially well recorded events from the lower crust was checked by means of a wavelet cross-correlation and inspection of the single-coverage field records, it was found that at least *some* reflections showed a negative reflection polarity (Fig. 6.40). These boundaries with inverse reflection coefficient R lie between reflectors of positive R, determined by both near-vertical and wide-angle events.

For comparison with data from other research areas a statistical evaluation of the frequency of occurrence of reflections was also carried out (Wever, 1984). Figure 6.41 shows the most extreme peak for any area investigated so

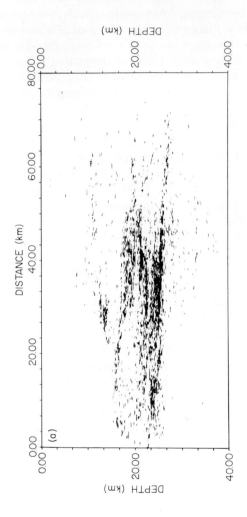

(a)

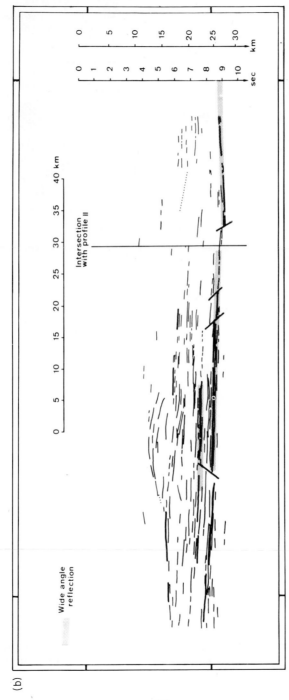

Fig. 6.38 (a) Composition of all single-record reflections along the line Urach I across a weak geothermal anomaly. Note reflection density and sub-Moho reflections below the anomaly. (b) Line drawing of profile Urach I (from 8-fold stacked records).

295

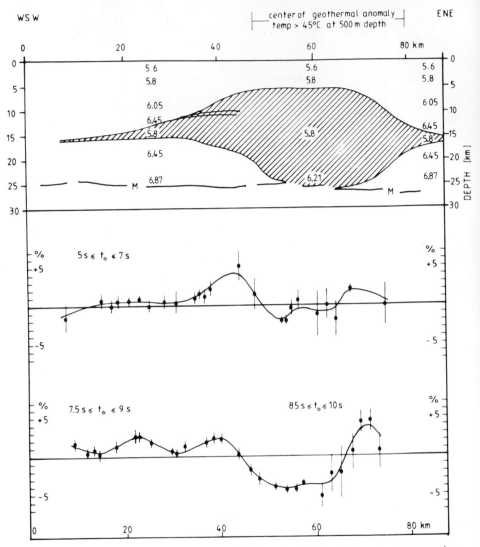

Fig. 6.39 Low-velocity body along U I with interval velocities in kilometers per second as obtained by a method of Krey (1976, 1980). Lower two diagrams show deviations from the average stacking velocities.

far. It is not yet known whether the good data quality or the existence of the LVB plays a greater role in producing this peak. Definitely, the lower crust is an extremely reflectophile zone with many subhorizontal lamellae, most probably a product of fractional crystallization with its horizontally oriented mafic cumulates.

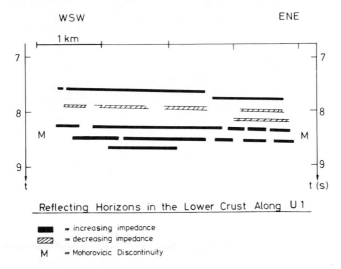

Fig. 6.40 Example of the polarity of reflecting boundaries along a 4-km section with an especially high signal/noise level along U I.

As mentioned before, an attempt has been made to connect some of the (comparatively short) reflection profiles in the Variscides with each other by DSS investigations. Figure 6.42 shows a line drawing from both kinds of investigations across the basic zones of the Variscan orogenies, from the North Variscan deformation front in the northwest to the Moldanubian area in the southeast.

In the northwest the detection of a thin-skinned nappe of Paleozoic sediments with its thrusting northwestward over the Variscan fore deep was one of the major achievements of the reflection surveys. The thrust fault seems to be similar in type, dimension, and age to that of the southern Appalachian zone of detachment, mentioned earlier. In Europe this Aachen thrust fault, also called Faille du Midi in Belgium and France, forms one of the most prominent features of Variscan origin. It can be traced through France and the southwest part of England into Ireland. The older London Brabant Massif must have acted as a ramp for the northwest and north-moving Variscan orogeny. For details see Meissner *et al.* (1983a).

Several commercial reflection surveys in Belgium, France, and Britain have also found this prominent fault zone, while farther to the east, where only the Variscan foreland basin was present, such a prominent thin-skinned nappe has not been found, and the North Variscan deformation front has moved farther to the north.

Regarding the nature of the thrust fault and its gouge zone, it was found by analysis of the field records that it consists of a 50–100-m thin low-velocity

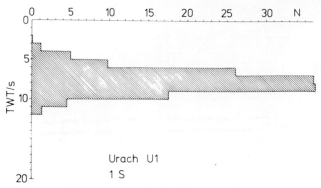

Fig. 6.41   Histogram of the occurrence of reflections per 1-s two-way travel time along U I. Compare with Fig. 6.36a–c. [From Wever (1984).]

layer producing a high-frequency, single-cycle reflection with negative polarity. Farther to the southeast the reflection survey across the Hunsrück boundary fault provided another—earlier—important result. It revealed, after an appropriate migration, two different crustal blocks with excellent data quality (although only a 3–4-fold coverage was applied). The northwest block has generally flat-lying reflectors while the southeast block in the Permian Saar–Nahe trough shows strong dips of reflectors, increasing with depth. Both blocks seem to be divided by a steep-angle fault zone of listric shape reaching down into the mantle. Direct evidence for this fault zone is provided by reflected refraction events in the section and by data from the mobile refraction stations emplaced in the extension of the reflection spread (Meissner *et al.*, 1980). The Saar–Nahe trough appears as a pronounced half-graben of postorogenic nature with a maximum thickness of sediments around 4 km. The question of why the northwest part of this graben was subsiding much faster than its southern part during the post-Variscan stretching period may be connected to the so-called mid-German crystalline high, which was overthrust over the Rhenoherzynian about 50–80 Myr before subsidence started. Old faults were probably reactivated, even under a different stress regime and a reverse direction of displacement.

The western Rhinegraben fault—in the center of Fig. 6.42—did not show continuous and reliable reflections. In contrast to the other areas of investigation, only small reflecting segments are observed at all crustal depths and even in the regime of the mantle. Possibly recent convective movements or an incomplete settlement in the form of a crust–mantle mix may be responsible for the disappointing results. The dip of the major western Rhinegraben fault could be determined by modeling of the first arrivals.

Urach profile II completes the section of Fig. 6.42. The Moho reflections and those of the lower crust dip in the southeast direction, a combined effect

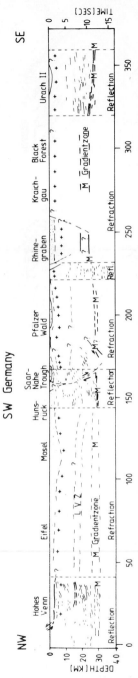

Fig. 6.42 Composite line drawing along a NNW–SSE line through the Variscan mountains of western Germany, combining reflection and refraction surveys. For situation see Figs. 6.37.

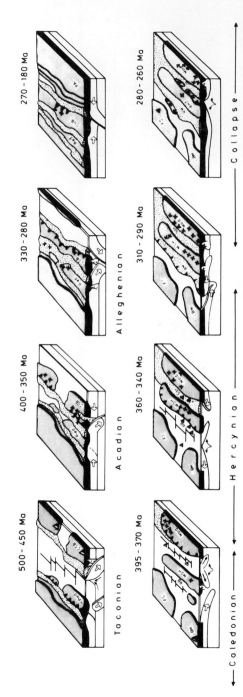

Fig. 6.43 Simplified kinematic model of main collisional events during the Appalachian and the European Variscan orogenies; Pied, Piedmont block; CSB, Charlotte Slate Belt; PA, Paleo-Africa; LBM, London Brabant Massif; FS, Fennoscandian shield; RB, Rhenish basin; MGH, Mid-German high; Mold, Moldanubicum; RM, Rhenish Massif; Sax, Saxothuringicum; RHZ, Rhenoherzynicum; +, granitic plutons; and ▲, volcanoes. [From Meissner *et al.* (1984).]

of the raised shoulders of the Rhinegraben and of the Bavarian Molasse trough in the southeast, where crustal depths are increasing toward the root of the Alps.

Both the near-vertical reflection surveys and the DSS investigations have supported the final outcome of a model for the origin of the European Variscides. The shifting and interstacking during crustal shortening became apparent from the NVR surveys, while the involvement of predominantly sialic platelets was made plausible by the velocity anomalies of DSS. Petrological arguments favor a continental rather than an oceanic environment. The abundant syn- and postorogenic granitic plutons emphasize the "bleeding out" of the crust and indicate a strong thermal event related to the different Variscan orogeneses. The approaching and later interstacked sialic platelets came from the southeast, like the continental pieces which were drifting toward the Grenville Province to form the Appalachians. Differences in the two orogenies of similar age lie in

(1) the thermal environment,
(2) the more oceanic regime in the case of the Appalachian platelets, and
(3) the sequence of belt-forming events: in the European Variscides orogenies moved from southwest to northwest, i.e., toward a continent, while in the southern Appalachians they started at the continent and moved away from it. Figure 6.43 shows a cartoon of the two different areas of Variscan collision.

6.3 CENOZOIC MOUNTAIN BELTS

The Cenozoic era from 65 Myr to the present has produced several remarkable features such as the Alpine–Himalayan orogenic belts, the major phases of the Rocky Mountains, the island arcs in the West Pacific, and the continental magmatic arcs of the Andes. Although there is apparently no difference between these and older processes or tectonic styles, the much better state of preservation of the previously mentioned structures makes them the preferred aim of tectonic studies. Simply speaking, erosion has not leveled the structures, sedimentation is not yet excessive, and crustal and subcrustal phenomena such as mountain roots are not yet annealed. Moreover, many tectonic processes can be observed in action and *in situ.*

Orogenic belts with large nappe displacements and crustal shortening are known from the Alps and other young structures, as well as from Variscan and Caledonian belts, as mentioned before. Unfortunately, so far no NVR studies have been performed in young mountain belts. Only in Cenozoic extensional areas have NVR studies been done by COCORP. Starting with the compressional structures, only DSS results can be presented in the following.

6.3.1   RESULTS FROM OROGENIC BELTS
IN CONTINENT–CONTINENT COLLISION ZONES

As seen in the situation map of Fig. 6.44, the Alps are covered by a very dense network of DSS lines. The shots have been fired from lakes, quarries, or specially drilled boreholes. All these investigations were carried out through international cooperation. Major tectonic features of the Alps are a pronounced, slightly asymmetric crustal root, a generally thick sialic part of its crust, and thin-skinned nappes, the latter two being a consequence of a considerable crustal shortening (Frisch, 1978, 1979, 1981). The African plate with its Adriatic and Italian peninsulas pushes northward toward the European continent. Other tectonic features of interest are the Insubric line, today a right-lateral strike–slip fault that might have started as a suture zone, and the Ivrea body, the famous span of the crust–mantle boundary with its gravity high, pierced through the crust as a consequence of a compressional stress regime (Berckhemer, 1968, 1969; Fountain, 1976; Christensen and Fountain, 1975; Fountain and Salisburg, 1981). Seismicity in the Alps is surprisingly low.

A number of review papers on the crustal structure and the tectonic features of the Alps have been written (Giese *et al.*, 1982; Mueller, 1982; Mueller *et al.*, 1980) as well as a book (Berkhemer and Hsue, 1982). Figure 6.45a shows a general cross-section along a W–E line through the Ivrea body, and Fig. 6.45b shows another W–E line which provides some more details on the crustal structure and velocities. We recognize the huge amount of low-velocity material making up nearly the whole crust. The displacement of the crustal root zone with regard to the maximum elevation is also seen.

Figure 6.46 shows a schematic N–S cross-section with some geologic information about the different tectonic units. It seems clear from these cross-sections that the southern or southeastern crustal block has been stacked on top of the northern one. It is not yet clear from these pictures where the major zone of detachment is. Considering the V-z structures along the profiles, the upper crust with some prominent LVZs and general velocities below 6.4 km/s, i.e., sialic material, covers 30 km of depth (!). A similar V-z structure has been found for the southern Rocky Mountains (Prodehl and Pakiser, 1980). This tremendous accumulation of upper crustal material can only mean that the zone of detachment must be somewhere in the lower crust. The lower crust in the Alps with velocities between 6.4 and 7.0 is only slightly thicker than in the adjacent continents. Part of the lower crust must have been sutured together with the uppermost mantle. The existence of a zone of detachment in the lower crust is certainly compatible with the postulation of a low-viscosity lower crust. An interesting observation on the crustal structure in the Alps and most other Cenozoic mountain belts is the lateral offset of the greatest crustal depth (root) against the area of highest elevation.

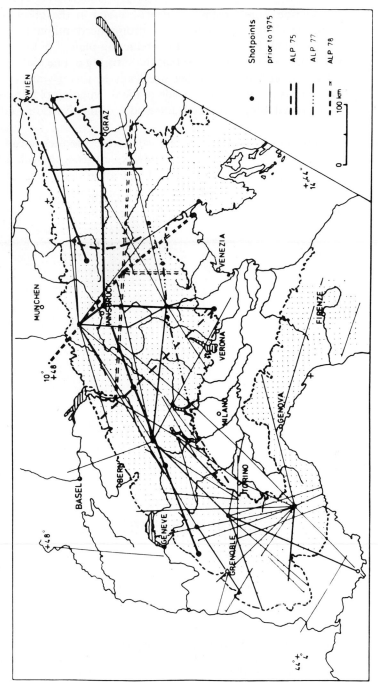

Fig. 6.44 DSS profile network in the Alpine area. [From Miller *et al.* (1982).]

This (slight) deviation from isostasy is also observed in the offset of the Bouguer anomalies (BAs), especially in a pronounced minimum, often followed by a maximum in the area of the obducting plate. Looking at the crustal cross-sections of Fig. 6.46 we see that the suturing of one plate (in the Alps, the northern one) below the other one leads to this asymmetry: the crustal root with a BA minimum at the side of the downgoing plate and the thin- and thick-skinned nappes on the leading edge of the overriding crustal section.

The other collisional mountain ranges north of the Mediterranean also have prominent crustal roots, although not always crusts with such a high percentage of sialic upper-crustal material. The Carpathians, for instance,

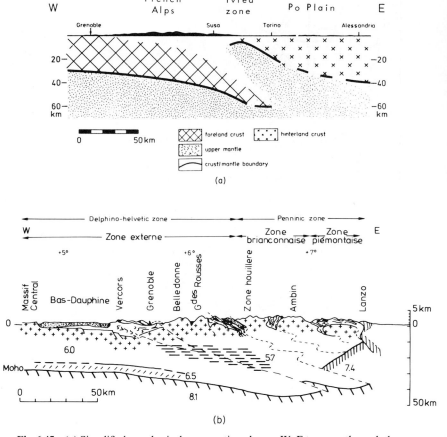

Fig. 6.45 (a) Simplified geophysical cross-section along a W–E traverse through the western Alps. [From Giese *et al.* (1982).] (b) Combined geologic–geophysical section along a similar cross-section. [From Miller *et al.* (1982).]

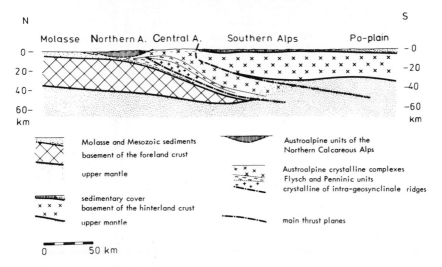

Fig. 6.46   Generalized N–S cross-section between the Bavarian Molasse and the Italian Po plain. [From Giese *et al.* (1982).]

bordering the Ukrainian platform, seem to have their lower-crustal part enlarged, as shown in the cross-section of Fig. 6.47 (Sollogub *et al.*, 1973; Horwath and Berckhemer, 1982). This figure also shows the Pannonian (Hungarian) back-arc basin with its hot and thin crust and elevated asthenosphere (Stegena *et al.*, 1975; Porsgay *et al.*, 1980). (See also the Moho map of Fig. 6.16.) Crustal structure in the Caucasus is still controversial.

The next Cenozoic mountain range of major interest is the large Himalayan collision zone, i.e., the indentation zone of the Indian plate into the Eurasian continent. After the Paleo-Tethys of the late Paleozoic age was closed in the Late Triassic by the collision between the Eurasian plate and the Tibetan plateau, another collision between the Tibetan plateau and the Indian plate produced the Indus–Tsangpo suture (Windley, 1983). The rather complicated collisional processes terminated about 38 Myr ago, when the northward movement of the Indian plate decreased from more than 10 cm/yr to about 5 cm/yr (Molnar and Tapponier, 1975, 1978) in the period from 70 to 40 Myr ago. Some 500–1000 km of crustal shortening have been proposed. This shortening took place in the overriding plate (Sengoer and Kidd, 1979), as observed clearly by a series of thrust planes with a southward-decreasing age. In addition, about 50 km of underplating (shallow subduction) of the Indian plate under Tibet along the Indus–Tsangpo suture may have taken place. Large ophiolite klippes are incorporated into the mountain belts. Major tectonic and metamorphic events have been dated by Maluski and Matte (1984).

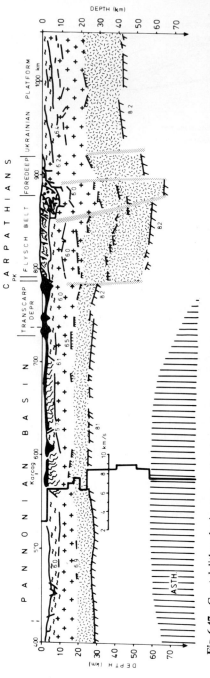

Fig. 6.47 Crustal–lithospheric cross-section from the Pannonian basin (Hungary) through the Carpathians to the Ukrainian platform; values in kilometers per second; Asth, asthenosphere. Note the deep Moho and the high-velocity lower crust below the Carpathians. [After Sollogub *et al.* (1973) and Horwarth and Berckhemer (1982).]

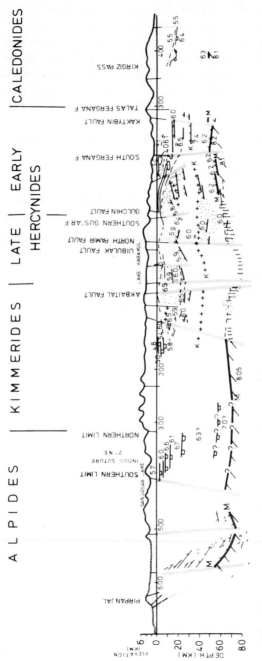

Fig. 6.48 Crustal cross-section through the Himalaya–Tibet area. Values in kilometers per second. [From Volvovsky *et al.* (1983).]

307

Between 1974 and 1979 DSS studies were carried out in the Pamir–Himalayan region in order to get some reliable estimate of the total thickness of the crust (Volvovsky *et al.*, 1983). A profile about 800 km in length crossed different orogenic belts in an S–N direction. A crustal cross-section is shown in Fig. 6.48. In the southernmost collisional belt crustal thickness reaches the highest values so far observed on Earth (about 80 km). The average crustal velocity is low, about 6.2–6.4 km/s according to Volvovsky *et al.* (1983). Other authors infer a 60-km crust with a continuously increasing velocity and a velocity jump at the Moho. According to the cross-section of Fig. 6.48 there is a thick layer at medium crustal depths of lower velocity north of the Indus suture zone, apparently continuing far to the north. The indicated fault zones are supposed to mark a block structure, as favored by the authors. Although they claim that these fault zones and the velocity structure argue against a doubling of the crust by lateral stacking and indentation, the observations presented are not too convincing.

The steep dip of the fault zones also seems to be debatable. The 70–80-km-thick crust with its low velocities is just double the amount of the undisturbed adjacent crustal thickness in the north and in the south. A schematic cross-section through the central Himalayas is presented by Windley (1983) (Fig. 6.49). The section is mainly based on geologic and petrological observations, but it certainly seems compatible with the DSS results. Also, an analysis of the strong gravity anomalies and the application of flexural rigidity models provide a model for underthrusting which basically agrees with Windley's model of Fig. 6.49. Only the crustal root is shifted slightly toward the north beneath the Indus–Tsangpo suture zone, thus producing an offset of the root with regard to the greater Himalayas, i.e., the highest elevation. As mentioned earlier, this seems to be a general phenomenon in zones of continental

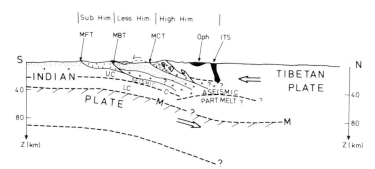

Fig. 6.49   Generalized geologic–geophysical cross-section through the Himalaya–Tibet area; MFT, main frontal thrust; MBT, main boundary thrust; UC, upper crust; LC, lower crust; M, Moho; Oph, ophiolites; ITS, Indus–Tsangpo suture; C, Conrad; and MCT, main central thrust. [After Windley (1983) and Seeber *et al.* (1981).]

collision where one plate (including some lower crust) underplates the other. Seeber *et al.* (1981) have performed teleseismic studies from a seismic network and focal plane solutions of great Himalayan earthquakes. They see evidence of one (or more?) detachment dipping at a shallow angle from the fore deep toward the Tibetan plateau.

In these collision zones much material of the gneissic–granitic upper crust and sediments are guided down to lower levels. Much of this material is transformed into a sialic–granulitic facies, which may make up the bulk of the lower crust in many areas of former collisions. Unfortunately, such granulites cannot be distinguished from mafic granulites or gabbros on the basis of seismic velocities (Christensen and Fountain, 1975).

Figure 6.50 shows a compilation of typical V–z curves of continent–continent collision zones (CCCs) (see also Table 6.5). It contains the CCCs mentioned in this section as well as some others that were not described. In addition, data from continental magmatic arcs (CMAs) are included, which will be treated in the next section. Regarding the young orogenic belts, we may summarize:

(1)   Maximum crustal thickness in the different belts varies between 40 and 80 km according to the intensity of collision. Prominent crustal roots are observed.

(2)   Crustal roots are laterally displaced along the suturing plate with regard to the maximum elevation. Nappe displacement takes place predominantly on the front of the overriding plate, i.e., on the opposite side of the root zone.

(3)   In general, a high accumulation of sialic material with $V < 6.4$ km/s forms a thick upper crust, suggesting a zone of detachment in the low-viscosity lower crust.

(4)   Wide-angle events like P_mP are generally very strong.

Table 6.5

Locations and References for the Standardized $V(z)$ Function of Fig. 6.50a,b (Continent–Continent Collision Areas)

Function	Location	Reference
01–02	Northern Alps	Will (1976)
03–06	Alps	Giese and Prodehl (1976)
07–010	Caucasus	Kondorskaya *et al.* (1981)
011	Caucasus	—
012–013	Himalayas	Mishra (1982)
014	Himalayas	Volvovsky *et al.* (1983)
015	Himalayas	—
016–020	Rocky Mountains	Prodehl and Pakiser (1980)

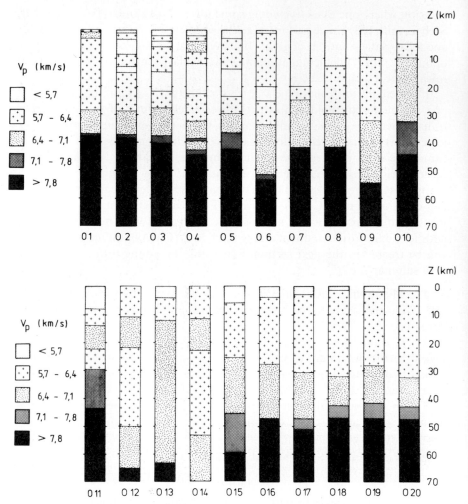

Fig. 6.50 Compilation of standardized velocity–depth functions of continent–continent collision zones (CCCs).

6.3.2 DSS AND OTHER GEOPHYSICAL STUDIES FROM CONTINENTAL MAGMATIC ARCS AND ISLAND ARCS

Relatively few DSS investigations have been performed in continental magmatic arcs. All arcs around the Pacific have been intensively investigated by seismicity studies, i.e., by carefully mapping hypocenters and determining focal plane solutions and the stress regime. The trenches have also been studied extensively by marine geophysical measurements, including NVR seismics, and sometimes the downgoing oceanic plate can be followed below the accreted wedge of sediments at the active margins. However, information on the deep structure of the arcs is scarce. Here, as in the Himalayas, gravity data often provide a powerful tool for constraining or establishing first crustal models. The reason for this good applicability lies in the very strong and often peculiar gravity anomalies from the trenches onto the continental margins.

Some examples will be given related to the crustal structure below the Andes mountains. The northern part of the Andes from the Central American isthmus down to Ecuador shows a strong positive Bouguer anomaly with a total length of 5000 km. It coincides with the so-called basic igneous complex. Parallel to the coast of Peru a similar gravity high is found. In general, the high is flanked by the BA minimum of the trench on the west side and the BA minimum of the central Andes, especially from their root, on the east side. Combined onshore–offshore DSS surveys were performed along several profiles across the continental margin and the Andes in Colombia. One of the crustal cross-sections obtained was shown in Fig. 3.5 together with the Bouguer anomaly. A prominent gravity high is located along the Western Cordillera, one of the three branches of the Andes in the northern section of South America, all along the Basic Igneous Complex. The DSS data show that it is formed by a mafic, probably rather young, oceanic crust of 6.8-km/s velocity which must have been obducted to the continent in a rather complicated oblique direction (Flueh et al., 1981; Flueh, 1982). (See also Section 3.15.)

The extension of one of the DSS profiles into the Central Andes revealed a very low velocity crust with velocities of 5, 1.0, and 5.6 km/s and a thickness of nearly 50 km (Fig. 6.51). These very low velocities may be related to a very hot crust. Widespread andesitic volcanism with a wide range from mafic to rhyolitic magmas leads to the suggestion that active magma chambers inside the crust might contribute to the extremely low crustal velocities (see Fig. 4.5). The feeding of such magma reservoirs by the melting of oceanic crust in the subduction zones below and possible contributions from the continental lithosphere and asthenosphere will be discussed in Section 6.5.

South of central Ecuador the structure of the Andes looks different: there are now only one or two mountain chains and no Basic Igneous Complex. According to gravity data, crustal depth in Bolivia/Peru should be about 75 km. The DSS surveys often failed to obtain reliable data from the huge Altiplano area. A very high absorption, possibly also related to an extremely hot crust, is invoked to explain the missing arrivals from the lower crust or the upper mantle. The DSS profile of Ocola and Meyer (1973) in the

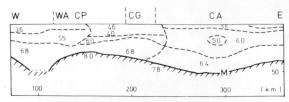

Fig. 6.51 Crustal cross-section along a DSS profile across the Andes in northern Columbia; values in kilometers per second; WA, West Andes; CG, Cauca Graben; CA, Central Andes; and M, Moho.

Altiplano parallel to the strike direction revealed a crustal thickness of 72 km. Giese's reinterpretation shows a layer with a 7.4 km/s velocity, i.e., a "transitional" velocity, between 43 and 70 km. Farther south, in the Chilean- -Bolivian part of the Andes, zones of high electric conductivity are found; a large zone in the lower crust below the Altiplano is surrounded by areas of very high conductivity at upper crustal levels (Giese, 1984, personal communication). Large intrusions of andesitic magma of slightly different chemistry at various crustal levels may be invoked for the seismic transitional velocities and the strong anomalies in electric conductivity.

In the central and southern Andes extreme differences in the age provinces, such as 1800 and 100 Myr appear along the crustal areas. This has led Nur and Ben-Avraham (1978, 1982) and Ben-Avraham *et al.* (1982) to suggest the accretion of different Pacific continental or oceanic platelets to the South American margin. Not only subduction of oceanic plates but also collision with incoming platelets may have contributed considerably to the formation of the Andes chains. The incorporation of older and younger crustal sediments into the subduction process also seems possible.

Another source of information on the crust in the magmatic arcs is the petrology and age dating of magmatic rocks and lavas. A schematic cross-section of the different types of rocks in the central Andes is shown in Fig. 6.52. Crustal thickening apparently is affected more by vertical additions than by horizontal compression and shortening. This is the basic difference from pure continent–continent collision with its *lateral* interstacking of crustal platelets. *Vertical* accretion with major contributions from the subducting oceanic crust, on the other hand, results in a true increase in crustal continental material (Brown and Musset, 1981).

The determination of the initial $^{87}Sr/^{86}Sr$ values (*I* values), as explained in Section 4.6.4, shows a significant correlation with the absolute age of the rocks. Figure 6.53 (McNutt *et al.*, 1975) shows that the oldest magmas had the lowest *I* values. They represent early melts of a simple, unmodified oceanic plate. A certain mixing with high-*I* material definitely appears in the

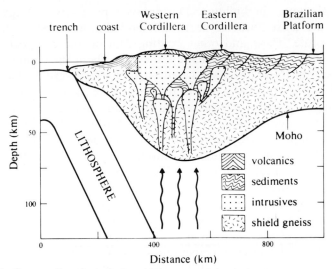

Fig. 6.52   Cross-section through the middle sector of the Andes. Example of a continental arc. [From Brown and Musset (1981).]

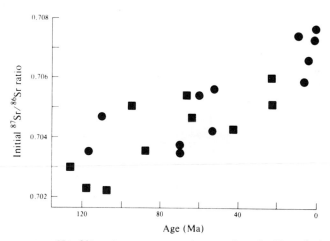

Fig. 6.53   Initial $^{87}Sr/^{86}Sr$ values versus age of magmatic rocks. Note the increase of the I values with time from "pure" mantle to enriched "contaminated" values; ■, intrusives; ●, extrusives. [From McNutt et al. (1975).]

process of mountain building. During such processes average *I* values increase and crust thickens, heats, and probably melts on its base at an increasing rate. These and other exciting data on the composition of the Andes cannot veil the lack of basic knowledge about the detailed structure and velocity in their subsurface based on appropriate seismic data.

Island arcs are generally considered an oceanic province. However, they are the major source of vertical crustal accretion, and they become attached to a continent during the final stage of the Wilson cycle. Their structure is generally complex. Deep-sea sediments are scraped off at the edge of the trench as the oceanic crust descends into the mantle. These sediment wedges may form a significant component of island arc crusts. Sometimes, as in the continental arc situation, a whole segment of oceanic crust including part of the peridotitic upper mantle is horizontally emplaced into the island arc structure. These are the famous ophiolite nappes or intrusions.

Similar to or even more pronounced than those in continental arcs are the high heat flow, high crustal temperatures, and mafic, andesitic, and rhyolitic magma complexes that generally form the core of the island arcs. Since 1950, many DSS studies have been made across the Japanese island arc. Velocity data from such studies are compiled in Fig. 6.54. Often a typical Moho with $V_P > 7.8$ km/s is not found. The large column of "sialic" velocities with $V_P \leq 6.4$ and crustal depths around 30 km may seem surprising and similar to Paleozoic or Mesozoic structures. Certainly, the high temperatures associated with intruding basaltic magma from below favor a reduction of velocity values and a thorough differentiation with a large thickness of sialic–rhyolitic material. The missing Moho, on the other hand, may indicate that in these cases the differentiation process is not yet complete and might

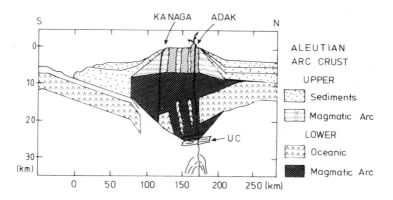

Fig. 6.54 Cross-section through the Aleutian island arc. Note the thick crust, new Moho and mafic and ultramafic cumulates (UC). [From Mahlburg-Kay and Kay (1985).]

still be in progress. As in continental graben structures the inability to find the Moho by DSS measurements has also been demonstrated for areas with recent volcanism in the Variscides of Germany (Mooney and Prodehl, 1978). Many new petrological data have recently been obtained from the Aleutian island arc (Grow, 1973; Kay, 1980). Kay *et al.* (1982, 1983) and Kay (1983) have shown that the tholeiitic volcanoes are large basaltic centers. Tholeiitic and calc–alkaline magmas can be clearly distinguished. The crust in the oceanic Aleutian arc, formed by huge intrusions and extrusions, reaches a thickness of about 25 km. Upper crustal rocks are differentiated from a parental magma of olivine tholeiitic composition. The lower crust is of basaltic composition and formed from complementary residual crystals, probably mixed with preexisting oceanic crust (Mahlburg–Kay and Kay, 1985). The uppermost part of the (newly formed) mantle may consist of ultramafic residues. Figure 6.54 shows a simplified cross-section of the Aleutian island arc. It may be considered a model for vertical accretion of island arcs which will be attached to a continent in a future collision.

Summarizing the V–z data from the magmatic arcs, we may state:

(1)   Continental arcs of the Cordillera type have a maximum crustal thickness between 40 and 75 km with prominent crustal roots.

(2)   Crustal velocities are still lower than in continent–continent collisional belts, especially in new and hot volcanic provinces ($V_P < 6$ km/s).

(3) $P_m P$ and P_n waves are often obscured or very weak, and the Moho is not always observable.

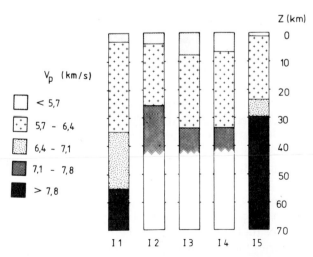

Fig. 6.55 Compilation of some standardized velocity–depth functions of island arcs. See Table 6.6 for location and references.

Table 6.6

Location and References for the Standardized $V(z)$ Functions of Fig. 6.55
(Island Arcs)

Function	Location	Reference
I1	Andes	Flüh (1982)
I2–I4	Japan	Research Group for Explosion Seismology (1966)
I5	Japan	Yamashina (1978)

(4)   Absorption of seismic waves is high.

(5) P_n velocities are generally lower than usual.

(6)   Seaward of the magmatic arc very long and narrow belts of ocean crust with strong gravity anomalies are occasionally found, apparently obducted onto the continent.

(7)   Island arcs have a maximum crustal thickness of about 30 km.

(8)   Crustal velocities are generally low.

(9)   Points (3)–(5) also hold for island arcs.

6.4   CONTINENTAL GRABENS, RIFTS, AND MARGINS

A graben is a simple, parallel-sided, downfaulted valley with a width between a few hundred meters and a few tens of kilometers (Mohr, 1982). At the larger end of this scale, the term graben overlaps with the terms rift and rift valley, generally a few tens of kilometers in width and a few hundred kilometers in length. In most cases a rift valley shows a complex of smaller-scale graben and horst blocks, i.e., a whole fault system along the rift floor. Most graben and rift valleys are asymmetric and some have developed into a half-graben. A rift system is understood as a set of rift valleys linked by parallel or en echelon faults, generally of a transform fault type. The development of rifts or rift systems is called taphrogenesis or just rifting. Many models of such developments exist; some of them are treated in Section 6.4.4.

Within the framework of plate tectonics continental rifts are understood as the beginning (sometimes as a rejuvenation) of a plate tectonic cycle. Heating and mobilization of tholeiitic magmas in the mantle result in a magmatic intrusion into the crust followed by volcanism and often doming and lifting of the shoulders, and later by breaking of the rigid upper crust in an extensional stage. The process might stop anywhere during its development, leading only to a geothermal anomaly or to a "failed rift" of various stages. Alternatively, if the forces from below are particularly strong and the tectonic conditions favorable, the rift might develop into a spreading stage, e.g., into a Red Sea feature and further into an Atlantic-type spreading ocean with margins that might still show many signs of the original rifting process. A remarkable review of rifts and rifting aspects is given in Pálmason (1982).

In the following, mainly geophysical data on continental rifts will be discussed and compared with each other. As young rifts or grabens are narrow, elongated structures, the limited resolution of seismic refraction studies can contribute only to a certain extent to the velocity–depth structure. On the other hand, seismic reflection studies have not always been successful, except for COCORP surveys in the extended rift area of the Basin and Range Province and in the Socorro rift. Common features of all continental rifts are:

(1)   negative Bouguer anomalies, mostly about 50 mgal, basically an effect of the sediment fill,

(2)   increased heat flow, and

(3)   increased seismicity, mostly being shallow and of tensional character.

In deriving depth values inside the rifts and grabens, not the absolute depth but the relative depth with respect to the surrounding area might be an important parameter for the assessment of graben development. This means that a continental rift must be judged by

(1)   its age,

(2)   its situation within the age province, and

(3)   its state of development.

Point (3) seems to be the most exciting and relevant feature of a continental rift. The following subdivision will be made on the basis of (3), keeping in mind also the age of the rifting process.

6.4.1   GRABENS AND NARROW RIFTS

One of the smaller but well-preserved rifts of Permo–Carboniferous age is the Oslo graben. It is located at the northwest termination of the Tesseyre–Thornquist line, a long lineament from the Ukrainian area through Poland and southern Scandinavia. This line appears as a narrow graben area in Poland (Guterch, 1977; Guterch *et al.*, 1976), but is a right-lateral transform fault from the north German plains to the Skagerak. Igneous activity started in the southern part of the Oslo graben and spread to the north (Sundvoll, 1978). This shows a fracture propagation, probably in response to transcurrent movements along the Tesseyre–Thornquist lineament, as suggested by

Ziegler (1978), who suspects that the graben is a pull-apart feature at the termination of the transcurrent lineament. It seems, however, that the Oslo graben extends to the Horn graben farther southwest, as shown in the situation map of Fig. 6.56. It is not clear whether doming or a large-scale subsidence in the N–S direction preceded the rifting. Some Proterozoic lineament seems to be related to rift development (Ramberg and Spjeldnas, 1978). The crust is thinned below the graben, but constraints from DSS and gravity data are poor. Similar in age and in volcanic eruptions is the half-graben-shaped Saar–Nahe trough, which was discussed in Section 6.2.4 as a postorogenic (post-Vasiscan) collapse structure. It belongs to the general fragmentation of the European Variscidcs which led to the initiation of many lineaments, one of which later developed into the Rhinegraben.

The Rhinegraben, although tectonically prepared during the post-Variscan collapse, came into evidence as an entirely new tectonic feature in Europe in the early Tertiary, together with many other Cenozoic volcanic rifts like the Limagne, Leine, and Eger graben. Subsidence began during the late Eeocene, immediately followed by a distinct culmination of volcanism (Illies, 1970, 1978). The Rhinegraben is about 30–40 km wide and about 300 km long, as seen in the map of Fig. 6.57 (Werner *et al.*, 1982).

There seems to be a relation between orogenic movements in the western Alps and the subsidence of the Rhinegraben. Compressional stresses exerted on the Alpine foreland seem to have initiated and enhanced sinistral shear motions, which even today dominate at the eastern Rhinegraben fault. They seem to have slowed down subsidence considerably (Illies, 1978; Illies and

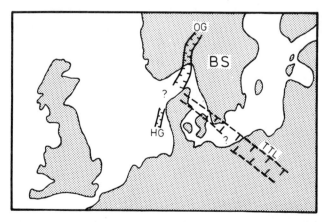

Fig. 6.56 Situation map of Paleozoic graben structures in the Scagerrak and the eastern North Sea; BS, Baltic shield; OG, Oslo graben; HG, Horn graben; and TTL, Tesseyre–Thornquist lineament.

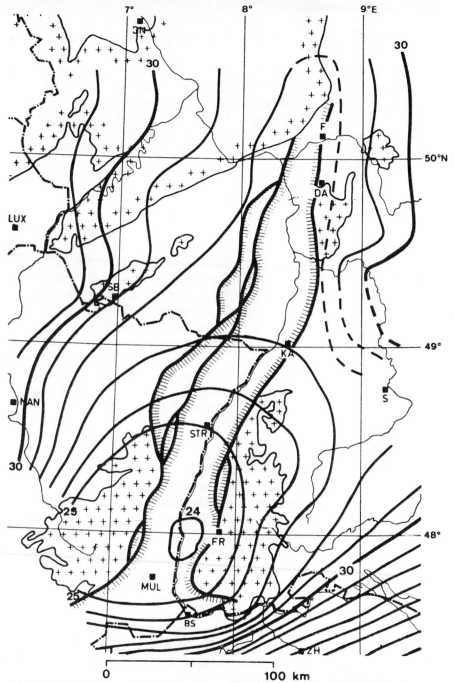

Fig. 6.57 Situation map of the Rhinegraben area and crustal thicknesses. Crosses represent Pre-Tertiary mostly uplifted basement units. [After Werner *et al.* (1982).]

Greiner, 1978; Ahorner, 1978). The whole tectonic history is complicated by episodic activity maxima.

The first evidence of an anomalously thin crust below the graben was given by Rothé and Peterschmitt (1950). After nearly 30 years of DSS studies with an especially dense network in the Rhinegraben's southern part, the crustal structure looks quite complicated. An unusually broad transition zone is found between crust and mantle below the southern part of the graben with velocities between 7.4 and 7.8 km/s, beginning at about 21 km depth (Prodehl, 1981). This structure was also termed "cushion" or "pillow" during the investigations. Outside the graben a rather well-defined first-order boundary between crust and mantle is determined, the Moho forming a large dome-shaped structure. A similar structure, although not constrained by as many observations as in the south, is reported for the northern part of the Rhinegraben (Meissner and Vetter, 1974; Meissner et al., 1976a,b; Mooney and Prodehl, 1978). Figure 6.57 shows a contour map of the Moho and in the Rhinegraben area. See also Fig. 6.42 for the graben's tectonic emplacement.

Similar but less pronounced rift structures in the vicinity of the Rhinegraben are the Rhonegraben and the Limagnegraben (Hirn and Perrier, 1974; Prodehl, 1981). Especially in the Limagnegraben a similarly large transition zone between crustal and typical mantle velocities is found between 22 and 40 km. It seems that a highly anomalous mantle or a kind of crust–mantle mix must be assumed for the structure of the transition zone, the differentiation process of intruding lavas or melting lower crust still being in progress.

The East African rift system is one branch of three, emerging from the Afar triple junction. The other two branches are the Red Sea rift and the Gulf of Aden rift, i.e., two oceanic developments, which will be treated in Section 6.4.3. Starting in the Afar triple junction the eastern branch of the rift system crosses a broad domal uplift in Kenya (Kenya rift); the western branch follows several lakes (see Fig. 6.58). At about 8°S both branches meet again and seem to cross each other. Although the width of the rifts is only between 30 and 60 km, they are the longest continental rift systems in the world. An elliptical area of 1200 × 2000 km was uplifted, but the uplift was moderate and did not start at the beginning of the rift process. While the western branch shows no signs of Tertiary volcanism, the eastern branch is very active (Barberi et al., 1982). There has been continuous volcanic activity since the Oligocene (35 Myr). Alkali enrichment is found in this branch, which is considered typical for continental magmatism. Only in the Kenya rift and around Addis Ababa has some DSS work been carried out (Griffith, 1972; Berckhemer et al., 1975). Interpretations included the observed negative BA. Teleseismic data were also applied to constrain crustal and upper-mantle structure (Guust and Hueller, 1982).

The V–z models show a decreased crustal depth in the western branch and

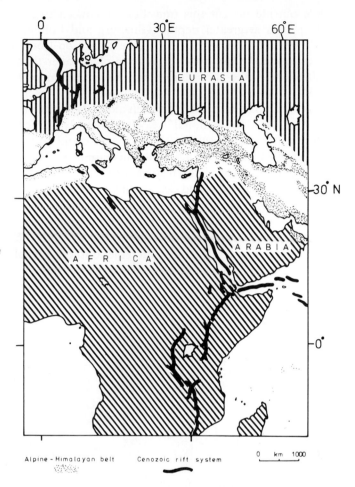

Fig. 6.58 Situation map of rift and graben structures around the Red Sea. [From Illies (1974).]

a broad transition zone between 20 and 60 km with an average velocity of 7.5 km/s in the Kenya rift. While the oceanic structure of the Red Sea is apparent in the south, magnetic lineaments and other typically oceanic features are obscured toward the north. On the northern side of the Red Sea there are the small Gulf of Suez rift and the Dead Sea rift. In both cases rift formation started in the late Oligocene. Both rifts are only about 35 km wide. The Dead Sea rift emerging at the northern end of the Red Sea, also at a kind of triple junction, shows considerable crustal thinning along the rift from NNE to SSW. Like other young rifts it shows a transitional zone from crust to mantle, while outside the rift this transition is reduced to a first-order boundary and appears at greater depths (Ginzburg and Folkmann, 1980; Ginzburg *et al.*, 1981; Makris *et al.*, 1983). The transition from crustal to mantle velocities inside the rift, however, is certainly smaller than in other rifts. Strong left-lateral strike–slip motions may have determined the recent development. The Gulf of Suez rift shows a more typical picture of rift structure. From two DSS lines parallel to the rift along its flanks a broad zone of 7.5 km/s velocity, i.e., a "transitional" velocity, is clearly defined (Makris, 1982). Inspection of the record section shows that it could equally well be interpreted as a gradient zone with a velocity increase from 7.0 to 8.0 km/s. There is also a reduction in crustal depth along the graben. Some selected *V–z* profiles from the rifts around the Red Sea are included in Fig. 6.59.

A rift with a much longer history and with mature rift features is the Baikal rift system (Kiselev *et al.*, 1978). Although on the average only 50 km in width, the whole system extends over 2500 km in the SW–NE direction. It can be considered as an isolated structure within the southern part of the East Siberian platform (Fig. 6.60). Volcanic activity may have started in the Late

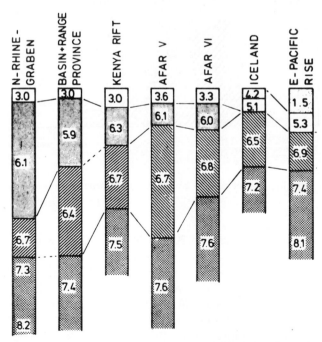

Fig. 6.59 Some typical velocity–depth functions from rifts and grabens. Numbers are V_p values in kilometers per second. [After Berckhemer *et al.* (1975).]

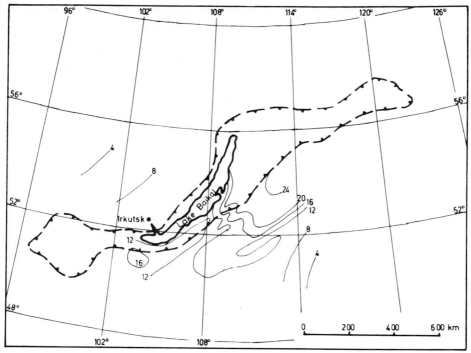

Fig. 6.60   Situation map of the Baikal rift system. Numbers are vertical movement rates in millimeters per year. [From Kolmogorov and Kolmogorova (1978).]

Cretaceous, but the peak activity occurred in the Miocene to early Pliocene. Subsidence seems to be concentrated in two major phases, and a tectonic relation to the orogenic phases in the Himalayan belt has been postulated from assessments of stress pattern and stres propagation (Tapponier and Molnar, 1977). As found at most other grabens (in particular around the Rhinegraben), a large area with a diameter of about 1000 km has been uplifted. Around the Baikal rift the area is still rising, with a reported rate of up to 26 mm/yr (!) (Kolmogorov and Kolmogorova, 1978). There is rather high seismicity with a tensional focal mechanism. Apparently, the rift is still active, in contrast to the rather inactive or distorted stress system around the Rhinegraben. In the rift zone proper there is a deep-going fault zone. There seems to be an LVZ between 12 and 17 km depth. Again, a P_n velocity of only 7.6–7.9 km/s is found in the transition zone between crust and mantle, and "normal" P_n velocities of 8–8.1 km/s are found at greater depths. It is interesting to note that the transitional zone with the transitional velocity extends to a much wider area than the rift. A hypothetical picture of the deeper part of the Baikal rift is presented in Fig. 6.61.

One of the few narrow rifts which have been intensively studied with NVR methods is the Rio Grande rift in New Mexico. Emplaced in a (rather young) Precambrian shield area (see Fig. 6.3), crustal extension started in the Oligocene, about 27–32 Myr ago, possibly also as a response to renewed tectonic activity in the southern Rocky Mountains in the west and a beginning uplift of the Colorado plateau. After 10 years of intensive multidisciplinary research the Rio Grande rift has become one of the best documented continental rifts in the world. The research activity is reviewed by Riecker (1979).

There are three different segments of the rift, each of which developed separately (Chapin, 1979) (Fig. 6.62). The largest extension with subsidence is in the southern segment, the Socorro branch. Near Socorro a bifurcation into two differently active branches appears, similar to structural patterns in the East African system. Rifting with subsidence had already started between 32 and 27 Myr ago. Volcanism began at 32 Myr in the Socorro rift. As in the Baikal rift, a pronounced episodicity is observed with a lull in the middle Eocene (20 to 13 Myr ago). Initial basaltic–andesitic volcanism changed to basaltic–rhyolitic suites after the lull, but strontium I values remained low. Epirogenic uplift is still observed today, but may have had its maximum rate between 7 and 4 Myr ago, certainly not in the beginning of the rift process.

There is ample geophysical evidence for an anomalous crustal and uppermantle structure. First, there are the usual anomalies in gravity, heat flow, and shallow seismicity. In addition, several DSS measurements have been carried out, the most important one consisting of a 350-km-long line along the axis of the Rio Grande rift and its extension to the north. The most detailed part of this profile is in the Socorro rift, where it crosses COCORP's NVR line, to be discussed later. Fig. 6.62 also shows the situation of the long

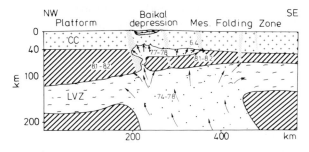

Fig. 6.61 Lithospheric–asthenospheric cross-section through the Baikal rift system. Values are in kilometers per second; CC, continental crust; LVZ, low-velocity zone (asthenosphere?). [After Zamarayev and Ruzhich (1978).]

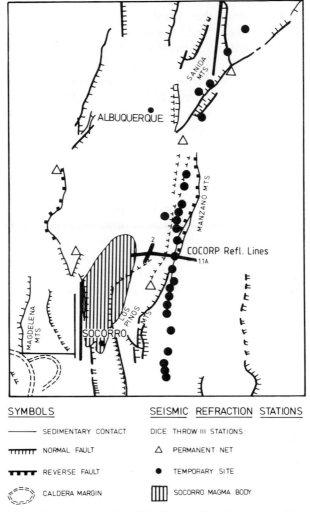

SYMBOLS   SEISMIC REFRACTION STATIONS

——— SEDIMENTARY CONTACT DICE THROW III STATIONS:

TTTTTT NORMAL FAULT   △ PERMANENT NET

▀▀▀▀ REVERSE FAULT   ● TEMPORARY SITE

CALDERA MARGIN   ‖‖‖ SOCORRO MAGMA BODY

Fig. 6.62 Situation map of the Rio Grande rift and seismic profiles.

N–S-striking DSS and COCORP lines 1 and 2, together with the basic geologic information. The long DSS profile was observed from a series of large explosions, as discussed by Olsen *et al.* (1978, 1979). Although no reverse shots are available, the *V–z* estimates seem reliable. There are good P_mP arrivals and another good reflected wave group from an intercrustal discontinuity, both branches being well observed also in the subcritical angle range. Comparison with DSS data from the flanks shows that crustal thinning is about 10–15 km. The P_n velocity in the central rift is between 7.6

and 7.8 km/s from about 33 km downward, indicating an anomalously low value similar to that below other rifts discussed before. Crustal velocities are low, but an LVZ is absent. It is interesting to note that the intracrustal reflector at 17 to 20 km depth correlates well with one found at COCORP's NVR lines.

In 1975 and 1976 COCORP's NVR survey was carried out with the objective of obtaining an improved picture of the crustal structure. Figure 6.63 shows line drawings of the traversing profiles 1 and 1a, perpendicular to the strike direction. As discussed by Brown *et al.* (1979), among the major features of these profiles are the pictures of the pronounced horst–graben structure inside the rift with sediment accumulations up to 7 km. The eastern rift boundary seems to be defined by a steeply dipping, seismically transparent zone. The same intracrustal reflector that came out so strongly in the DSS profile is also very clearly observed in the record sections of NVR lines. It is actually—as in the DSS profile—the most prominent reflector in the whole crust. It was predicted by Sanford *et al.* (1973, 1977) on the basis of microearthquake studies that there should be a large magma body at midcrustal depths. In the area of the NVR and DSS cross-point it should be at a depth of about 20 km, as shown by a special investigation by Rinehard *et al.* (1979), who mapped S-wave reflection points from more than 200 microearthquakes.

It seems, therefore, that this thin and flat-lying magma body is responsible

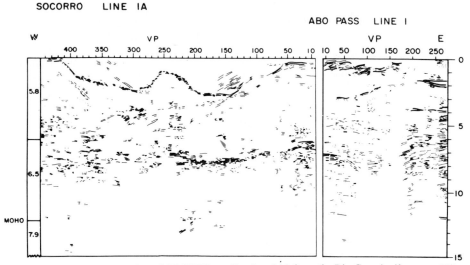

Fig. 6.63   Line drawing of COCORP profiles 1 and 1A from the Rio Grande rift survey. VP, Vibrator point. [From Brown *et al.* (1979).]

for the strong reflectors in both kinds of seismic sections. Unfortunately, the polarity of the reflections could not be determined. A layered structure of this body must be assumed in order to explain the strong reflected events. The crust–mantle transition is observed in the form of rather discontinuous reflections. These are much stronger along profile 1A (in the strike direction of the rift) than along profile 1. The depth and the structure of the Moho are in agreement with those from the DSS data and also compatible with a "weak" transition to the anomalous mantle.

In an effort to summarize the DSS data from the less developed rift standardized V-z profiles are compiled in Fig. 6.64 (see also Table 6.7). The observed transition zone at the crust–mantle boundary is indicated only in the average high velocities of 7.2–7.8 km/s and does not reflect the thinning of the crust inside the rifts properly. We may summarize the characteristic features:

(1)   Generally the crust is thinner inside a rift area than in its surrounding.

(2)   In nearly all rifts a smooth transition zone from crustal to mantle velocities reaching up to 40 km in thickness has been found by appropriate interpretation. If no transition zones are mapped by the individual interpreters at least an unusual thickness of the 7.2–7.8 km/s layer is given.

(3)   These transitional velocities are considered anomalous mantle velocities.

(4) $P_m P$ is mostly strong in the supercritical range, in accordance with gradient-zone models.

Table 6.7

Locations and References for the Standardized
$V(z)$ Functions of Fig. 6.64 (Rifts)

Function	Location	Reference
R1	Rhinegraben	Ansorge et al. (1970)
R2	Rhinegraben	Meissner et al. (1970)
R3	Rhinegraben	Edel et al. (1975)
R4	Rhone Valley	Sapin and Hirn (1974)
R5	Oslo graben	Cassell et al. (1983)
R6	Baikal rift	Puzyrev et al. (1973)
R7	Baikal rift	Puzyrev et al. (1978)
R8	Dead Sea	Ginzburg and Folkman (1980)
R9	Dead Sea	Ginzburg et al. (1979a,b)
R10	Dead Sea	Ginzburg et al. (1981)
R11–R12	Red Sea	Makris et al. (1983)
R13	Afar	Makris (1975)
R14–R17	Afar	Pilger and Rösler (1975)
R18–R20	Afar	Ruegg (1975b)

(5)   The shoulders of rifts including the Moho are mostly elevated. In addition, there is often a regional doming up to 1000 km in diameter.

(6)   Some rifts show evidence of intrusions of magma.

(7)   Other geophysical anomalies include negative BAs, high heat flow values, increased, generally very shallow, seismicity, and layers of increased electric conductivity at various depths.

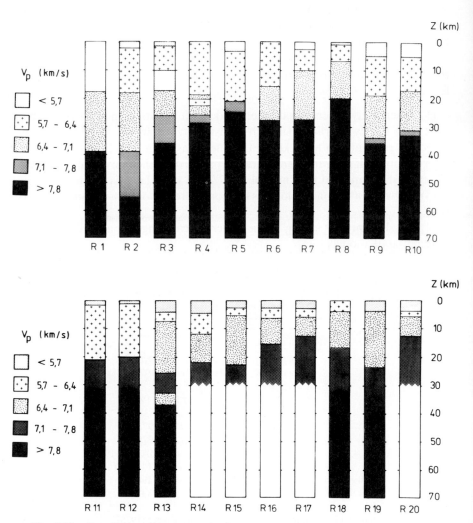

Fig. 6.64. Compilation of some standardized velocity–depth functions of continental rift structures. See Table 6.7 for locations and references.

6.4.2 ADVANCED RIFT STRUCTURES AND SPREADING CENTERS

According to most rift models, a development to more mature structures with broad extensional features and finally to a spreading branch of an ocean with continental margins seems possible and logical. An example of an advanced stage of a rift structure is the Basin and Range Province in Utah, which has been cited as "the largest intracontinental rift in the world" (Allmendinger *et al.*, 1983). Late Cretaceous to early Cenozoic thrusting was followed by middle to late Cenozoic extension. This later extension in the last 15 Myr has formed the typical basin and range (B & R) topography with high-angle faults and blocks, separated by basins. According to geologic field exploration, the normal faults account for most of the extension of the *upper* (brittle) crust, which has been estimated as up to 100% in the northern B & R area.

Some DSS studies together with interpretation of a large *positive* BA resulted in the model of a shallow crustal depth with the Moho at 25–30 km (Smith, 1978). Again, an "anomalous P_n velocity" of 7.4 km/s underlies the thin crust. This typical transition velocity is similar to the situation in the smaller rifts. The relatively thin crust, high heat flow, low crustal velocities, shallow seismicity, and some high-conductivity layers all fit into the picture of a (highly developed) rift system. Some reflection profiles from exploration surveys had already shown subhorizontal layering in the crystalline part of the crust—although no seismic slip on low-angle faults had been reported from studies of the shallow seismicity—when a 170-km east–west line and some additional cross-lines were observed by a COCORP survey (Fig. 6.65). The large section presented in Fig. 6.66 shows continuous shallow-angle reflectors, correlating with faults detected at the surface. Apparently three major zones of detachment, much deeper and more pronounced, underlie the B & R complex. Normal faults either bend downward in a listric way or seem to be terminated by the subhorizontal master faults. These faults apparently can be followed downward to about 12–14 km, which is the depth of the postulated transition zone from the rigid upper crust to the ductile lower crust. As expected, some weak NVRs but in increased numbers are observed at Moho depths at 25–30 km, where a gradient layer, possibly without a first-order boundary, is supposed to begin.

As mentioned by Allmendinger *et al.* (1983), the mechanism of the Cenozoic extension remains enigmatic. The structural similarity between this extensional thin-skinned tectonic pattern and those of compressional regimes is striking. It has been speculated that the faults were created during the Cretaceous–early Cenozoic compressional period when a convergent plate boundary was present in the western part of North America. When the stress

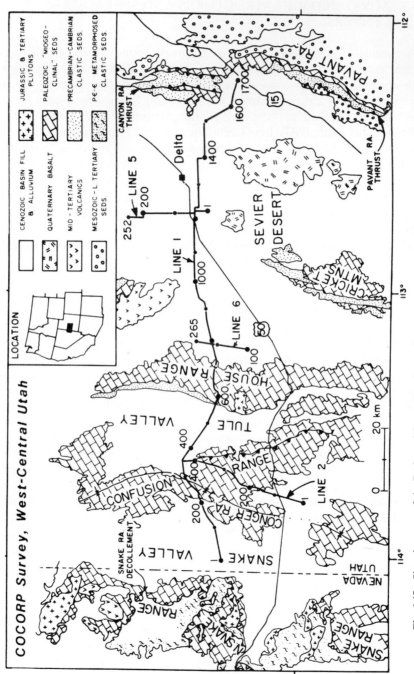

Fig. 6.65 Situation map of the Basin and Range Province and COCORP reflection profiles. [From Allmendinger *et al.* (1983).]

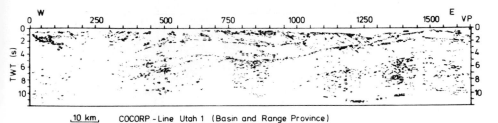

Fig. 6.66   Line drawings of profile 1 in the Basin and Range Province. Note reflections of zones of detachment (and metasediments?) in the upper part, while at lower crust levels, especially in the east, intrusions with a lamellation character are supposed. VP, Vibrator point; TWT, two way time. [From Allmendinger *et al.* (1983).]

pattern changed, possibly in connection with a continental back-arc structure or a subducted ridge, the same old fault systems were used again. In this way the B & R province may not present a "typical" picture of rift developments. However, similar mechanisms related to spreading asthenospheric flow patterns or intrusions of magmas into the crust and its modification may be general features of rifts.

Another example of the development of a large rift system is from the Variscan continental margin in the Gulf of Biscay (Fig. 6.67). Although not much is known about the deeper crust, extensive NVR data collected by marine measurements demonstrate the tectonic development in this area. After an earlier extensional phase in the Triassic, major rifting and extension began during the Early Cretaceous at about 140 Myr and ended in Aptian time at about 110 Myr. Blocks were rotated during this period by up to 20–30° along listric normal faults (Bally *et al.*, 1981). Rifting occurred on a preexisting marine Mesozoic basin of the Variscan continental shelf and was related to the rotation of the Iberian Peninsula. Rifting (and spreading) stopped when rotation stopped. The DSS surveys in this area (Avedik and Howard, 1979) had revealed an upper (brittle) crust with $V_P < 5$ km/s, underlain by a (ductile) lower crust with the (continental) velocity of only 6.3 km/s and a Moho at 12 km with a normal P_n velocity. Hence, a crust has developed in this area which is oceanic in terms of today's sea level and Moho depth, but still continental and sialic, consisting of upper-crustal rocks, in terms of seismic velocity and crustal material. Figure 6.68 shows a SSW–NNE directed migrated line drawing after Montadert *et al.* (1979a,b). The zone of detachment is (again) at about 10 km depth, where the listric faults turn horizontally. Here, a major zone of detachment must have developed during the rifting period. This zone (again) marks the transition from brittle to ductile behavior in a hot and wet crust during the rifting stage. In the Biscay area this zone seems to coincide with a change from (Triassic?)

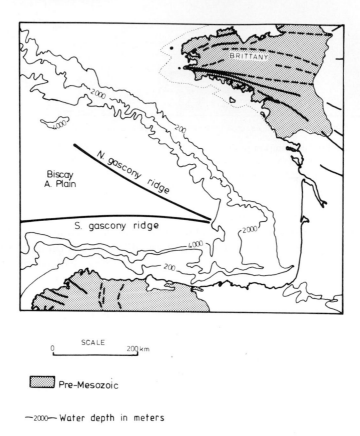

SCALE

0 _____ 200 km

▨ Pre-Mesozoic

—2000— Water depth in meters

Fig. 6.67 Situation map of the Gulf of Biscay with major structures. Depth of sea bottom shown in meters.

sediments to Variscan?) basement. It should be emphasized that in this zone of rifting and subsidence velocities (and material) typical of lower crust are completely missing. (See the discussion in Section 6.4.4.) Hence, this example of rifting may not be typical either and must be connected with the margin development. Cooling and subsidence since 110 Myr have brought P_n velocities back to normal.

A third and last example of rifts in an advanced stage is the "classical" Red Sea area and Afar triple junction. Compared to the Gulf of Biscay, this is still a young and active rift. The whole rift system may be considered a continentward extension of the mid-ocean ridge system of the Indian Ocean. Small ages and low rates of plate movement compared to oceanic rifts may be considered responsible for the limited development. Spreading rates in the last 2 Myr have been estimated to be 0.75 cm/yr in the northern part and

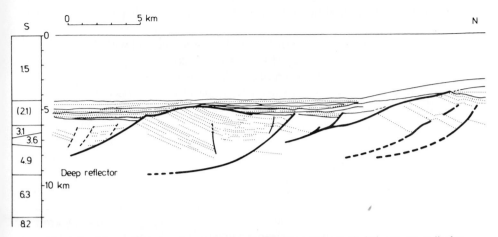

Fig. 6.68 Migrated line drawing across the northern part of the Gulf of Biscay perpendicular to the geologic lineaments. Dotted lines are NV reflections, and thick lines are fault zones. [From Montadert *et al.* (1979b).]

0.6 cm/yr in the south, and they were possibly never much larger than 1 cm/yr. The initiation of the Afar triangle dates back to the early Miocene (25–23 Myr) (Barberi and Varet, 1975). This is slightly later than the appearance of the first volcanics in the eastern branch of the East African rift system, which dates back to 35 Myr, or in the southern Red Sea (41 Myr). There is agreement among all researchers that since at least 3.5 Myr ago oceanic crust has been created in the southern part of the Red Sea, in the Gulf of Aden, and apparently in the Afar triangle. Le Pichon *et al.* (1973) discuss several approaches to the earlier development. They emphasize the contrasting models of a kinematic approach (based on a best fit of the coastlines, except for Afar) and a continental approach (which assumes that only during the last 4.5 Myr has "real" oceanic crust been created and that the rest of the Red Sea, including Afar, is a subsiding and thinned continental crust). In the kinematic approach the whole 300–350-km width of the Red Sea is considered oceanic; in the continental approach only the central 70 km near the ridge axis is considered oceanic. It seems that this discrepancy is only an apparent one. The arguments of the continental approach lose weight when some additional spreading inside the Afar triangle is allowed (and there is ample evidence for a rotation and shifting of the "Danakil Alps") and the DSS data are not uncritically described as being from "continental basement." Moreover, there is a zone of magnetic lineation in the southern Red Sea, i.e., south of 19° latitude, which dates back at least 5 Myr and covers a width of at least 200 km. In the northern part of the Red Sea magnetic lineation becomes obscured, and the topography of the sea floor is characterized by many small (additional?) strike–slip motions nearly perpendicular to the strike direction.

Dealing first with the southern sector of the Red Sea and with Afar, results of DSS measurements of Ruegg (1975a) and Berckhemer *et al.* (1975) helped to clarify the situation. The discrepancy mentioned before is again visible in relating Ruegg's upper 6.4 km/s layer to a "basaltic" (i.e., oceanic) layer, or Berckhemer's equivalent 6.2–6.3-km/s layer to an upper continental layer. However, this discrepancy is also a fictitious one. Although the upper crust seems to be sialic (i.e., continental), with velocities less than 6.4 km/s prevailing, it is thinner than in the adjacent continents and is underlain by velocities around 6.8 km/s, compatible with an oceanic affinity. The P_n velocity between 7.4 and 7.6 km/s mentioned by both groups of authors is considered typical for hot oceanic provinces. However, we must remember that it is also typical for hot continental rifts. The relation of the 6.8-km/s lower crustal velocity to an "oceanic" lower crust is certainly not unique either. As shown in the V-z sections in Fig. 6.59, crustal thicknesses and velocities in the Afar depression are similar to those of continental rifts. A certain "basification," which might be considered the first step to "oceanization," may be seen in the relatively thin upper crust and thicker lower crust. It may be important to state that a thorough differentiation such as that on top of plumes or island areas has not (yet?) taken place.

On the other side of the Red Sea, slightly to the north, the southwest part of the DSS profile through the Arabian Peninsula had its starting point at the margin of the rift area. There seem to be two Mohos at the flank of the shield, the lower one being a continuation of the Moho under the shield area and the upper one representing that of the rift area (Flüh and Milkereit, 1984). Both Mohos seem to be separated by an LVZ. We will show in Section 6.4.4 that such a structure is to be expected if basaltic magma intrudes the low-viscosity, heated lower crust and has some time for differentiation into an upper mafic or even sialic light part and lower ultramafic cumulates, forming a new Moho (see Fig. 6.9c).

It is not possible to arrive at a uniform picture of the advanced stages of rifting. Structures, including those shown in the examples, are really very different from each other and are controlled by the timing and the intensity of upcoming magmas from the mantle and by the host area in which rifting takes place. An attempt to summarize such advanced stages is as follows:

(1)   Even more than in the less advanced rifts, crustal thinning is observed in advanced stages of rift development. It seems that the more advanced a rift is, the thinner is the crust until the small oceanic depths are approached.

(2)   A transitional layer is observed at mature rifts that are still in the rifting and spreading stage. If resolution in $P_m P$ is insufficient to detect a

possible gradient zone, at least a reduced P_n velocity is observed in the uppermost mantle.

(3) This reduced transitional velocity can be related to a hot mantle regime with rising and not yet differentiated basaltic magmas and to extensional features.

(4) Usually, P_mP is strong only in the supercritical range, which is compatible with a gradient layer in the uppermost mantle.

(5) Cooling after the termination (or migration) of the rifting process has the following consequences: (i) further subsidence, (ii) P_n velocities returning to normal, and (iii) development of a higher Moho (still shallower crust) by fractional crystallization of the formerly intruded magmas.

6.4.3 PASSIVE CONTINENTAL MARGINS

Passive continental margins are considered a consequence of an earlier rifting process. They belong to a rather advanced stage of the plate tectonic cycle and may be transformed into active margins once a critical stage of negative buoyancy has been reached.

In contrast to the continental rifts, the margins, especially the passive ones, have been intensively studied by marine reflection seismology. They are among the best observed continental structures on Earth, although research has often been restricted to the sediment accumulation. However, even from the sediments a tremendous amount of information can be obtained on the subsidence pattern, sea level changes with transgressions and regressions, tectonic events, and the whole stratigraphic sequence. Because of their high coverage, long streamers, and homogeneous surface conditions, marine near-vertical reflection measurements provide higher data quality than do land surveys. The increasing resolution of the seismic data acquired has shown that sediment accumulation (and hence subsidence) is much greater along nearly all passive margins than was previously thought. Unfortunately, the limited length of streamers (up to 4 km) and limited offset usually do not permit calculation of velocities for depths greater than 10 km, and the penetration into basement is often very limited. Hence, Phinney and Odom (1983) state that "old fashioned marine refraction data" provide important constraints for the interpretation of the multichannel reflection sections. These constraints mainly concern the velocities in the lower parts of the sediments, in the basement, and in the uppermost mantle.

Sediment structure and the development of continental margins have been extensively described by Talwani *et al.* (1979), Blanchet and Montadert (1981), Keen *et al.* (1981), Pitman III (1983), Eldholm and Sundvor (1979), and Phinney and Odom (1983). Only three examples of passive margins will be discussed here. They were selected according to their different ages, indicating different stages of development. A young margin is found in the Red Sea, the Gulf of Aden, and the Gulf of California; the northern Atlantic is considered of intermediate age; and the west coast of Africa, the east coast of America, and the Gulf of Mexico typify old margins (Eldholm and Montadert, 1981). At least four important structural features must be analyzed:

(1)   the landward boundary of the magnetic lineaments as an indication of where a "truly" oceanic crust terminates,
(2)   the boundary of (mostly listric) faults or fault concentrations marking the (original) rift boundary,
(3)   the subhorizontal boundary between postrift, quiet sedimentation and prerift, faulted-block structure, and
(4)   the subhorizontal and deepest part of listric faults indicating the boundary between brittle and ductile layers.

In the Red Sea and the Gulf of Aden the pattern of opening has been essentially stable since the early Miocene, i.e., since about 25 Myr ago. True sea floor spreading seems to have started 10 Myr ago as estimated from the width of the central zone with magnetic lineations, which is about 100 km wide adjacent to a magnetic quiet margin of 40 km on either side (Cochran, 1981). This "pre-sea floor spreading" zone was subsiding at least 15 Myr before spreading set in. The Gulf of Aden is characterized by a SW–NE spreading with many fracture zones approximately in this direction. As there is a lack of sediments, subsidence must have occurred only by cooling after the spreading set in. Before spreading, lateral intrusions into the lower crust and vertical dyke injections must have modified the crust. Not much information is available on the crystalline basement in this area. Fig. 6.69 presents a picture of the sea floor topography and some model calculations from Cochran (1981) for the Gulf of Aden; the development of a double Moho in the lower part of the Red Sea margin, as determined by DSS studies, was the subject of Fig. 6.9c.

Another young margin of similar age is the Gulf of California. The San Andreas transform system joins the divergent plate boundary in the Gulf. Preexisting continental rocks have been rifted, sliced, and rotated. Although toward the end of the Miocene seas flooded part of the subsiding basin, true—slightly oblique—spreading began in the southeast about 4.5 Myr ago, as shown by the arrangement of magnetic anomalies (Crowell, 1981). High heat flow, seismic activity, and volcanism indicate the spreading center. Information about the deeper structure obtained by seismic methods is still missing.

As an example of a middle-aged margin, the Norwegian area is discussed. The Norwegian Sea opened in the early Tertiary, about 60 Myr ago, and the

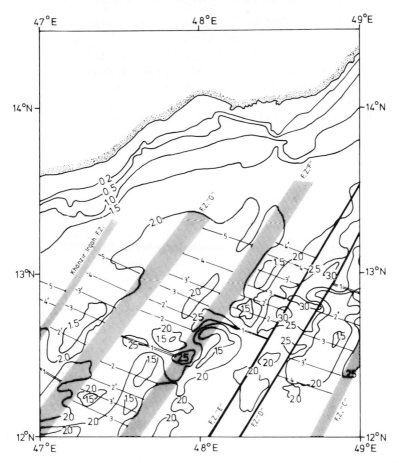

Fig. 6.69.   Sea floor topography in the Gulf of Aden. Note the oblique fracture zones; depth values in kilometers. [From Cochran (1981).]

sediment cover is not yet very thick (Talwani *et al.*, 1980). The rate of spreading decreased from about 2 cm/yr to only 0.6 cm/yr according to the analysis of the magnetic anomalies. Extensive multichannel NVR studies were performed all along the margin, and a DSS profile, the Blue Norma, covered the whole shelf area. Unpublished NVR data along the Blue Norma line show several listric faults in the Mesozoic sediments. This means that extension and subsidence were active before spreading set in. Looking back to Fig. 6.21, even an individual basin (with rifting?) had developed, the Westfjord basin, between the shore and the old Lofoten horst. It may be one of the many failed (or starved) rifts and basins in the Atlantic area which terminated when the spreading at the present center took over (Meissner,

1979). On top of the faulted block structure the postrifting Neozoic sediments show a subhorizontal and undisturbed character; they were apparently deposited during a stage of quiet, thermally controlled, subsidence. The Moho dips from 20 km at the outer shelf to 40 km below the Caledonian mountain range over a distance of about 350 km. A layer of 7.1–7.2-km/s velocity marks the bottom of the crust on top of an uppermost mantle with $P_n = 8.0$ km/s. Northwest of the Blue Norma line there is the Vøring plateau, a rather unusual shallow feature. There is an escarpment across it at the landward side which was defined as the ocean–continent boundary by Talwani and Eldholm (1972), but this suggestion was highly debated (Hinz and Weber, 1976). Indeed, slightly seaward of the Vøring plateau escarpment magnetic anomaly 24 is present, the last landward anomaly which can be followed from the northern Lofoten basin onto the plateau. Seaward of the plateau a huge sequence of seaward-dipping reflectors in a wedge-shaped configuration was found by the multichannel NVR surveys (Figs. 6.70 and 6.71) (Mutter *et al.*, 1982, 1984). Talwani *et al.* (1980) suggest that the wedge is composed of Tertiary volcanic products, a combination of lava flows and volcanogenic sediments. This would fit the velocity picture, the velocities strongly increasing with depth due to increasing compaction. A huge, rather isolated volcano or a series of volcanoes, active between about 54 and 60 Myr ago, could have produced this enormous amount of volcanic material. The Vøring plateau itself may be considered as a volcanic plateau created during

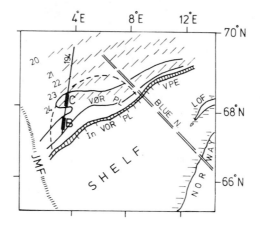

Fig. 6.70   The Norwegian continental margin in the area of the Vøring plateau. Situation map of one (out of 12) multichannel seismic reflection profiles (194); JMF, Jan Mayen fracture zone; 21–24, magnetic anomalies; VPE, Vøring plateau escarpment; OU, outer Vøring plateaux; IN, inner Vøring plateau; BLUE N, Blue Norma onshore–offshore DSS profile (see Fig. 6.19); and LOF, Lofoten area. Sea floor spreading started shortly after the time of magnetic anomaly 25 (63 Myr).

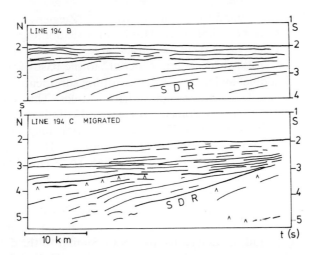

Fig. 6.71 Line 194, part B, as a line drawing with seaward-dipping reflectors (SDR); ∧ represents possible beginning of acoustic basement. Line 194, part C, as a migrated line drawing with seaward-dipping reflectors (SDR); ∧ is acoustic basement (?); SDRs are thought to originate as a combination of large subaerial volcanic flows and isostatic adjustment and occur between the area of magnetic anomalies 24 and 25. [From Mutter *et al.* (1982).]

the time when spreading started and the whole North Atlantic was in its peak tectonic activity. Hinz *et al.* (1984) emphasized the importance of seaward-dipping reflectors as being a special kind of passive margin.

The Bay of Biscay margin, where rifting was active between 140 and 110 Myr ago, is slightly older than the Norwegian margin. As mentioned in Section 6.4.2, the well-preserved listric normal faults partly involve the basement and are intimately related to the rifting process (Fig. 6.68). The low-angle faults reach a depth of about 10 km, and the folded and faulted prerift sediments are from about 5 to 9 km. They are overlain by unfaulted postrift sediments, apparently also deposited during an era of quiet subsidence. The quiet, undisturbed subsidence, after the rifting process had stopped and/or jumped to another position, is certainly a thermal effect and belongs to the general picture of margin evolution. Numerous models for this subsidence caused by thermal contraction have been calculated, especially for the mature margins (see Keen *et al.*, 1981).

Following the age sequence and the marginal basins of failed rifts of the European Atlantic area, there is the North Sea basin, where maximum extension took place between the Middle Jurassic and Early Cretaceous, i.e., between about 160 and 120 Myr ago. Both multichannel NVR and some DSS data are available. At least down to the Zechstein level, and probably extending much deeper, listric normal faults have been observed, especially concentrated at and around the "Central Graben" (Ziegler and Louveraus, 1979). There seems to be a short-lived Middle Jurassic doming phase before rift formation. Ziegler (1984) suspects considerable subcrustal erosion and a diapiric ascent of "hot asthenospheric material" to the base of the crust.

Again, there is a quiet and undisturbed postrifting subsidence of the North Sea rift and the whole North Sea and adjacent areas. This subsidence is greatest in the area that was domed up before, also indicating a thermal contraction mechanism. Figure 6.72 shows a situation map and Fig. 6.73 provides a cross-section through the North Sea, based on DSS and NVR data (Barton and Matthews, 1984; Barton *et al.*, 1984). The Central Graben still seems to show the shape of the crustal thinning below, the effect of subcrustal erosion, and/or formation of a new (shallower) Moho, although P_n and crustal velocities are back to normal. The whole North Sea basin has developed like a huge continental margin, although its asymmetry must be attributed to some additional tectonic processes. Similar to the development at the younger Norwegian margin in the north and the Gulf of Biscay in the southwest, extension ceased after rifting and spreading moved to the mid-Atlantic rift.

Most extensive NVR surveys and other geophysical studies have been performed along the large old margins of the middle and southern Atlantic, where continental rifting and crustal thinning took place during the Late Triassic–Early Jurassic, i.e., about 200 Myr ago, and sea floor spreading began in the Early to Middle Jurassic.

The older the margins and the thicker the sediments, the more difficult and complex is the picture of crustal structure. Very dense networks of multichannel NVR surveys and some refraction data are available along the broad

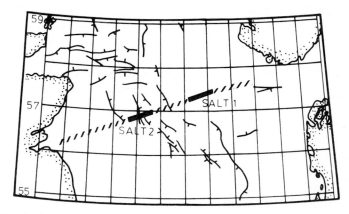

Fig. 6.72 Situation map of DSS observations and some NVR data (solid lines) through the North Sea. [From Barton *et al.* (1984).]

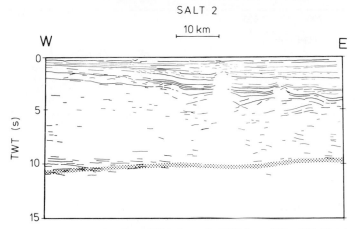

Fig. 6.73 Crustal cross-section from NVR along the DSS line of Fig. 6.72. Hatched line is Moho from DSS. [From Barton *et al.* (1984).]

margin of the eastern United States. Very rapid and highly variable postrifting subsidence had already started during the Jurassic (Grow and Sheridan, 1981). Transverse fracture zones divide the margin into four sedimentary basins. The only one of them which will be shown here is the Baltimore Canyon trough beneath the outer continental shelf between Virginia and New Jersey, which shows at least some control on the deep velocity structure. It has a maximum width of about 150 km. Subhorizontal reflectors, presumably of postrift, quiet sedimentation, are present down to 13 km depth. Again, below these strata there are the deformed pre- and synrift sedimentary rocks, which reach down to a maximum depth of 18 km (!) (Grow, 1980).

Figure 6.74 shows a situation map and Fig. 6.75 presents a geological–geophysical section through the Baltimore Canyon trough. The East Coast Magnetic Anomaly, present along the whole North American margin mostly as a strong feature situated near the continental slope, is supposed to mark the boundary between the sediment-covered continental and the sediment-covered oceanic basement, as seen in the figure. It should be stressed, however, that the basement below the "continental section" of the shelf is not well constrained, especially not the lower crust below the faulted prerift sediments. The other three basins along the eastern U.S. margin are slightly more complicated, with different salt intrusions, volcanic instrusions, and ashes (Sheridan, 1974). Figure 6.76 shows V–z data from a large DSS survey along the continental, i.e., the inner, quiet zone of the New Jersey shelf (Sheridan *et al.*, 1979). It demonstrates the heterogeneity and the rugged appearance of the margin in its lower crustal–upper mantle structure. There

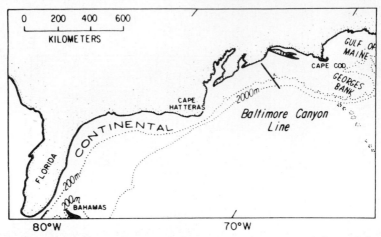

Fig. 6.74 Situation map of the U.S.—Atlantic margin with bathymetric lines and situation of the Baltimore Canyon line.

is no convincing indication of prerift doming. The Blake plateau is about 350 km wide with water depths of only 500–1500 m and a transitional, ill-constrained crust of more than 30 km thickness, probably related to a failed (Triassic) graben and much igneous activity.

Some profiles, well constrained by DSS work, were obtained in the *Gulf of Mexico* (summarized by Antoine *et al.*, 1974), where subsidence of the continent and crustal thinning is comparatively strong. The adjacent coastal plains in the Gulf Coast of Texas have crustal thicknesses of 30–33 km and very low velocities in the whole crust, similar to the area of the North German plains near the North Sea. Sediments in the northwest part of the . Gulf reach thicknesses of 15 km. The whole northern part is also dominated by extensive salt diapirism. In all these cross-sections and models of the passive margins a large part of the shelf area *and* the adjacent continent show

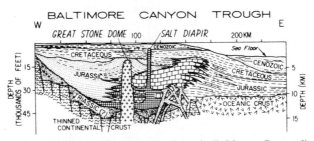

Fig. 6.75 Geologic–geophysical cross-section along the Baltimore Canyon line through the Baltimore Canyon trough, an example of an advanced passive margin. [From Grow (1980).]

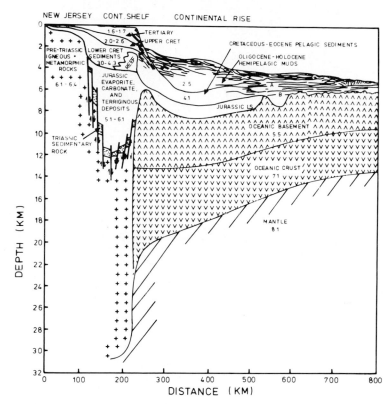

Fig. 6.76 Depth structure and $V(z)$ data along the "continental" zone of the passive Atlantic margin near New Jersey. [From Sheridan *et al.* (1979).]

considerable amounts of postrift subsidence. While sedimentary layers and the upper crystalline crust dip seaward, the Moho dips landward and the whole crust thins out seaward. This observation certainly does not agree with a simple loading and bending model. It seems much more plausible that the newly created oceanic lithosphere in the rifts (including the crust and Moho) has somehow also modified the adjacent continental structure. The lower crust of the adjacent continents especially seems much thinner or even completely missing, as in the case of the Gulf of Biscay or the North German plains.

The European and West African Atlantic margins have also been investigated by both marine NVR studies and DSS networks, some of them being observed along onshore, offshore profiles (Weigel *et al.*, 1982). Figure 6.77 shows a situation map of two DSS profiles with some additional bathymetric

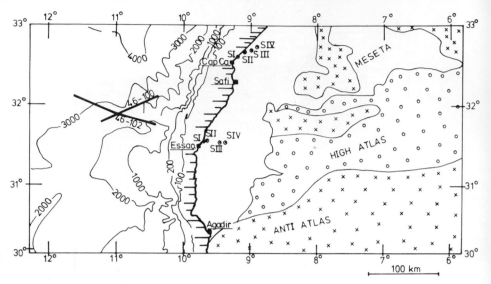

Fig. 6.77 Situation map of two DSS profiles in the African Atlantic margin and some bathymetric and geologic data; S, shot points on land; contours represent sea bottom depth in meters. [From Thiessen (1979).]

and geologic data included. Both profiles are presented in the form of a simplified cross-section in Fig. 6.78. There seems to be material of the upper continental crust present below slope and shelf areas, but the lower crust certainly looks very different in both profiles. The 7.1 km/s layer is often found below shelves (see also Fig. 6.19). It seems to be well constrained and may indicate mafic material left over from the rift process when it intruded the lower continental crust. Summarizing the data for the passive margins, we may state that their heterogeneity is even greater than that of the advanced rift structures. They have all developed individually according to their age, their rate of development, their sediment supply, and their host rocks.

We may summarize:

(1) The older the continental margins, the more subsidence, sediment accumulation, and individual development have taken place. Subsidence also affects the adjacent continents.

(2) The boundary between oceanic and continental crust is generally marked by the termination of the oldest oceanic magnetic lineation; in the old margins it is a prominent positive anomaly, e.g., the East Coast Magnetic Anomaly of the United States.

(3) Unfaulted, quiet sedimentation on top of older faulted sediments indicates postrift, quiet subsidence. Only in the old, mature margins is the quiet sedimentation disturbed by diapirism and intrusions.

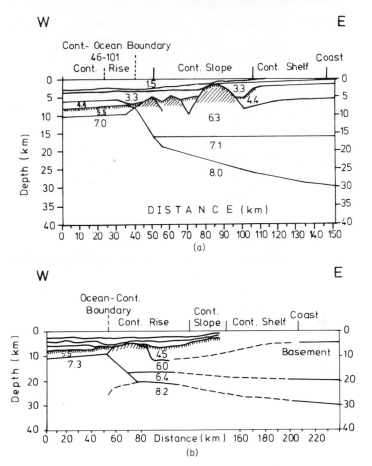

Fig. 6.78   (a) and (b) Simplified crustal cross-sections along the two profiles of Fig. 6.77. [From Thiessen (1979).]

(4)   At middle-aged and old margins listric faults mark the prerift sediments (if prerift subsidence with sedimentation existed). High-angle faults sometimes end at subhorizontal zones of detachment.

(5)   During the rifting stage individual large volcanoes and increased igneous activity have modified the margins locally and left their traces in the form of plateaus and a wedge-shaped layering around them.

(6)   Abundant failed (or starved) rifts are found in the prespreading stage of the Atlantic Ocean. They generally terminated after the new spreading center was established. They contribute considerably to the heterogeneity of the lower part of the crust in the margins.

(7)   The transition from oceanic to continental crust, mostly hidden below huge sediment thicknesses, is often difficult to observe in the old margins.

(8)   Resolution of DSS surveys for the lower crust is often insufficient.

6.4.4   SOME MODELS FOR THE DEVELOPMENT FROM RIFTS TO MARGINS

So far, observations from rifts and margins at several stages have been discussed. All these observations should fit into a physically and geologically reasonable model. Among the models there are the simple physical ones such as the McKenzie (1978) stretching model and the Royden intrusion model (Royden *et al.*, 1980). Both models can account for the observed postrifting heat flow as a function of age. McKenzie's model results in a thinned crust after stretching, as seen schematically in a summary by Cochran (1981) depicted in Fig. 6.79. The model of Royden *et al.* (1980) is a modification of earlier models of Karig (1971) and Scholz *et al.* (1971) in which interarc spreading or mantle diapirism was considered the primary source while the horizontal extension was seen as a secondary process. The simplified (Royden

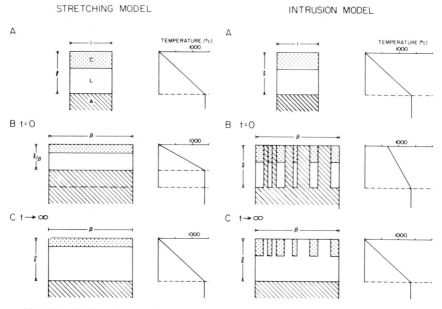

Fig. 6.79   McKenzie's stretching and Royden's intrusion model, as seen schematically in the review of Cochran (1981). Note the assumption of sudden start of stretching (resp., intrusion) at time $t = 0$. Immediately after stretching length 1 will be 1-β, and thickness (= C + L) will be l/β C, crust; L, lithosphere (subcrustal); and A, asthenosphere.

et al., 1980) model may explain crustal modifications by magmatism but fails to take proper account of crustal thinning.

The two simple models shown in Fig. 6.79 neglect geologic observations of prerift uplift or subsidence, magma chemistry, crustal structure, and viscosity. They suffer from the assumption of a sudden stretching at the beginning (McKenzie), a sudden intrusion of dykes from the asthenosphere (Royden *et al.*, 1980), or a sudden heating event (Bott, 1980). The assumption of any sudden event like a "big bang" is certainly a gross oversimplification and is not justified, at least in those cases where the prerift situation can be reconstructed and the reasons for rifting may be found. Hence, the initial period of the preparation of rifting should also be considered.

Actually, there seems to be a consensus among most geologists and geophysicists that there are basically two main periods in the development of rifts and grabens: a prerift and often extended period of crustal thinning, block faulting, and stretching and a second postrift period of spreading with quiet subsidence and sedimentation of the margins. As mentioned earlier, this development might stop anywhere along its way. Widespread evidence of the two main periods, mainly based on seismic stratigraphy, has been mentioned in the previous sections. Only a few investigations deal with the physical background for the first period. Heiskanen and Vening-Meinesz (1958), Girdler (1964), Artemyev (1966), Karig (1971), and Scholz *et al.* (1971) associate the stretching of the crust with "diverging convective flows in the mantle," the rising part coming from some depth below the asthenosphere. The second, postrift stage is well explained by:

(1)   thermal cooling and contraction of the growing lithosphere (Parson and Sclater, 1977; Watts and Ryan, 1976),

(2)   the load of sediments in connection with (1) (Bott, 1973; Sleep, 1971), and

(3)   phase transitions, for instance, gabbro to eclogite, in connection with (1) and (2) (Spohn and Neugebauer, 1978; Neugebauer and Spohn, 1978; Artyushkov, 1981).

The contribution from both (1) and (2) is estimated by Steckler and Watts (1978) for Atlantic-type margins. Explanation (3), involving deep crustal metamorphism, provides an attractive additional force in connection with (1) and (2) and may induce a step in subsidence which may contribute to the final transformation from passive to active margins.

Analyzing the observed data for the subsidence of margins, important corrections must be applied. There is first the "back-strip" correction, taking care of the effect of sediment loading, i.e., compression, densification, and additional apparent subsidence (Sclater and Christie, 1980). Rather simple corrections seem adequate for this purpose (Crough, 1983). Seismic data and borehole information generally supply the necessary parameters. After back-stripping (deloading), model calculations for stretching are applied in order to calculate the stretching factor β (see Fig. 6.79) for the explanation of the thinned crust below the margins. At the Atlantic type of margins the amount of thinning is least on the continent side and greatest on the seaward side, as deduced from the observed Moho depth and dip. However, a uniform stretching factor for the whole lithosphere is generally applied. Jarvis and McKenzie (1980) also calculate the effect of cooling during the rifting stage. Turcotte and Ahorn (1977) use a cooling half-space model for computing subsidence and heat flow. Calculations of cooling of the lithosphere of the margins are similar to earlier calculations of growing sea floor depth and lithospheric thickness in oceanic areas. With the right values for volume expansion, thermal conductivity, and heat supply, the famous dependence of sea floor depth on age ($\Delta h \sim t^{1/2}$), the observed heat flow and the seismicity could be well explained by such "thermal contraction" models (Parson and Sclater, 1977).

In some models flexural rigidity, increasing with increasing age, is introduced (Watts, 1982). This concept makes an important contribution to all loading and deloading processes and also explains many gravity anomalies. See Section 3.1.1. Vetter and Meissner (1977, 1979) calculate stress differences between a number of continent–ocean transitions. In the crust between 10 and 20 km stress differences up to nearly 100 MPa are directed toward the ocean, resulting in a stretching in the continental part and a possible flow of crustal material toward the oceanic crust. In the upper mantle between 30 and 80 km, however, stress differences up to 50 MPa are directed toward the continent, leading to a certain flow of oceanic mantle material toward the continent.

All these forces related to density and pressure differences along boundaries are only of a secondary nature, however. The primary force for initiating stretching and rifting and finally spreading must come from below. Convection patterns in the asthenosphere and the rise of plumes from below are incorporated in the model of Meissner (1981b), which also takes into account the viscosity layering in mantle and crust. Stretching by rising plumes is a part of analytical solutions of cylindrical or two-dimensional plumes by Li *et al.* (1983). These plume models may explain the prerift development by diverging flows in the mantle. In particular, the rising of plumes in an area of a strong viscosity gradient results in strongly diverging motions in the upper part of the plume. As a consequence, horizontal extensional stresses are exerted at the subcrustal lithosphere. These stresses favor faulting and the beginning of volcanism as a forerunner of rifting. The rising plume *might* lead to a regional updoming if the divergence in the area of the upper part of the plume is greater than zero. Alternatively, updoming might be negligible if the divergence is equal to zero and the same quantity of mass is transported sideward as the quantity that is rising. Even subsidence might be generated on top of a rising plume if lateral movements dominate, which might be enhanced by subduction somewhere else or by an in-line propagation of adjacent rifts.

Figure 6.80a depicts the initial deformation of a rising plume that intrudes the asthenosphere, causing uplift if it is under a certain overpressure or causing subsidence if the lateral intrusion into the low-viscosity astheno-sphere dominates. In any case, horizontal extensional stresses of the lateral plume movements will cause a stretching and a certain subcrustal erosion. Volcanism and magmatism start and the first (high-angle) faults and narrow grabens appear at the surface.

Figure 6.80b shows the next stage of rift development. Plume material has entered the low-viscosity lower crust at several locations. It fills the stretched parts of the lower crust, where it might differentiate, mix, and form a pillow of intermediate (crust–mantle) velocities with gradient zones and smooth seismic boundaries. The shoulders of the rift structure become elevated, and abundant normal faults develop, resulting in a horst–graben structure with extended volcanism. Huge volcanoes develop at this stage. In addition to the lateral intrusions in the lower crust, lateral emplacement of plume material in the asthenosphere continues and keeps up extensional stresses and regional stretching.

Figure 6.80c depicts the first postrift stage of sinking and spreading. The rift has extended considerably, and new lithosphere (including oceanic crust, Moho, and uppermost mantle) is created at the center of the rift, where the first magnetic lineaments appear. The plume has developed into a convection cell. The new (oceanic) Moho is on top of the old continental one at the margins of the rift, creating an LVZ between them. Failed (or starved) rifts terminate their development. Lateral stresses below the new margins and block faulting also terminate because, after the total breaking of the continental lithosphere, the two split continental segments ride on top of the convection cell. Below the margins regional cooling and contraction set in, resulting in a quiet subsidence with unfaulted sedimentation.

Figure 6.80d shows the situation of young to middle-aged margins. Spreading at the center of the rift continues. Cooling and regional subsidence of shelf areas and adjacent continents also continues as does quiet un-disturbed sedimentation of the margins. The new (oceanic) shallow Moho may join the continental one. Resistance to further spreading might develop and lead to compressive stresses in the adjacent continents. Large teleseismic delays around the margin bear evidence of the large asthenospheric intru-sions.

The model shown in Fig. 6.80 is based on the viscosity structure below

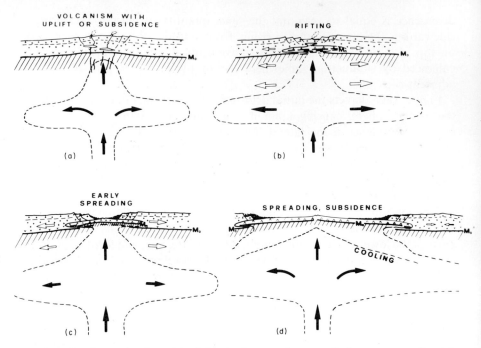

Fig. 6.80 Conceptional model of the development from prerift features to continental margins: (a) prerift stage, (b) rifting stage, (c) first postrift stage, and (d) young continental margin.

continents with a broad minimum in the asthenosphere and a smaller one in the lower crust (Meissner, 1981b). A rising plume or some loosely connected rising plumes in a variable-viscosity mantle and crust lead to plume structures and features that agree with most geological and geophysical observations. There are, for example:

(1)   the general, regional uplift (or subsidence) in the preparatory stage with initial volcanism and a limited stretching,

(2)   the maximum of continental stretching associated with block faulting, rising of shoulders, intrusion of magmas, huge volcanoes, and crustal pillows,

(3)   the postrift, quiet subsidence with beginning unfaulted sedimentation and concentration of rifting at a central spreading center, and

(4)   the continued subsidence of shelf areas and adjacent continents as a consequence of cooling and shrinking of the formerly intruded hot plume material and hot lithosphere.

Mature margins like the ones described in Section 6.4.3 with large sediment accumulation and a possible gabbro–eclogite transition at depth must show a continued tendency for sinking. At the old Atlantic margin it seems to have developed into a small "silent subduction." It is believed that such margins may easily be transformed into active ones, a process that might be further supported by a downward-directed component of a convection cell. Such a development is not only compatible with the old concept of Stille's pattern of orogenies (Stille, 1944), it may also be considered a part of the Wilson cycle.

6.5 ACTIVE MARGINS

The active margins, also called convergent plate boundaries or areas of subduction, are considered a part of the oceanic regime. Hence, only their most important features will be described. Generally, deep trenches are observed along the strike of the subducting oceanic plate. The plate subducts landward of the trench as deduced from the signature of seismicity, which can be followed down to a maximum depth of 700 km along so-called Benioff zones. While these Benioff zones, with their inclined zones of hypocenters and their focal plane solutions, provide the seismological constraints for subduction, simple extrapolation of sea floor spreading velocities, based on magnetic lineaments and deep drilling, provides the long-time evidence for subduction. The most prominent active margins today are found around the Pacific Ocean.

6.5.1 ISLAND ARC STRUCTURES

At the western margin of the Pacific Ocean is the Japanese island arc, where the old and cold Pacific plate subducts seaward of the continent, a process which most probably is related to the fact that the density of the oceanic lithospheric plate has become greater than that of the underlying asthenosphere. Flexural rigidity and possible convective movements may lead to a subduction angle of about 45° on the average. In many cases a buckling of the plate with a gravity and topographic high is seen seaward of the trenches. Scraping off of sediments may lead to a fore arc in front of the occanic magmatic arc which is mainly built of basaltic to optiolitic lavas and the related intrusions at depth. As mentioned in Section 6.3.2, basaltic magmas seem to originate from the top of the subducting plate with its oceanic crust and may run through various stages of contamination and differentiation until they have reached the surface or the upper crust. Figure 6.81 shows a schematic picture of the crust through the Japanese island arc

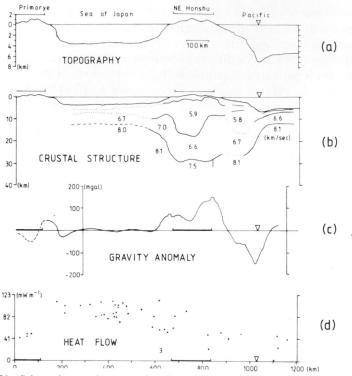

Fig. 6.81 Schematic crustal cross-section through the Japanese island arc. (a) Topography, (b) seismic velocities, (c) gravity, and (d) heat flow. [From Yoshi (1979).]

with some geophysical observations included. It should be stressed again that the island arcs (and the continental arcs) are sites of vertical crustal accretion. The main contribution to the growth of the crust seems to come from the differentiation process beneath these arcs. Island arcs will later be accreted to a continent during another part of the tectonic cycle.

Behind an island arc there is generally a back-arc basin. This is an area where sometimes rifting and generally quiet subsidence are observed without magnetic lineaments and with an increased heat flow. It is certainly an area of extension. It may be caused by the seaward-retreating subduction and by some secondary convection pattern induced by the retreat or by secondary rising of magmas from the subducting plate. Numerous models have been suggested for an explanation of the back-arc basins where a transitional and strongly modified continental crust prevails. This might be considered an extended zone of slow extension where divergent motions at depth and subcrustal erosion dominate over an actual breaking of the still rigid upper crust.

6.5.2   CONTINENTAL ARC STRUCTURES

Active margins without island arcs have deep sea trenches immediately in front of the continent. Landward of the trench a generally narrow coastal plain is found in front of mountain ranges, which often are called "continental arcs" or "cordillera-type mountains." While the North American margin of the Pacific was a zone of subduction until the early Cenozoic, it now shows a complicated pattern with remnants of a subducted ridge, pull-apart basins, and large transform faults (e.g., the long San Andreas fault). The South American Pacific margin also consists of different units, although its gross structure seems so simple and uniform. The margin consists of a northern sector, running from Central America to Ecuador, where relatively young oceanic lithosphere subducts with shallow angles and in a slightly oblique direction. A large amount of oceanic lithosphere was obducted onto the continent in the Early Cretaceous–Late Tertiary in this area and formed one of the three mountain ranges of the Andes, as discussed in Section 6.3.2.

Separated by a transform fault, the old central chain of the Andes is the dominant mountain range with old and recent volcanism. It may be considered the continental arc *sensu stricto*. Its chemical and magmatic signature does not differ very much from that of island arcs, but being on truly continental crust it is much higher and has a deeper root (Section 6.3.2). Farther to the south, the South American margin is by no means uniform. At many places oceanic plates subduct under rather shallow angles. The pattern of volcanoes is not continuous and is probably related to the present accretion of oceanic aseismic ridges or platelets of oceanic or even continental affinity. If these units arrive at the trench they are not easily incorporated into the margin. In fact, trenches disappear, subduction and volcanism seem to stop at these sites, and shallow earthquakes increase in number and intensity. Figure 6.82 gives an impression of the indentation of oceanic platelets into the South American Pacific margin and their relation to volcanic chains (Ben-Avraham *et al.*, 1981). The different radiometric ages and the different rocks along the South American margin suggest that accretion of such anomalous oceanic platelets has played a major role through the long history of that margin. Occasionally, some subduction of edges of margins also takes place.

6.5.3   ACCRETED TERRANES

Accreted terranes along present or formerly active margins have gained increased attention in recent years. Accreted terranes belong to the margins that are active in a certain sense, although sometimes the mechanism of such

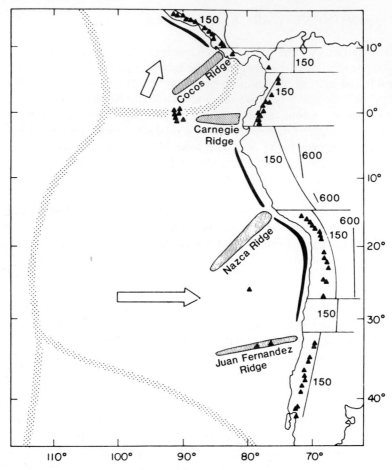

Fig. 6.82   Zones of volcanism and shallow subduction along the Pacific coast of South America. Thick black lines represent trenches, and black triangles represent recent volcanoes. Numbers give maximum depth of Benioff zone. Open arrows show direction of motion. [From Ben-Avraham *et al.* (1981). Copyright 1981 by the American Associative for the Advancement of Science.]

accretion is not well known and real subduction may not take place, as mentioned in the previous section. Evidence for such accretion in the past comes from fossil stratigraphic records and paleomagnetic studies, which show that some "exotic" parts of an old or young margin have paleodirections completely different from those of the rest of the continent. Some units have crossed many thousands of kilometers and suffered various rotations before they reach a continent (Ben-Avraham *et al.*, 1981, 1982; Nur, 1983).

Anomalous oceanic features such as island arcs and aseismic ridges, but also fragments of continents, hot spots, or even oceanic ridges, are all approaching the continent and must be digested. During the collisional stage this may be considered real continent–continent collision, thus creating mountain ranges, nappes, and crustal thickening, often associated with ophiolite belts.

Over 50 such terranes have been identified in western North America; many of them originated at equatorial or southern latitudes. Most of these terranes are composed of rocks that are oceanic in nature, including gabbros, basalts, cherts, and shales. Some others show clear continental affinities. Today there are about 100 rises in the oceans which might be viewed as future allochthonous terranes that are still in migration. The ones that look like microcontinents have crustal depths between 25 and 42 km (Nur and Ben-Avraham, 1978, 1982). Missing magnetic lineations show that they are at least not part of a typical oceanic crust. Traces of hot spots, leaky transform faults, or plume activity have been invoked for the oceanic terranes. Breaking apart of pieces of continents or a massive differentiation of basaltic oceanic units has been postulated for terranes of continental affinity. Throughout the entire Alpine–Himalayan chain, but also in older mountain ranges such as the Appalachians, there is a multitude of continental terranes, ranging in size from island arcs to subcontinents like India or the Arabian Peninsula; such "slivers" often have very different tectonic, stratigraphic, and paleomagnetic histories. As mentioned before and shown in Fig. 6.82, gaps in volcanism along the eastern South American margin mark the present-day process of digesting small anomalous oceanic terranes. The emplacement of ophiolites shows the traces of the consumption of such oceanic plateaus. The greater buoyancy of mature, larger, and more continental units must result in collisional belts such as those which predate the closing of the Iapetus in the Caledonian and Variscan structures, or those in the Himalayas which were formed by many fragments that collided with the Eurasian continents long before India finally arrived at the southern margin of Eurasia.

6.6 SPECIFIC CRUSTAL STRUCTURES

In this final part of the discussion of crustal structures some specific features are selected which have attracted special attention in recent years. Geothermally anomalous (hot) crust may be considered a preparatory stage of rift formation, which, as mentioned several times, might stop anywhere during its development. Such areas of hot crust inevitably generate an additional differentiation and often show an LVZ or an LVB. However, not all LVZs are related to goethermal areas, as will be shown in Section 6.6.2. Fault zones of various ages and their dips, depth range, and composition are selected as a special topic. Their detection provides a challenge for geophysical investigations. They are discussed in Section 6.6.3.

6.6.1   GEOTHERMAL AREAS

Geothermal areas are areas of high heat flow. Strong anomalies are accompanied by volcanic and magmatic activity and all kinds of upward-moving materials and related phenomena such as mantle plumes, mantle crests, diapirs, hot spots, mud volcanoes, geysers and hot water springs, and salt diapirism. Geothermal power plants are all situated in areas connected with volcanism or magmatism in some form, using dry steam for their turbines. District heating may be performed with moderate ground-water temperatures (Ungemach, 1980; Meissner, 1981a).

As shown in Fig. 6.2, heat flow depends on the age of the last essential tectonism of the area under investigation. Various books and reports are written on general problems of geothermics and geothermal energy (e.g., Ungemach, 1982; Buntebarth, 1980; Cermak and Haenel, 1982; Rinehart, 1980) or with specific geothermal areas like the (weak) geothermal anomaly in Urach (Haenel, 1982) or the strong anomaly in Yellowstone (Smith, 1982). In this section only those topics that are strongly related to the crustal structure will be reviewed. Plumes or hot spots generally induce a geothermal anomaly. Basaltic magmas, chemically similar to those of oceanic rifts or islands, are formed in the upper mantle. Magmas rising to the crust may stop anywhere, preferably in the low-viscosity lower crust, where they are still lighter than their surroundings and might generate partial melting. Depending on the tectonic setting, on the material of the lower crust, and on the overpressure of the reservoir, volcanic activity may assume a basic to ultrabasic chemistry of tuffs ejected to the surface by a large number of pipes, as in the Urach area, or may form huge masses of basaltic to rhyolitic suites, as in the Yellowstone area. The effects of crustal heating by intrusions are manyfold:

(1)   The velocities (and densities) of the heated region are reduced, compared to adjacent areas. In the Geysers area a 25% reduction in velocity is reported (Iher *et al.*, 1979).

(2)   A number of related phenomena occur, such as a minimum in the gravity and the magnetic field as well as in the electric resistivity.

(3)   Very high temperatures may result in a partial melting of crustal and intruded material and give rise to rhyolitic magmas and a further differentiation of the crust.

(4)   Seismicity is generally shallower in a geothermal anomaly than in its vicinity, and seismic noise generally shows a maximum.

Based on the different boundary conditions such as crustal material and setting, overpressure, and general magmatic energy, crustal structures also become very different. Two areas, the Urach example, a weak geothermal anomaly in a 30-km sialic Variscan crust, and the Yellowstone area, a strong anomaly in a 40-km basic Precambrian crust, will be discussed.

Seismic data from the Urach anomaly were already discussed in Section 6.2.4 and illustrated in Figs. 6.38 and 6.39. The center of the geothermal anomaly shows a heat flow of only about 90 mW/m². This is only 50–70% higher than found in its surroundings. The center is situated in the middle of a great number of volcanic pipes, filled with olivine nephelinitic tuffs of Tertiary age. Some dykes and some maar-type structures are present. Magmas are apparently from great depth, i.e., about 80 km (Maeusnest, 1982), but mantle fragments or any granulite–gabbroic rocks are missing. Indeed, ejected fragments of granite, granodiorite, and gneisses are found, occasionally showing evidence of partial fusion (Wimmenauer, 1982). This means that there must have been some additional magma chamber in the crust, where upper crustal material was melted and/or incorporated in the explosion-like volcanism.

However, the small volume of massive volcanic material erupted and the low degree of differentiation observed favor the conclusion that there was a rather short epoch of volcanic activity, probably between 11 and 14 Myr ago. The most significant geophysical feature below the Urach anomaly is the pronounced LVB shown in Fig. 6.39.

If the whole crustal velocity anomaly of up to 10% is explained by a temperature effect, a temperature difference of more than 400° could be expected (Bartelsen et al., 1982). This is in contrast to the observed moderate heat flow values, the small positive (!) gravity anomaly, and the moderate anomaly in electric resistivity. The distribution of hypocenters has a minimum, i.e., it shows shallower depth in the area of the LVB, thus indicating the effect of elevated temperatures (Meissner and Strehlau, 1982). It is suggested, hence, that the LVB is still a zone of elevated temperature, but that additional modifications of a former melting process such as thermal cracks, fissures, and individual differentiation into small alternating velocity lamellae contribute significantly to the LVB structure. Enhanced lamellation is definitely present in the lower part of the LVB, as shown in Fig. 6.38. Thermal cracks might have been induced by cooling since about 10 Myr ago. Figure 6.83 shows the lateral extension of the LVB as determined by NVR and DSS (Trappe, 1983). It may be an example of limited thermal activity in space and time acting on a shallow and rather sialic crust.

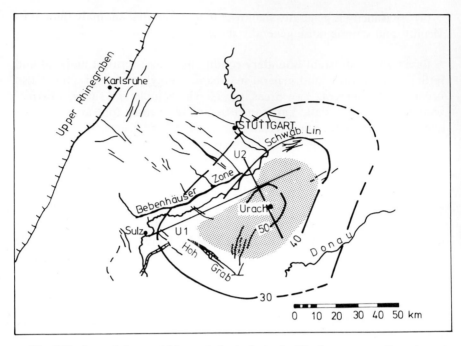

Fig. 6.83 Area of the crustal low-velocity body in the Urach area, a small geothermal anomaly. Compare also Fig. 6.37; Hoh Grab, Hohenzollerngraben. U1, U_2, NVR-surveys. Numbers are temperatures at 1000 m depth. [From Trappe (1983).]

The Yellowstone Snake River plain has also been extensively investigated by geophysical methods, although NVR investigations are missing. The heat flow within the Yellowstone area is about 1600 mW/m², i.e., more than 15–40 times higher than in the surrounding region. The area of the strong heat flow anomaly is rather circular and concentrated around the Yellowstone caldera. It is not much wider than that of the Urach anomaly, i.e., about 70 km in diameter. The volcanic activity, on the other hand, was much higher and covered a much larger area, as seen in Fig. 6.84, where lines and shot points from DSS investigations are shown.

Early Tertiary (?) volcanism was basaltic and produced a wide range of volcanic landforms, including shield volcanoes, plateaus, and numerous small cones and domes, similar in structure and chemistry to those of the diverse Icelandic volcanic types (Greeley, 1982). Silicic volcanism began at about 12 Myr in the southwest and moved to the Yellowstone area, where continuous eruptions of rhyolites and ash flow tuffs were beginning 2.2 Myr ago, interrupted by caldera-forming collapses, the last one at 600,000 years forming the present Yellowstone caldera.

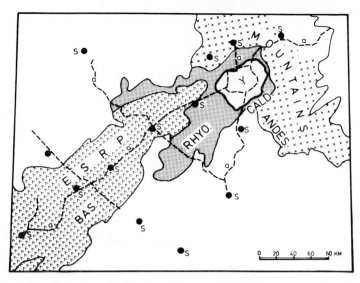

Fig. 6.84 Situation map and volcanic activity around the Yellowstone geothermal anomaly; ESRP, eastern Snake River plain; BAS, basaltic; ANDES, andesitic; CALD, caldera; RHYO, rhyolitic; S, shot points; and — —, DSS lines. [From Smith *et al.* (1982).]

The large volume of silicic volcanic material in an old crust suggests the presence of magma chambers in the uppermost (sialic) crust or a profound melting and differentiation of the deeper crust. Uplift is still going on at a maximum rate of 14 mm/yr.

First indications of a large LVZ and an anomalous mantle came from large teleseismic P-wave delays, indicating a 25% reduction in crustal velocity and a 5% reduction of mantle velocities in the upper 250 km (Iher *et al.*, 1979). For the crust these data could not be confirmed by the DSS studies, the outcome of which was really a surprise. Completely in contrast to the Urach example, a velocity increase was found below 10 km depth relative to the surrounding area (Smith *et al.*, 1982; Braile *et al.*, 1982). Even the Moho shows comparable depth, although its P_n velocity is slightly lower (7.9 km/s). As shown in Fig. 6.85, an anomalous, high-velocity body is seen between about 10 and 20 km below the caldera area. Here, only the upper 10 km and possibly only the upper 5 km show pronounced LVZs with laterally strongly varying values. It is interesting to note that a delay time analysis from shot points as near as 20–60 km (Lehman *et al.*, 1982) shows a very detailed map of LVBs down to 10 km.

Low-velocity zones in the crust below 20 km are incorporated in the *V–z* models of Priestley and Orcutt (1982). However, these LVZs are not very

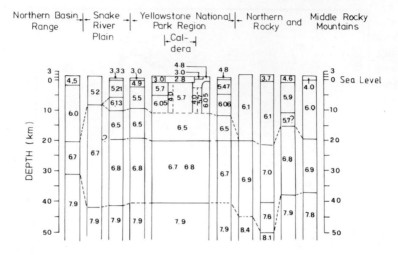

Fig. 6.85   Velocity cross-section below the Yellowstone geothermal anomaly. [From Braile *et al.* (1982).]

pronounced and cannot be clearly defined within the resolution capacity of DSS studies. The P_mP is of excellent quality and P_n also appears as a strong arrival in the record sections.

The gravity picture so far is controversial. A strong positive simple BA of +60 mgal is present within the generally negative background, in accordance with the high-velocity body in the center (Sparlin *et al.*, 1982). The plotted residual BA (Blank and Gettings, 1974; Lehman *et al.*, 1982) shows a −60-mgal minimum within the caldera, apparently in accordance with the LVZs in the upper 10 km with especially low V_P and low-density zones at the caldera margin. These are attributed to zones of intense fracture and/or partial melt. Strong geyser and steam eruptions are related to these areas.

Magnetotelluric soundings in the Yellowstone area revealed a high-conductivity layer in the middle or deeper crust, its upper part ranging from about 10 to 15 km in the southwest to about 20 km in the northeast with rather poor depth constraints. In summary, the Yellowstone geothermal anomaly represents one of the major continental volcanic–tectonic provinces on Earth. There are many apparently controversial observations. Against the interpretation of the anomaly as that of a cooling granitic batholite near the surface, there is the evidence of large teleseismic residuals and high uplift rates. The large volumes of silicic volcanic material seem to originate from the uppermost 10 km, where average P_g velocities are about 5.7 km/s, compared to 6.1 km/s in the surrounding area. The heat for large-scale melting of this layer may be provided by a domelike rise of strongly mafic material (the

6.5-km/s layer) below the center of the anomaly. Alternatively, the 6.5-km/s high-velocity body might be a product of melting of the uppermost 20 km, leaving a mafic body as a residue of a differentiation process. Whatever the reasons are, the Yellowstone area is an example of a very strong thermal anomaly that is now in the upper crust, while the lower crust, the Moho, and the uppermost mantle are not much modified or have already come back to normal conditions. Certainly, more detailed investigations with NVR surveys would shed more light on this unusual anomaly.

6.6.2 LOW-VELOCITY ZONES

It has been mentioned several times that crustal LVZs can be created in different ways. Eight possibilities are mentioned in the following. It appears that the simplest physical interpretation for an extensive crustal LVZ is the temperature effect. As deduced in Section 3.8.4 [Eq. (3.80)], the critical thermal gradient that makes $dV/dz < 0$, i.e., for the appearance of a low-velocity layer, is

$$(dT/dz)_c = \rho g[-(\partial V/\partial p)_T /(\partial V/\partial T)_P].$$ (6.1)

The values $(\partial V/\partial p)_T$ and $(\partial V/\partial T)_P$ have been determined by a number of laboratory experiments (see Section 3.8.4). Christensen (1979) shows that in areas of high heat flow the true gradient is often larger than the critical one, i.e.,

$$(dT/dz) > (dT/dz)_c.$$ (6.2)

This seems to be the main reason for LVZs and LVBs in hot crusts. They are mostly stronger in V_S than in V_P. They appear in young mountain ranges, rifts, and most geothermal areas. Up to about 13% reduction of velocities can be obtained by this effect. If velocity anomalies in hot regions are larger and reach -25% (as in the case of the magmatic Geysers area), almost necessarily partial melt is involved, which might reduce V_S and V_P considerably.

In such hot areas some changes of mineral structure might also take place, such as that of α- to β-quartz, as postulated by Gutenberg (1959). Extensive investigations of this transition have been carried out by Kern (1982a,b), as mentioned in Section 4.1. It should be stressed that the last two reasons for an LVZ—partial melt and the α–β-quartz transition—take place only in extremely hot crusts with very strong dT/dz gradients.

Another reason for an LVZ may stem from areas of high pore pressure. High pore pressure might arise from dehydration reactions at certain depths and might reduce the effective pressure p_{eff} in such a way that an LVZ is created (again see Fig. 4.4). In general, the effective pressure is expressed as

$$p_{eff} = p_L - p_P, \qquad (6.3)$$

where p_P is the pore pressure and p_L the lithostatic pressure. Furthermore,

$$(dV/dz) = (\partial V/\partial p)(dp/dz). \qquad (6.4)$$

If $dp_L/dz < dp_P/dz$, then

$$\frac{dp}{dz} = \frac{dp_L}{dz} - \frac{dp_P}{dz} < 0 \qquad (6.5)$$

and $dV/dz < 0$ because of Eq. (6.4), where $\partial V/\partial p > 0$. Hence a certain increase of pore pressure $dp_P/dz \geq g\rho$ will lead to a critical $dV/dz \leq 0$.

Dehydration reactions generally produce denser mineral phases, thus leading to an increase in pore space and a reduction of bulk volume. Minerals like amphibole, mica, and serpentine are all common in the crust, especially amphibole in the "amphibolite layers" at medium crustal depths. All these minerals dehydrate continuously over a wide range of p-T conditions (Kern, 1982a,b).

Granitic plutons or other rising diapirs with their inherent V_S or V_P velocity being smaller than that in their surroundings might create an LVB. When they intrude sideward like magmas or plume material, they might well create an LVZ.

In zones of extension and subsequent cooling, cracks, fissures, and small faults might lead to a reduction of velocities, as known by the phenomena of dilatancy.

Overthrusting or interstacking of high-velocity nappes or wedges over low- or normal-velocity material may create an LVZ of considerable dimensions. This is a widespread phenomenon in all kinds of compressional zones. It might appear in the upper crust as well as in the mantle.

A final possibility is provided by the traces of major fault zones. They may constitute a *narrow* LVZ themselves, as shown in the example of the Hohes Venn area (Section 6.2.4). This will be mentioned again in the next section. It should be stressed again that not all mapped LVZs seem to be real. Any small delay in travel time branches, which might be caused by a lateral detour of the ray in a slightly inhomogeneous area, is often attributed to a fictitious LVZ. Certainly, confirmation from reversed or overlapping travel time branches or from NVR surveys is needed to delineate a real LVZ.

6.6.3   FAULT ZONES

Fault zones have been observed in old shields as well as in young tectonic areas, where earthquakes concentrate along their trace. Some fault zones cross the whole crust although they are aseismic at greater depth. Unfortunately, fault zones are often hard to observe by seismic or other geophysical methods. First, their limited width, ranging from meters to a few hundred meters, shows that they are generally smaller than, or just comparable to, the seismic wavelength. Steeply dipping faults, like many normal faults near the surface, are observed in DSS studies only if their offset is larger than the dominant wavelength, i.e., at least about $\frac{1}{2}$ km. Reversed or overlapping travel time branches should certainly be used in order to distinguish fault zones from spurious effects of near-surface layers. Steeply dipping faults are also not easily seen in NVR surveys. They often appear in NVR sections as reflected refraction, i.e., as a steeply dipping event with an apparent velocity $V_{app} = V_1/\sin i$ being equal to the velocity in the crustal layers horizontally traversed, e.g., $V_{app} = 6.5$ km/s. The use of geophone and shot patterns, stacking, or even velocity filtering might reduce the amplitudes of reflected refractions considerably. In many cases near-vertical fault zones such as transform faults can be detected in NVR sections by their diffraction pattern. Diffractions are related to the small-scale inhomogeneities of the fault zone itself. An offset of adjacent crustal blocks is another indirect sign of a steep-angle fault.

Shallow dipping and subhorizontal fault zones are excellent candidates for reflectors to be observed by NVR surveys. They are generally not seen in DSS studies. It seems that the NVR method is the only way to detect such fault zones, and ample evidence has been collected from many parts of the world. Figure 6.86 shows the fault zone system of the Wind River overthrust, which

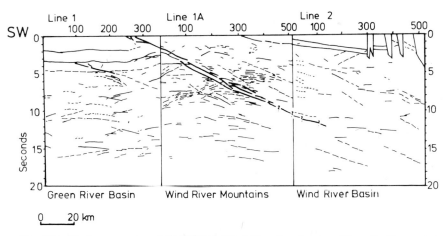

Fig. 6.86   Fault zone system of COCORP's Wind River thrust system. [From Smithson *et al.* (1979).]

can be followed down into the lower crust. It is interesting to note that the older sediments in the east do not seem to be affected by the compressional regime that led to the large overthrust structure. This means that the fault plane itself has transformed most of the horizontal compressional stress into strain rates. It might be well lubricated and show high pore pressures. Figure 6.87 shows an example of the fault zone system from the Caledonian area as found by the WINCH lines of BIRPS.

The Great Glen fault, a remarkable syn- and post-Caledonian transform fault, and some important overthrust faults can be clearly seen, the first one by its diffraction pattern and the others by their direct reflections. Note the appearance of these reflections in the upper and middle crust and again in the mantle, thus leaving the low-viscosity lower crust relatively free of fault-trace reflectors.

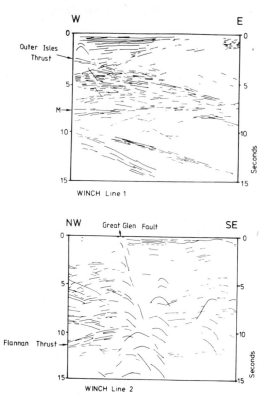

Fig. 6.87   Three fault zones from BIRPS WINCH lines: the Flannan thrust and Outer Isles thrust (top) and the Great Glen fault (bottom). Note the direct detection of low-angle faults (top) and the more indirect signs of steep-angle faults (bottom). [From Brewer *et al.* (1983).]

Figure 6.88 shows a simplified picture of fault zones of different lengths and dips as found by a variety of NVR studies. There are the steep-angle normal faults of young or narrow continental rifts or half-grabens, the shallow-angle overthrust faults, and also the shallow-angle extensional faults of extended and mature rift structures. Whether these follow traces of preceding over-thrust zones—as in the Basin and Range Province—or whether they are expressions of enhanced subhorizontal movements in the low-viscosity lower crust of warm areas is still an open question. Another problem is the nature of the fault zones at great depth and the reason for the reflections, which are often strong and continuous. Petrological observations and *in situ* studies of exposed fault zones show strong mylonitization, i.e., a dramatic disturbance and alteration of the original rock structure. Such mylonite zones occur in the middle or lower crusts. They are sometimes only a few meters wide but may also extend over several hundred meters. Sibson (1984) and Etheridge and Wilkie (1979) have analyzed the shape and the microstructure of such mylonite zones. Their reduced viscosity and other observations were listed in Section 4.5.3. The low-viscosity gouge material, a loose rock flour, gradually turns to so-called cataclasite with increasing depth. Cataclasite is a more compact rock conglomerate with generally smaller grain sizes. Below the generally narrow cataclasite zone the true mylonites appear, apparently created by quasi-plastic processes (Etheridge and Wilke, 1979; Anderson *et al.*, 1983).

Mylonitization generally alters the rock structure and also the density ρ and velocity V. In particular, a stronger orientation of grains must lead to a seismic anisotropy. Such a limited zone of enhanced anisotropy might result in a seismic reflection, even if V and ρ were the same in the average. Lamellation formed by several branches of mylonite zones might further enhance the reflection coefficient.

Jones and Nur (1982) during their laboratory investigations found only some mylonites with an anisotropy greater than 10%. In these samples V_p was found to be slowest perpendicular to the foliation. Most mylonites did not differ remarkably from their host rocks. It seems that from the structural point of view only the presence of phyllosilicates like micas with their sheetlike structures can explain a good reflectivity. In addition, environmental conditions such as high pore pressure or the sandwich structure of several branches of a mylonite zone may lead to the observed continuously strong reflectivity of many fault zones.

Another critical parameter of the fault zone is its width. Anderson *et al.* (1983), investigating the petrology, grain size, and chemistry of the San Andreas fault and the Gabriel fault, distinguish between the upper 3 km, where rather loose gouge material and breccia in a rather wide fault zone are found, a second, rather narrow part between about 3 and 8 km depth, where

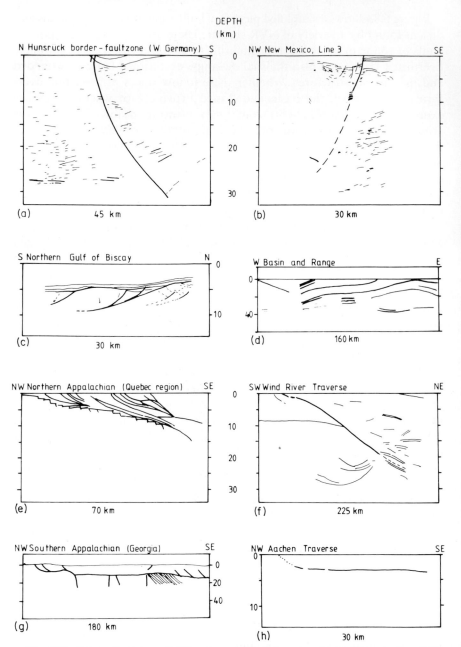

Fig. 6.88 Schematic picture of different crustal fault zones as obtained from various NVR studies. (a), (b), (c), and (d) are extensional stress system ("normal" faulting) and, (e), (f), (g), and (h) are compressional system (thin-skinned or skinned faulting).

cataclasite with a generally brittle behavior dominates, and a third, lower part from 8 km to greater crustal depth, in which true mylonites appear in a widening fault zone where ductile processes prevail.

Figure 6.89 shows such an idealized cross-section of the San Andreas fault. Whether the change in width is really a general phenonemon must be investigated by detailed NVR studies. For the southern boundary of the Rhenish Massif (Fig. 6.42) there is at least an indication of a change in thickness in the previously mentioned way.

Fault zones may contain key information on the stress pattern, structure, and rheology of the crust. Their behavior should be studied through a concentrated effort.

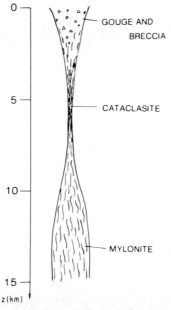

Fig. 6.89 Schematic cross-section through a strike–slip fault as derived by Anderson *et al.* (1983) for the San Andreas fault zone.

Chapter 7 | The Evolution of the Continental Crust

7.0 INTRODUCTION

Several observations of crustal structure and thickness, age provinces, and petrology have been discussed in the previous chapters. In particular, in Sections 6.1–6.3 an age pattern for major geologic provinces was established. Continents were introduced as being composed of progressively younger belts around some old nuclei. A difference in composition also seemed to emerge when regarding velocity–depth structure of crusts of different age: the older units being rather mafic (with higher average V_P velocities), the younger ones being more sialic with lower V_P velocities and a smaller crustal thickness. This is certainly a point to be investigated. While crustal thickness apparently *decreases* during evolution, average lithospheric thickness certainly *increases*. Another interesting subject is the rate of crustal growth. As mentioned in Chapter 4, crustal growth is related to the depletion of the mantle. However, in order to arrive at a more qualitative picture, processes of vertical crustal accretion in island arcs and continental arcs must be judged in terms of new crustal mass produced versus recycling processes at suture zones. Whether lateral accretion of platelets and microcontinents enlarges the total crustal volume must also be checked. Certainly it reduces the crustal surface in crustal shortening processes.

Many books and special issues of journals are devoted to processes of crustal evolution, most of them emphasizing the geologic–petrologic side of the process (e.g., Windley, 1977; Tarling, 1978; Strangway, 1980). O'Connell and Fyfe (1981) and many articles by Wasserburg and De Paolo in various journals review physical and geochemical constraints. Geophysical treatments are found in Mueller (1977, 1978) and Meissner (1983).

There is no doubt that real crustal growth is based on extraction of material from the mantle by melting and magmatic extrusions and intrusions. This global and even planetary differentiation and accretion process must be weighed against the process of remixing caused by the early meteorite bombardment and by vigorous convection. The presence of any convection cells means that vertical and horizontal movements took place from the beginning, even before differentiation. Before a model for crustal evolution is presented, some of the well-established and less well-established preconditions of crustal evolution will be introduced.

7.1   PHYSICAL AND SOME PETROLOGICAL BOUNDARY CONDITIONS

As mentioned in Section 6.1 when we dealt with the crustal structure in the Archean, one of the most fundamental and well-established differences between the early Earth and the present Earth is the heat production (see Fig. 6.2). Unfortunately, some of the heat-producing radioactive elements were still in the mantle or in a state of vigorous mixing. In addition to radioactive heat production in the early Earth, there was more heat left from the planetary accretional peak and core formation, although a quantitative assessment of those sources depends heavily on model calculations. It might suffice in the present context to state that the Earth was certainly hotter in its early times and lost its heat mainly in the beginning by a variety of convective and volcanic processes.

In the following, various consequences of an increased temperature in the outer layers of the Earth will be discussed. Because the temperature at the surface was about the same as today, the first consequence of the increased heat regime is a stronger average temperature gradient, i.e., a higher heat flow. More complicated models of thermal history also show the increased heat flow of the early Earth (Sharpe and Peltier, 1979). The next consequence of the higher temperature is that any plates that were formed by cooling of the outer layers must have been thin. The average higher thermal gradient must have led to a shallow lithosphere–asthenosphere boundary, hence to a small plate thickness. However, the high heat of the mantle necessarily led to a vigorous convection. Such a motion will start as soon as a critical parameter, the nondimensional critical Rayleigh number R_c, is surpassed. In convection processes the Rayleigh number R is always greater than R_c:

$$R = g\alpha\beta\rho d^4/h\eta > R_c, \tag{7.1}$$

where g is gravity, α the coefficient of thermal expansion, $\beta = dT/dZ$ in excess of the adiabatic value, d the thickness of the convecting layer, h the thermal diffusivity, ρ the density, and η the viscosity.

Inserting reasonable values for these parameters, convection is certainly possible, even today and even if the thickness of the convecting layer is only 300 km. The value of R_c is about 1700 for fixed boundaries (Cathless, 1975) and goes down to 700 in the models of various authors. All model calculations for convection in the early Earth produce a strongly reduced thickness of the thermal boundary layer (plate) and a very strong dT/dz value for the upper 30–50 km with a large deviation on top of rising and descending parts of convection cells. The stronger temperature gradient of the early Earth and especially the lower viscosity (η) values must have led to higher Rayleigh numbers and a vigorous convection with a greater irregularity and a greater number of convection cells. Widespread melting on top of rising parts of convection cells, dynamic plumes or hot spots, created many patches of differentiated mantle material: the first crusts. In the earliest part of the Earth's history all of these patches must have been recycled back into the mantle because of the much higher drag forces, related to faster movement than today. Apparently some of these crustal patches survived from 3.8 Gyr ago. In this time period crustal growth (CG), understood as the difference between crustal accretion (CA) and crustal recycling (CR), was still small:

$$CG = CA - CR \gtrsim 0 \tag{7.2}$$

and

$$CG = d(CV)/dt \tag{7.3}$$

where CV is the crustal volume, although both CA and CR were much greater than today. A schematic picture of Eqs. (7.2) and (7.3) is shown in Fig. 7.1. Plates carrying the first crusts must have been thin and may have been restricted to upper crustal levels in the beginning.

Downgoing parts of the convection cells dragged the plates down, back into the hot mantle, a process that in a way was resumed later by modern subduction, where, however, not the drag forces but the negative buoyancy of the cooling and shrinking plate provides the force for recycling.

Convection in the early Earth certainly was more irregular. Hot areas on top of rising cells and cooler parts on top of sinking cells may also explain the different temperature gradients that have been postulated for early Archean areas. Jarvis and Campbell (1983) emphasize the high temperature gradients obtained from the high MgO content of the strongly mafic "komatiites" in

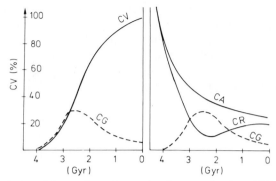

Fig. 7.1 Evolution of the continental crust showing crustal volume (CV) and crustal growth (CG); $CG = d(CV)/dt$; $CG = CA - CR$, where CA is the crustal accretion and CR the crustal recycling.

the Archean, which is related to a high liquidus temperature T_m on the basis of high-pressure melting experiments. The liquidus is estimated to be about 1650°C (Bickle, 1978).

On the other hand, geothermal gradients calculated from Archean metamorphic assemblages indicate only slightly larger temperature gradients than today (Burke and Kidd, 1978). This discrepancy may easily be solved by assuming that komatiites are created on top of rising convective motions, which may have come from great depth, the first continental patches accumulating much nearer to downward-directed cells (as do modern magmatic arcs). The beginning of crustal differentiation and the appearance of felsic patches in the upper part may have contributed to an increasing buoyancy and a certain resistance toward crustal recycling.

Another consequence of the generally higher temperature of the early Earth is related to the amount of (crustal) material extracted in a certain melting process. Magmas were first strongly mafic, like the primitive magnesium-rich komatiites. A much larger percentage of partial melt was extracted from the mantle and led to huge gabbroic crustal thicknesses, which are still observed in some old nuclei on the earth and on the terrae of the Moon and the terrestrial planets. The density of primitive magmas was higher than today. It was also higher than that of the first primitive crust in most places and could only penetrate the lowermost part of these units.

Subsequent heating events have created an advanced differentiation of crustal material on Earth, with light upper crusts, mafic lower crusts, and shallower mantles whose uppermost part is made of the ultramafic residues of the melting process. With the decrease of the drag forces of convective movements and with an average increase in the regularity and radii of convection cells, crustal recycling decreased. The horizontal part of a convective cell does not contribute to crustal growth, but it must have led to first collisions, underplating, and interstacking. The gradual crustal growth with time [Eq. (7.2)] is reflected in the overall mantle depletion of those volatile elements that enter the crust in differentiation processes, as discussed in Section 4.6.4. The initial values (I values) of various radioactive elements can discriminate between magmas derived from the mantle and those obtained by crustal fusion at depth. In particular, ε values of the Sm–Nd and Rb–Sr methods, introduced in Section 4.6.5, give clear evidence that crustal growth is still going on, although the growth rate is decreasing. This will be discussed further in Section 7.3, after a review of the geological observations.

7.2   GEOLOGICAL OBSERVATIONS

The geological observations on rock units of different ages, together with the physical constraints, provide the basis for models of crustal evolution. A mafic-to-felsic compositional progression of magmatic units, the decreasing proportion of volcanic to sedimentary rocks, and the increasing appearance of oceanic ophiolites are only some examples of the changes in crustal composition and structure. In the following, geological observations will be discussed for four selected age groups.

7.2.1   THE FORMATIVE PHASE (4.6–3.9 Gyr)

For the pre-Archean period there are reliable geological–planetological observations only from lunar studies, mainly those of the Apollo Program. Only those features of lunar crustal research will be mentioned which can be transferred to the development of the early Earth. The heavy bombardment between 4.6 and 3.9 Gyr ago must have hit the early Earth with even greater force than the Moon because of the Earth's stronger gravity, although the Earth's early atmosphere may have filtered out the small meteoroids and the Earth's hydrosphere reduced some of the impacting power. This bombardment, also termed the "accretional tail" of planetary accretion, caused severe mixing of the upper layers and shifts the radiometrically determined ages of the lunar terra crust to a peak at about 4 Gyr. The terra crust, a thick feldspar-rich layer covering about 80% of the lunar surface, is commonly explained as a result of cooling and fractional crystallization of one or more large magma bodies. Flotation of crystallizing feldspars in a large magma ocean of several hundred kilometers thickness would result in the formation of an anorthositic-gabbroic crust such as that observed in the terra region. A similar process is supposed to have acted in the early Earth, where high degrees of melting and high temperatures, greatly exceeding the solidus, must also be assumed for the first several hundred million years. However, on Earth cooling was slower, plates stayed much thinner, convection power was greater, and the meteoritic impacts were stronger. All these processes led to strong mixing and recycling of any differentiated material. On the Moon, cooling and thickening of surface plates occurred so rapidly that any plate tectonic movements must have been restricted to the first stage of heavy bombardment. For a detailed study of lunar and planetary tectonics see Head and Solomon (1981), Guest and Greeley (1977), or Meissner (1983). As mentioned in Section 7.1, no pre-Archean terrain survived on Earth. Recycling of very thin and unstable first plates with some crustal patches matched crustal accretion in this time of most vigorous convection and heavy bombardment (see again Fig. 7.1).

7.2.2 THE ARCHEAN (3.8–2.5 Gyr)

The strong decrease in meteoritic impact density observed at 3.9 Gyr on the Moon nearly coincides with the time of the first appearance of crustal patches on Earth—which, by some lucky coincidence, have survived until today. "Survived" in this sense means that no recycling and no complete melting have affected these early domains. From the small scatter and the good coincidence of the $^{143}Nd/^{144}Nd$ I values with the CHUR curve (Section 4.6.4) it is clear that the mantle was still rather homogeneous and not depleted. This means that it cannot have formed large areas of continental crust which were not incorporated into the convecting mantle again, an observation which coincides with the small patches of old crusts that survived from the early Archean. Two major periods of Archean crustal activity can be recognized: 3.8–3.5 Gyr and 2.9–2.5 Gyr. The first continental segments that can be recognized today consist of granite–greenstone associations and gneissic terrains (Kroener, 1979, 1981). They are widely dispersed over the Earth, and most of them consist of tonalitic orthogneisses with I values of Sr and Nd close to those of an undepleted mantle evolution. There is a large variety of mafic to felsic metavolcanic remnants. At least the mafic and ultramafic complexes like the Mg-rich komatiites must have been derived directly from the mantle; their I values are all very close to the undepleted mantle curve. Repeated melting would leave behind a depleted granulite facies rock, as is actually observed in many deeply croded granite–gneiss terrains.

It is most interesting to see that the high-grade metamorphic gneisses are surrounded by low-grade volcano–sedimentary sequences, the so-called greenstone belts (Brown and Musset, 1981); see Fig. 7.2b. Greenstone

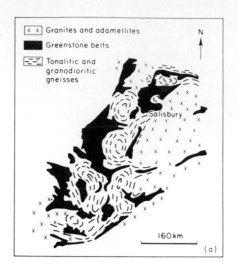

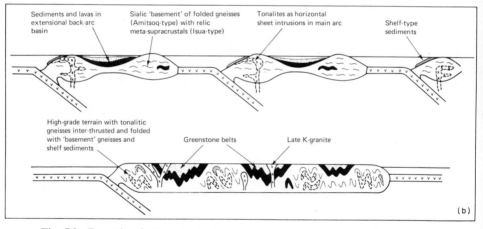

Fig. 7.2 Example of (a) early Archean high-grade gneisses surrounded by low-grade greenstone belts and (b) growth of continents in Windley's (1977) mobilistic view.

sequences vary considerably in thickness, composition, and metamorphic grade. They are long, narrow belts and might have been created in narrow and rather shallow basins. Windley (1977) considers these basins to represent the equivalent of modern back-arc structures which were later compressed in a collisional stage to form the first microcontinents (Fig. 7.2b). There is no doubt that the erosional cycle was also present. Granitic magmas and plutons, some of them coming from a hot lower crust, entered the high-grade and low-grade areas at several locations. A more static model was presented by Fyfe (1974, 1978). According to this view, granitic domes form on top of hot spots. Formation of greenstones by the eruption of recycled crustal material takes place between the granitic domes in areas of crustal thinning on top of descending convection cells. However, the different geothermal grades in the Archean, which have been derived by petrological arguments (see the previous section), and the inavoidable postulation of strong convective motions favor Windley's mobilistic concept. The highly mobile microcontinents could easily collide with each other and thicken considerably, leading to deformation and metamorphism, in agreement with the geological observations of orogenic reworking of preexisting crust. From petrological reasoning, based on the grade of metamorphism and on present-day seismic observations, the continental Archean "crust" was about 50 km thick on the average and certainly comparable to plate "thicknesses."

Fyfe's concept of a thickening of crustal parts by a strong and continuing melting of dense underlying hot spots contains the assumption of a considerable thickening of gabbroic crusts by continuous magmatic "underplating" in the first place, similar to processes in the Moon. On Earth, however, various melting events differentiated the crust later and hot spots would more easily generate crustal thinning. Kroener (1977) emphasizes the lack of rocks of ophiolite affinity such as blueshists in the Archean nuclei of Africa and postulates ensialic orogenies. It seems, however, that blueshists, which require a high p–low T environment, could not have been created when *shallow* underplating and indentation of the first sialic platelets took place and Benioff-type subduction did not occur. Ensialic orogenies, on the other hand, may well have taken place in some nuclei of the Archean. Crustal ensialic basins and crustal thinning may develop, e.g., on top of diverging movements below. Whether large parts of oceanic crust are formed depends on the degree and the time period of convective movements. Later compressions may really involve thinned, predominantly sialic, crust, as possibly in some part of the Variscides. A significant proportion of sediments are chemical precipitates of silica and iron, deposited in shallow water and free of clastic deposits.

It follows from the different and diverging geological findings and from the theories that the early Archean crustal segments were places of highly different development, both tectonically and laterally. A common feature is the occurrence together of high-grade metamorphic gneisses and surrounding low-grade volcano-sedimentary greenstone belts, apparently both within the blocks and around the block margins, both so intimately related to each other that each must have influenced the other. Thermotectonic events have produced such diverse phenomena as linear mountain belts and even nappe displacements (platelet collisions?) on the one hand, and broad, thermally affected domains magmatically intruded, yet tectonically not much deformed, on the other hand. In the late Archean, at around 2.8 Gyr, a more systematic pattern of old cratons surrounded by initially mobile belts develops, which then seems to dominate crustal accretion over most of the Earth's history (see again Figs. 6.3 and 6.7).

According to Veizer and Jansen (1979) crustal growth was slow until 3 Gyr, and only about 12% of the present-day continental area was covered by crustal segments at that time. That is, 12% of the finally surviving crustal segments can be made out, a figure that is compatible with the mantle depletion curve. Actual crustal coverage may have been much higher, although toward the end of the Archean resistance of the growing microcontinents to recycling may have critically increased. Cratonization occurred at an accelerated rate between 3 and 2 Gyr, i.e., from the late Archean to the early Proterozoic (Goodwin, 1981), and broad shelves could develop. The overall tectonic style of the Archean is determined mainly by thin and small plates and still vigorous convection below (Goodwin's "interplate tectonics").

7.2.3   THE PROTEROZOIC (2.5–0.7 Gyr)

A widespread phase of stabilization with massive granitoid emplacements led to the division between the Archean and Proterozoic at 2.5 Gyr. Crustal growth apparently reached its peak during the early Proterozoic. Separate peaks of orogenic activity can be recognized, such as those depicted in Figs. 6.3 and 6.7, where apparently large mobile belts were attached to the old nuclei. Whether these belts were created on top of convection-driven subduction zones or formed by some other large-scale, linearly arranged melting events is still an open question. In any case, these belts or microcontinents were later driven toward one of the nuclei and finally attached to them in a number of distinctly episodic orogenic events. The episodicity might be a function of a changing convection pattern. The increased growth rate is compatible with the deviation of the Nd and Sr initial values from the undepleted mantle curve, showing that by now an appreciable part of the mantle is involved in quasi-permanent crustal segregation. With progressive cooling of the mantle, geotherms would migrate downward and the peridotite solidus would move upward, partly by degassing of the mantle (Piper, 1978). There is no more mantle-wide melting with small-scale circulation, but a transition to larger convection cells with slower movements. The first biological explosion could take place because of the generation of large shallow marginal basins and shelves, greatly enhancing the photosynthetic production of O_2. While in the Archean the first banded iron formation (BIF) bears evidence of only a small O_2 supply to shallow sediments, the onset of massive red beds at the beginning of the Proterozoic marks a changed O_2 level. The time period from 2.5 to 2 Gyr is the time of greatest accumulation of sedimentary iron in the Earth's history. Great BIFs appear worldwide (Goodwin, 1981), possibly in response to a greatly improved exchange between deeper Fe^{2+}-bearing ocean waters from anaerobic zones and the upper oxidizing water of the new shelves. Strongly increased carbon segregation takes place around 2.0 Gyr.

In the early Proterozoic, i.e., from 2.5 to 1.5 Gyr, thick sedimentary–volcanic accumulations in ensialic basins and along mobile belts are a typical feature. In addition to the massive granitoid emplacements, large layered mafic complexes are also emplaced in the stabilized continental crust. Numerous diabase dyke swarms pierce the old cratons. In general, the tectonic pattern changes from the Archean microplate tectonics to "intraplate tectonics." Blueschists and ophiolites are still missing. Benioff-type subduction apparently has not yet taken place. Drag forces had decreased considerably and gravitationally powered subduction could not yet take place because oceanic plates must have been still too hot, thin, and buoyant. Any subduction of growing oceanic plates must have been shallower than today, but because of the still high temperature regime magmatic arcs and partly ensialic back-arc basins could easily have been developed. Activation of the asthenosphere at depth by either steeply dipping Benioff zones or strong vertical convective movements might have had a minimum during the early and middle Proterozoic. Figure 7.3 shows a hypothetical model of Proterozoic active margins, compared to the present average situation.

Tectonic patterns slowly seem to change toward the beginning of the late Proterozoic, around 1.5 to 1.2 Gyr. Lithospheric plate thicknesses approach present dimensions and densities. On the continents episodic events with huge anorthositic–granitic volcanic associations or some of basaltic provenance continue, some of them in a truly nonorogenic environment. The anorthositic and granitic plutonic-volcanic activity especially reached a peak between 1.6 and 1.2 Gyr. Large I values of Sr and Nd observed in many places indicate that quite a number of these magmas were derived from reworking of crustal material. For a long time middle Proterozoic continents resisted fragmentation, but in the late Proterozoic genuine geosynclinal belts appear at the continental margins. Extensive basins developed, filled with coarse continental sandstones, flood basalts, clastic deposits, and conglomerates. At other places rifting of old cratons can be observed for the first time. Paleomagnetic evidence is compatible with the assumption of a huge middle Proterozoic supercontinent (Piper, 1978), which was probably kept together by the continuous underplating of oceanic plates on its margin. However, such compressive forces must have decreased toward the end of the Proterozoic, and the opening of the Iapetus (the northern Pre-Atlantic) may have

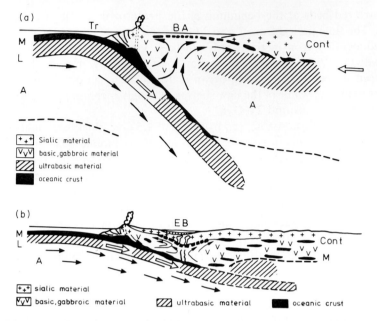

Fig. 7.3 Comparison of conceptual models of (a) today's active margins with (b) an early Proterozoic environment (similar to today's subduction of some young oceanic sections). Oceanic plates were thinner and faster and did not show negative buoyancy; Tr, trench; A, asthenosphere; BA, back-arc basin; EB, ensialic basin; Cont, continent; M, Moho; and L, Lithosphere.

started as early as 1000 ($\pm$200) Myr ago. Toward the end of the Proterozoic the first ophiolites and blueschists also appear, signs of a true Benioff-type subduction process, see Fig. 7.3a. The downgoing oceanic plates mobilized the asthenosphere. Different tectonic styles are found during this transitional period. However, the change to modern plate tectonics with its *whole* spectrum of rifting, marginal basins, trenches, Benioff-type subduction, island arcs, collision, and paired metamorphic belts does not become evident before the end of the Proterozoic period.

7.2.4 THE PHANEROZOIC (0.7 Gyr TO PRESENT)

In the early Phanerozoic several mobile continental plates are present which accumulate to (another?) supercontinent—Pangea—at around 0.3 to 0.4 Gyr. This supercontinent, which must have blocked mantle heat flow and heat release over more than one-third of the Earth's surface, became fragmented (again?) at around 0.2 Gyr. Although the radiometric heat production further declined, a very intense thrust tectonic took place in the last 0.6 Gyr, marked by huge collisions of plates, interstacking of crustal units, subduction, suturing, and even obduction of oceanic plates onto and into continental margins. Such processes of the Phanerozoic era have already been described in various sections of this book. Many of these processes, like accretion of small patches of oceanic and continental affinity to the continents in mild or stronger collisions, might also have occurred in the Proterozoic, but are obscured. Growing oceanic plate thicknesses led to subduction of old, cold, and dense oceanic plates, often already in front of continents, creating the island arcs and possibly also dragging down some sediments. Increased rigidity of plates and especially of continental crusts led to the transitory huge sialic crustal thicknesses in the young collisional belts. Splitting of huge continental masses by rifting is another expression of still continuing convective motions in the mantle. Smaller plates exhibit stronger horizontal velocities than larger ones. The largest ones seem to be highly immobile. The ε_{Nd} and ε_{Sr} curves of mantle-derived magmas are still increasing, though at a smaller rate. Apparently the growth of continental crust has slowed down considerably. As observed by the decreasing crustal thicknesses and the increasing percentage of (low-velocity) sialic components, crustal differentiation processes are still going on in the Phanerozoic: crusts become shallower and lighter, while plate thicknesses still increase.

7.3 THE MECHANISM OF CRUSTAL GROWTH

The gradual increase of the initial $^{143}Nd/^{144}Nd$ and $^{87}Sr/^{86}Sr$ ratios (I values) of mantle-derived magmas seem to be a general indication that the mantle is still depleting and crust is still being formed, although at a reduced rate compared to the growth curve for the Archean–Proterozoic boundary. Crustal growth in this context is always understood as the increase of volume of crustal material as expressed by Eq. (7.2), i.e., crustal accretion minus crustal recycling. It has been argued that recycling of seawater and of highly enriched sedimentary or granitic material might disturb the isotope ratio of the mantle considerably. For a discussion of this subject see Brown and Musset (1981). The systematic changes of I values with time along many magmatic arcs over time periods of 0.1 Gyr seem to contradict the hypothesis of a large contribution of recycling (see again Fig. 6.53). It appears that the most reasonable explanation for the variation of the ε_{Nd} and ε_{Sr} values is a

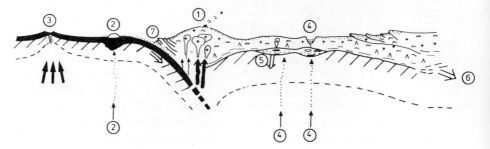

Fig. 7.4 Schematic picture of today's locations of accretion (↑) and recycling (⇓). Numbers are discussed in Sections 7.3.1 and 7.3.2.

quasi-continuous, though episodic, crustal growth over the Earth's history, an irreversible process that has not yet terminated. Nevertheless, there are different types of recycling and also different types of crustal accretion, such as horizontal or vertical contributions. The processes of accretion and recycling may be considered in terms of leaking circles with certain time constants, as will be shown in Section 7.3.3.

Figure 7.4 gives an overview of locations of accretion and recycling. In this simplified picture filled arrows show the localities of vertical (or horizontal) accretion and open arrows show areas of recycling. The various processes will be discussed below.

7.3.1 VERTICAL CRUSTAL ACCRETION

It was mentioned in Section 7.2.2 that early Archean crust in nearly all cases came entirely from the mantle (Moorbath, 1976). The few continental nuclei did not have much chance to collide with each other. Being quickly transported to the relatively cold parts near downgoing convection cells, they probably had only a poor chance to differentiate. Moreover, any upper differentiated part would have long since eroded away. Today, vertical crustal accretion still plays a major role in regional and global growth of the continental crust. As indicated in Fig. 7.4, vertical crustal accretion is envisaged for the following sites:

(1) continental arcs and island arcs (later to be attached to the continental crust),

(2) oceanic islands (later to be attached to the continental crust),

(3) oceanic rifts, creating a basaltic crust, most of it subducted again and converted to melt and eclogite,

(4) continental hot spots, e.g., geothermal anomalies and rifts.

Most of these features were discussed in Chapter 6. Their contribution to vertical crustal accretion in the present context will be explained here. As shown in Fig. 7.4, magmatic arcs, situated along active continental margins, generally develop (1) andesitic volcanism, (2) large granitic batholiths, parallel to the margin, and (3) various amounts of clastic sediments (Brown and Musset, 1981). Continental arcs differ from island arcs mainly in crustal thickness and in the ratio of intrusions to extrusions. In island arcs the formation of newly created magmatic material takes place on top of oceanic crust, and extrusive rocks dominate; in continental arcs magmatic material is added on top of and inside existing continental crust, i.e., in the form of extrusive andesites and—more important—dioritic-granodioritic intrusives, with generally only a small contribution from crustal melting (see Figs. 6.52 and 6.81).

Both island arcs and continental arcs receive their magmatic material from dehydration and melting of the downgoing oceanic lithosphere, probably mainly from the oceanic crust (Ringwood, 1974). The subducting H_2O-rich oceanic crust provides the magmas and the volatiles, which in turn might reduce the melting point of the continental or oceanic asthenosphere.

In terms of major element composition, arc basalts can be most easily produced by direct partial melting of quartz eclogite, the high-pressure form of subducted basaltic oceanic crust. Also, the trace elements, e.g., the pattern of the rare earth elements, can be matched with subducted oceanic crust (Toksöz and Hsui, 1978a,b). How the high temperatures for melting can be produced seems problematic, however, if the melting point is not lowered by H_2O. Diapirism requires about 1000 to 1300°C, values that are hard to obtain without introducing additional heat sources. Refined model calculations by Hsui *et al.* (1983) show a subduction-induced convective current which constantly brings hot mantle material in contact with the subducting surface of the slab, i.e., the oceanic crust, at about 100 to 130 km depth. The effects of both shear heating and dehydration will also contribute to the melting of the oceanic crust at these depths. All three heat sources might be responsible for the sharp volcanic fronts of many arcs, whose generally narrow band of volcanoes seldom exceeds a width of 20 km.

Depending on the buoyancy and the velocity of rising magmas as well as on the thickness of the overlying magmatic crust, magmas might stay hot and erupt, or cool and form teardrop-shaped intrusions. See Fig. 7.5 and compare it with Fig. 6.52. It should be emphasized that island arcs and continental arcs along active margins form the major additions to the continental crust. The continental arcs receive this new material in a direct way; the island arcs suffer a future compressional phase in a shrinking ocean in a mild or strong horizontal accretion. The material itself is *leaking* from the rift–subduction cycle, resulting in a growing depletion of the remaining (oceanic) mantle; this

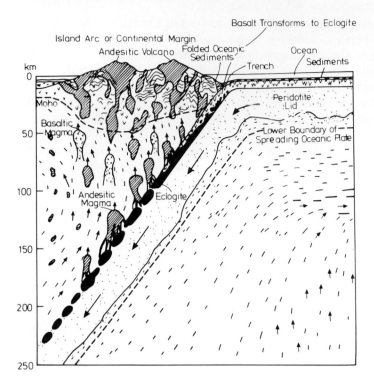

Fig. 7.5 Generalized picture of rising magmas above a subducting oceanic plate. [From Ringwood (1969).]

explains the positive ε_{Nd} and negative ε_{Sr} values of MORBs at oceanic ridges, where the rotating material might appear again after 0.1 to 0.2 Gyr. Brown and Musset (1981) estimate the rate of vertical accretion in the magmatic island and continental arcs to be 0.5 km^3/yr. That is less than half the rate of annual crustal growth needed to produce the present-day volume of continental crust during the last 4 Gyr. As shown in Fig. 7.1 and discussed again in Section 7.3.3, the growth rate was much greater between 3 and 2 Gyr.

Another important consequences of magmatic arc formation is its impact on the episodic growth of continents during all geologic ages. The average chemical composition of arcs with mafic as well as granitic–rhyolitic contributions is similar to those of the successive belts of different ages around the old nuclei. It suggests that these belts of successive age provinces, with time intervals of more than 1 Gyr at places, were all created by direct or indirect vertical accretion. This would mean that magmatic material leaking from a downward-directed part of the oceanic lithospheric crust—transported downward by convection or present-day subduction—always formed the major part of crustal accretion and crustal growth. The ε_{Nd} and ε_{Sr} values of such magmatic arcs can be used for an assessment of the material that has really been extracted as *new* material from the mantle compared to that of *preexisting* crust. Investigating a Proterozoic batholith about 1.6 Gyr old on a bordering 1.8-Gyr-old craton, it seems that it is really a mixture of mantle-derived magmatic material and preexisting sediments or metasediments from the border zone. Figure 7.6 shows that about 50% of the ε_{Sr} and ε_{Nd} values cluster at the average island arc values. Other data points are more scattered and show the typical negative ε_{Nd} and positive ε_{Sr} values of continental crusts and sediments (De Paolo, 1980a, 1981a) with an age of 1.6 Gyr.

Compared to the incorporation of magmatic arcs into the continents, the contribution from oceanic or continental hot spots is relatively small. As mentioned in Section 6.5.3, there are many aseismic ridges, i.e., volcanic oceanic island and other fragments "swimming" in the oceanic lithosphere. According to Ben-Avraham *et al.* (1981) they make up about 10% of the area of today's ocean. They were also extracted from the mantle, possibly from deeper layers than the MORBs, as seen by their overall calc-alkaline composition and their lower depletion compared to the MORBs. Some of them (already?) acquired a continental affinity, with sialic upper crusts. They are eventually carried to an active continental margin and incorporated into it in a collision zone or possibly in a mild accretional process, especially if accretion is not perpendicular to the continental margin and a more lateral attachment takes place. It seems that vertical accretion by hot-spot volcanism and segregation of crustal material on top of irregular plume patterns was much more important in the hot mantles of the Archean and possibly in the present Venusian mantle (Meissner, 1983).

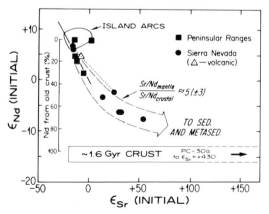

Fig. 7.6   Display of the ε_{Nd} and ε_{Sr} initial values from a Proterozoic batholith. The transition of island arc values to those of crustal origin is clearly defined. [From De Paolo (1980b).]

Hot spots and anomalous thermal regions are also found inside continental plates. A minor contribution to crustal accretion certainly comes from these continental hot spots. It was mentioned when discussing continental rifts and geothermal areas that new material from the mantle is intruded directly into the continental crust. Basaltic plateaus, huge volcanoes, or "only" intrusions of variable dimensions are all added to the crust to form a crustal accretion. Most important may be mantle heating and the intrusions of basaltic magmas into the low-viscosity lower continental crust, causing uplift and later possibly a geothermal area or a rift. The many "pillows" with seismic velocities around 7.5 km/s and gradual crust–mantle transitions in the young and narrow rifts (Section 6.4.1) together with the huge extrusions provide clear evidence for vertical crustal accretion. Many of the tholeiitic magmas are denser than the material of the upper and middle crust and will tend to accumulate in the lower crust, possibly in the form of ponds which later differentiate.

Herzberg *et al.* (1983) have investigated the density of magmas from the mantle and compared them with those of crustal material. They arrive at the conclusion that the (primitive) komatiites and the MORB tholeiites are denser than the average material in the lower crust. The crust, hence, acts as a density filter, accumulating the dense mantle-derived magmas at its lowermost level but permitting andesites and other light lavas to arrive at upper crustal levels; see Fig. 7.7. This process leads to an additional differentiation of the crust with each magmatic event, while ultramafic residues are returned (again) into the mantle. Additional differentiation and cooling of such areas, on the other hand, lead to subsidence.

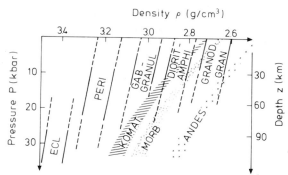

Fig. 7.7 Comparison of the density of various magma types to those of upper-mantle and crustal rocks; ECL, eclogite; PERI, peridotite; GAB–GRANUL, gabbroic-granulites; DIORIT—AMPHI, diorite-amphibolite; ANDES, andesite; GRANOD-GRAN, granodiorite-granite; KOMAT, komatiites; and MORB, mid-oceanic ridge basalts. [After Herzberg *et al.* (1983).]

7.3.2 HORIZONTAL CRUSTAL ACCRETION

Horizontal accretion may contribute to crustal growth only on a local scale. The general horizontal accretion in collisional zones results in a thickening and shortening of the crust and does not increase the existing crustal volume. Today, horizontal crustal accretion takes place at locations such as:

(1) continent–continent collisional belts,
(2) leading edges of active margins with sedimentary wedges and island arcs and other oceanic structures (7 in Fig. 7.4).

Crustal shortening up to 100% is reported for the large continent–continent collisions, where the crust of one of the colliding continents generally follows the downward motion of the previous oceanic subduction, as was shown in Fig. 6.49 for the Himalayan collisional belt. The main problem of these simplified pictures of compressional tectonics is getting rid of the subcrustal lithosphere of the overriding plate. As observed for the Himalayas and Tibet, the crust is doubled but there is only *one* subcrustal lithosphere below. A possible solution for the doubling of the crust is delamination of one crust within the weak and low-viscosity lower continental crust. The subcrustal lithosphere of one plate with its rigid uppermost part is delaminated and led downward toward the asthenosphere, where it might be resorbed. Such a hypothetical model, i.e., a pattern of "digestion," is shown in Fig. 7.8.

Some smaller collisions preceding the main Indian–Eurasian collision, on the other hand, have thermally weakened the whole lithosphere in the area and mobilized the asthenosphere so much that an incorporation of the subcrustal regions into downgoing convection cells might also be possible. However complicated such a collisional process may be, the orogenies are areas of crustal digestion, i.e., locations of negative growth or crustal recycling, as shown in Section 7.3.3. Local and regional accretion may certainly take place at the leading edges of continents in places where huge sedimentary wedges are scraped off the subducting oceanic crust. However, these sediments come from the continent and, hence, do not contribute to net crustal growth on a global scale.

Accretion of oceanic platelets like island arcs, volcanic islands, aseismic ridges, and other oceanic structures also does not contribute to crustal growth. These structures have already been created by vertical accretion, as shown in Section 7.3.1, and are now only transported horizontally to the continental margins. Because of their buoyancy, they are not easily subductible. They produce a limited continent–continent collision, which is strong if the relative velocities are high, if the platelets are large, and if their direction is about perpendicular to the margins. If their approach is more oblique they

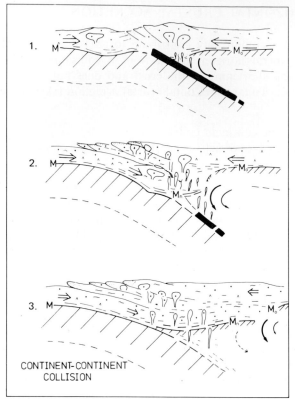

Fig. 7.8   Conceptual model of a Himalaya-type collision belt as an example of interstacking and crustal shortening. M_o, old Moho; M_n, new Moho.

are a bit more "silently" attached to the continent, creating the long strike–slip faults observed along many of the continental margins.

7.3.3   CRUSTAL DIFFERENTIATION AND RECYCLING

Looking back at Fig. 7.4, we see that there are basically three candidates for recycling:

(1)   a general crustal differentiation of hot, but cooling, crusts at and after thermotectonic events (5 in Fig. 7.4);

(2)   subduction zones, where not only oceanic crust but also some sediments or gneisses from the continental crust (CC) are being subducted;

(3)   suture zones in compressional belts, where crust is led to greater depths and differentiation processes (6 in Fig. 7.4).

As demonstrated in Fig. 7.1, recycling of crust limits crustal growth and was most effective in the formative phase and the Archean. The decrease of drag forces with time and the increasing buoyancy of the growing and differentiating CC also must have limited the amount of recycling. Nevertheless, as will be shown at the end of this section, *some* recycling of continental crust takes place on a global scale. Crustal differentiation occurs if part of the crust is heated by intruding magmas or just by conduction from underlying hot parts of the mantle. A light granitic–rhyolitic melt fraction will separate from a mafic or ultramafic residue. Even the lower continental crust consisting of granulites or amphibolites might melt, especially if some volatiles, e.g., H_2O, come up with a basaltic magma from below, e.g., from a subduction zone or another anomalous region in the underlying mantle. By means of several or multiple melting events the crust becomes increasingly light and sialic. The granitic–gneissic material of the upper crust increases in volume and the granulitic–gabbroic material of the lower crust decreases in thickness. The residues may form a new, shallow Moho on top of the old one.

This process of renewed crustal melting and differentiation is shown schematically in Fig. 7.9, which illustrates the effect of differentiation by heat supply only. Differentiation upon such a thermal event may transform the typical crust of a shield into a more differentiated crust with a thickness of only 30 km. More quantitative examples are shown in Fig. 7.10. The whole process might develop completely isostatically. (It was mentioned in Section 6.2 that the newly created shallow crust west of the Russian platform is isostatically balanced with regard to the shield crust.) Introducing the boundaries of Fig. 7.10, isostasy between the two blocks is achieved if

$$\rho_2 h_2 = \rho_2 h_2^* + \rho_1 \, \Delta h_1 + \rho_3 \, \Delta h_3 \tag{7.4}$$

with $h_2^* = h_2 - \Delta h_1 - \Delta h_3$. This equation transforms into

$$\Delta h_3 = (\rho_2 - \rho_1)/(\rho_3 - \rho_2) \, \Delta h_1, \tag{7.5}$$

where ρ is the density of different layers, h the thickness of different layers, and Δh the change of thickness. This means that by melting and differentiation of the (granulitic–gabbroic) h_2 layer, an additional sialic column (Δh_1) and an additional ultrabasic column (Δh_3) are created, Δh_1 increasing the sialic upper part. Therefore, the thickness of the uppermost mantle also increases and the total crustal thickness decreases. In the example of Fig. 7.10 $\rho_3 - \rho_2$ and $\rho_2 - \rho_1$ are taken as 0.3 g/cm^3 in accordance with density values of 2.7 g/cm^3 for the upper crust, 3.0 g/cm^3 for the lower crust, and 3.3 g/cm^3 for the uppermost mantle. The V-z profiles presented in this figure are comparable with those of shield areas, e.g., in the Russian platform, and those of the young areas accumulated later like the west European continent (see

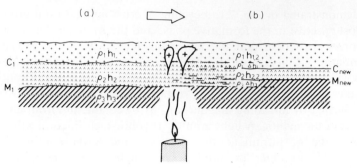

Fig. 7.9 Differentiation of an ancient mafic lower crust (a) into a light, sialic melt fraction and an ultramafic residue by supply of heat only. The outcome is a differentiated lighter and shallower crust (b); ρ, density; h, thickness of layers; C, Conrad; and M, Moho.

Fig. 3.5). As mentioned in Section 3.1.4, the nearly linear relationship between ρ and V results in equal areas for the difference between the two V–z curves. Some V–z profiles presented in Fig. 6.27, e.g., V17–V20, V25, V26, and V28, do not show any convincing contribution of "typical" velocities of the lower crust ($V_P > 6.4$ km/s). The disappearance of these layers in some areas is a mystery, but one explanation is the complete melting of lower-crustal

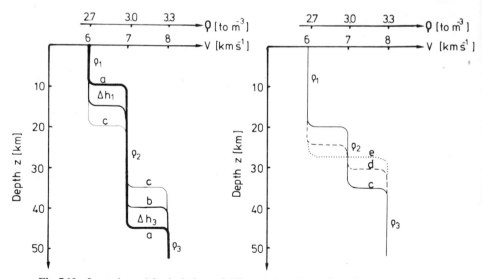

Fig. 7.10 Isostatic model calculations of differentiation of a mafic ancient lower crust into a low-density (low-velocity) sialic (upper) part and a high-density (high-velocity) ultramafic residue, to be added to the mantle as Δh_3. For the diagram $\rho_2 - \rho_1 = \rho_3 - \rho_2$ and hence $\Delta h_1 = \Delta h_3$ [Eq. (7.5)]. A *complete* melting of the lower crust with density may transform a shield crust (curve a) to curves d and e.

material, thus creating the thin crust (less than 30 km) and its totally sialic composition.

Concerning recycling, it is interesting to note that the ultramafic residue of the melting and differentiation process is incorporated, i.e., recycled, into the mantle. It is depleted of most volatiles and not easily remelted again. The observed velocity structure of the uppermost mantle in DSS studies of many areas may be attributed to such a layering, the upper part being caused by young cumulates from the crustal remelting process.

In general, remelting of the crust with its necessary high temperatures cannot be performed without magmas from the mantle, which enter at least the lower low-viscosity crust. This process was recently investigated by Kushiro (1980), who found that tholeiitic basaltic melts are heavier and denser than plagioclase at deeper crustal levels and lighter than plagioclase at upper crustal levels. As mentioned in Section 7.3.1, the consequence of this behavior would be the creation of ultramafic cumulates at the bottom of a magma chamber in the lower crust and feldspar-rich material at its top. Hence, the crustal thickness might even decrease although material is added to it. A part of the newly intruded magma is incorporated into the cumulates, which may then form the uppermost part of the mantle. The crust-mantle pillow during the heating maximum of rift processes also may differentiate into a lighter and shallower crust and an ultramafic residue if the rift process terminates at an early stage.

Locations of possibly major recycling are the oceanic trenches, where subduction may drag sediments down. This process is facilitated by the fact that the approaching oceanic plate is broken by bending stresses. This breaking into graben (and remaining horst) structures starts seaward of the trenches in the topographic and gravity high, where bending also starts. The grabens, being parallel to the line of subconduction, may easily be filled by sediments from the nearby continent or island arc. There is no doubt that all around the Pacific sediments are trapped in these grabens; this has been especially well observed by multichannel reflection profiles across those active margins where sediment thickness is small or moderate and fore-arc sediment accretion is limited (Isacks, *et al.*, 1968; Hilde, 1983). In the case of a large sediment thickness on top of the approaching oceanic plate the major part is scraped off at the foreland. Figure 7.11 shows two pictures with different sediment thicknesses, different subduction angles, and different segmentation into horst-graben patterns.

The amount of recycled material in these cases is certainly debatable. If sediments in grabens and other traps are subducted, their density might increase by diagenetic and metamorphic processes. If so, they may stick to the downgoing plate and carry Sr-rich, Sm-rich material down into the continental asthenosphere, where it may be melted and remixed again. As a

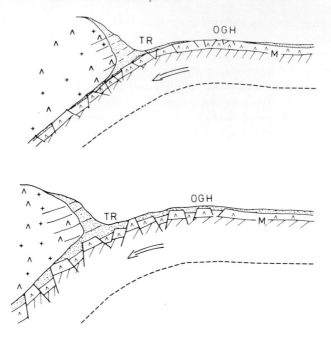

Fig. 7.11 Principle of sediment trapping along bending parts of descending oceanic lithospheres (possible sites of major recycling of sediments). According to many reflection data, the forming of horst–graben structures in the oceanic crust starts at the outer gravity high (OGH), and the grabens are filled with sediments. [After Hilde (1983).]

consequence of the possible remixing, the isotope ratio of erupting lavas may be enriched or at least become of "mature" nature. However, as shown in Fig. 6.53, magmas with such "crustal contamination" mostly occur in regions of thickened crust and may also acquire their crustal component during its ascent to the surface.

Compact continental crust is guided to great depths in the suture zones of continent–continent collisions. As indicated in Fig. 7.8, the lower part of the downgoing crust is easily melted in the high T–high p environment. In some "ponds" considerable ultramafic residues may be created and even recycled to greater depth with the downgoing subcrustal lithosphere of one of the colliding plates, which might also carry some material of the lower crust down and back into the mantle. Such processes seem plausible but have not yet been investigated in detail. Even the asthenosphere may be mobilized and involved in these processes.

In addition, much material of the gneissic–granitic upper crust and sediments are guided down to lower levels. Much of this material is transformed into a sialic–granulitic facies, which may make up the bulk of the lower crust in many areas of former collisions. Unfortunately, such granulites cannot be distinguished from mafic granulites or gabbros on the basis of seismic velocities (Christensen and Fountain, 1975).

De Paolo (1983) tries to estimate the amount of globally recycled material by means of the different isotopic development of Nd and Hf, i.e., ε_{Nd} and ε_{Hf}, for magmas from island arcs and those from oceanic ridges. Assuming that MORBs are similar to average mantle composition, he finds that the actual rate of increase of both ε values is much smaller for the last 2 Gyr than the slopes to be expected from the estimated mother/daughter ratio of the mantle. His estimates for recycling are 35% of crustal mass per 1 Gyr or 2.5 km^3/yr. This would be a factor of 5 *higher* than the estimated rate of accretion in the magmatic arcs and is certainly not compatible with the global depletion curve of the mantle or with any known process of crustal recycling. Compared to this extreme assessment of recycling, most other estimates indicate smaller recycling rates (Reymer and Schubert, 1984). Summarizing all the available data, it seems that recycling of crustal material today is probably still smaller than crustal accretion, as indicated in Fig. 7.1.

A simplified picture of the global oceanic leaking cycle is shown in Fig. 7.12. In addition to the pattern of the oceanic rift–subduction cycle with its leaking branches at the magmatic arcs (accretion), there is the sedimentary

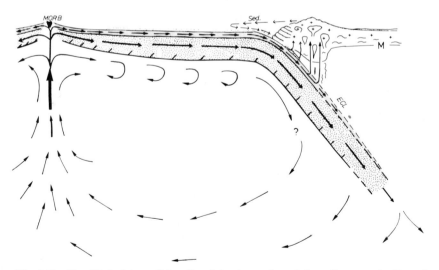

Fig. 7.12 Simplified picture of the rift–subduction cycle and the sediment cycle. Note the leaking sites at the magmatic arcs (accretion) and the subduction; MORB, mid-oceanic ridge basalts; ECL, quartz eclogite; and Sed, sediments.

cycle with its possible leaking branch in the subduction zones (recycling).
Figure 7.13 demonstrates the complicated, predominantly vertical and more
local (or regional) accretion and recycling processes in a section of a "typical"
differentiated, young continental crust.

The final outcome of these processes is the crustal section at the left-hand
side of Fig. 7.13. The more often crustal melting and differentiation take
place, the thicker will become the upper sialic and granitic part, the thinner
the lower mafic and granulitic part, the smaller the crustal thickness, the
lower its seismic velocities and the stronger the reflected seismic waves from
the lower crust and the crust–mantle boundary. In such a mature, e.g.
Variscan, crust reflective lamellae mostly start at the Conrad level and always
end at the Moho. The creation and the subhorizontal ordering of the

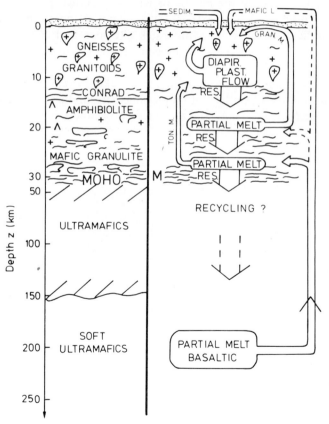

Fig. 7.13 Section of a strongly differentiated, i.e., young, continental crust (left) and
accretional and recycling processes which resulted in its origin (right); based on a concept of
West (1980); RES, residue; L, lavas; and M, melt.

reflecting lamellae take place by gravitational differentiation in a low viscosity environment with a possible contribution of shearing. Lamellae are preserved over a certain time period, probably freezing in by cooling. Their final distortion, as observed in old shield areas, might be controlled by a general time-dependent homogenization, involving slow creep processes. The evolution of crustal thickness is demonstrated in Fig. 7.14. It is based on a statistical evaluation of wide angle and refraction profiles such as those compiled in Fig. 6.13 for shields (SH), Fig. 6.23 for Caledonian structures (CA), Fig. 6.27 for Variscan areas (VA), Fig. 6.50 for present orogenic belts (OB) and other well-known data sets for the Moon, passive margins (MA), and oceanic crust (OC) (see also Figure 1.1). Except for the transient root zones of orogenic belts, which melt away in post-orogenic creep processes, the other crustal units show a definite age dependence as mentioned in various sections of this book, especially in Section 6.1. We may fit a straight line through the data points in a z versus $\log t$ diagram giving

$$Z = 25 \log t - 32 \tag{7.6}$$

with Z in kilometers and t in megayears. The geophysical base for this relationship lies in the multistage development of the younger crusts with their repeated melting and recycling processes and in the more primitive but large extraction of crustal material from the mantle of the old nuclei.

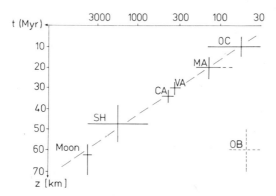

Fig. 7.14   Crustal thickness versus age; OB, orogenic belts; OC, oceanic crust; MA, passive margins; VA, Variscan areas; CA, Caledonian areas; and SH, shields.

References

Adám, A. (1980). *J. Geomagn. Geoelectr.* **31**, 1–46.

Adám, A., and Wallner, A. (1981). *Earth Evol. Sci.* **1**, 280–285.

Adám, A., Horwáth, F., and Stegena, L. (1977). *Acta Geol. Acad. Sci. Hung.* **21**, 251–260.

Adám, A., Vanyan, L. L., Hjelt, S. E., Kaikkonen, P., Shilovsky, P. P., and Palshin, N. A. (1983). *J. Geophys.* **54**, 73–75.

Ahorner, L. (1970). *In* "Graben Problems," (H. Illies and St. Mueller, eds.), pp. 155–156. Schweizerbart, Stuttgart.

Ahorner, L. (1975). *Tectonophysics* **29**, 233–245.

Ahorner, L. (1978). *In* "Alps, Appenines, Hellenides," (H. Closs, D. Roeder, and K. Schmidt, eds.), *Int. Union Comm. Geodyn. Sci. Rep.* **38**, 17–19. Schweizerbart, Stuttgart.

Ahrens, L. H. (1965). "Distribution of the Elements in Our Planet." McGraw-Hill, New York.

Aki, K. (1969). *J. Geophys. Res.* **74**, 615–631.

Aki, K. (1979). *J. Geophys. Res.* **84**, 6140–6148.

Aki, K., and Richards, P. G. (1980). "Quantitative Seismology." Vols. 1 and 2. Freeman, San Francisco.

Alekseev, A. S., Belonosova, A. V., Burmakov, I. A., Krasnopetertseva, G. V., Matveeva, N. N., Pavlenkova, N. I., Romanov, V. G., and Ryaboy, V. Z. (1973). *Tectonophysics* **20**, 47–56.

Alexander, L. G. (1960). "Third Australian–New Zealand Conference on Soil Mechanics," pp. 161–168.

Allègre, C. G., Dupré, B., Richard, P., and Rousseau, D. (1982). *Earth Planet. Sci. Lett.* **57**, 25–34.

Allenby, R. J., and Schnetzler, C. C. (1983). *Tectonophysics* **93**, 13–31.

Allmendinger, R. W., Sharp, J. W., Von Tisch, D., Serpa, L., Brown, L., Kaufman, S., and Oliver, J. (1983). *Geology* **11**, 532–536.

Anderson, D. L. (1981). *Science* **213**, 82–88.

Anderson, D. L. (1982). *Earth Planet. Sci. Lett.* **57**, 13–24.

Anderson, D. L., and Minster, J. B. (1979). *Geophys. J. R. Astron. Soc.* **58**, 431–440.

Anderson, D. L., and Minster, J. B. (1980). *In* "Source Mechanism and Earthquake Prediction," pp. 13–22. American Geophysical Union, Washington, D. C.

Anderson, D. L., Osborne, R. H., and Palmer, D. F. (1983). *Tectonophysics* **98**, 209–251.

Ando, C. J., Czuchra, B., Klemperer, S., Brown, L. D., Cheadle, M., Cook, F. A., Oliver, J. E., Kaufman, S., Walsh, T., Thompson, J. B., Lyons, J. B., and Rosenfeld, J. L. (1985). *Tectonics,* in press.

Angenheister, G., and Pohl, J. (1976). *In* "Explosion Seismology in Central Europe," (P. Giese, C. Prodehl, and A. Stein, eds.), pp. 290–302. Springer-Verlag, Berlin and New York.

Ansorge, J., Emter, D., Fuchs, K., Lauer, J., Mueller, St., and Peterschmitt, E. (1970). *In* "Graben Problems," (H. Illies and St. Mueller, eds.), pp. 190–197. Schweizerbart, Stuttgart.

Antoine, J. W., Martin, R. G., Pyle, T. G., and Bryant, W. R. (1974). *In* "Geology of Continental Margins," (C. A. Burke and C. L. Drake, eds.), pp. 683–694. Springer-Verlag, Berlin and New York.

Artemyev, M. E. (1966). "Isostatic Gravity Anomalies," (in Russian). Nauka, Moscow.

Artemyev, M. E., and Artyushkov, E. V. (1967). *Geotektonika* **5,** 41–57.

Artyushkov, E. V. (1981). *Tectonophysics* **73,** 9–14.

Arvidson, R. E., Guinness, E. A., Strebeck, J. W., Davies, G. F., and Schulz, K. J. (1982). *EOS* **63,** 261–265.

Ashby, M. F., and Verall, R. A. (1978). *Philos. Trans. R. Soc. London A* **288,** 59–95.

Assumpção, M., and Bamford, D. (1978). *Geophys. J. R. Astron. Soc.* **54,** 61–73.

Avedik, F., and Howard, D. (1979). *In* "Initial Report of the DSDP," (L. Montadert and D. G. Roberts, eds.), Vol. 48, pp. 1015–1024. U.S. Govt. Printing Office, Washington, D.C.

Avedik, F., Fucke, H., Goldflam, S., Hirschleber, H., Meissner, R., Sellevoll, M., and Weinrebe, W. (1984). *Ann. Geophys.* **2,** 571–578.

Babcock, E. A. (1978). *Am. Assoc. Petrol. Geol. Bull.* **62,** 1111–1126.

Backus, G. E., and Gilbert, J. F. (1968). *Geophys. J. R. Astron. Soc.* **16,** 169–188.

Backus, G. E., and Gilbert, J. F. (1970). *Philos. Trans. R. Soc. A* **266,** 123–145.

Bakun, W. H., Bufe, C. G., and Stewart, R. M. (1976). *Bull. Seism. Soc. Am.* **66,** 363–384.

Bally, A. W., Bernoulli, D., Davis, G. A., and Montadert, L. (1981). *Oceanol. Acta* **4,** Suppl. C3, 87–102.

Balsley, J. R., and Buddington, A. F. (1958). *Econ. Geology* **53,** 777–805.

Bamford, D. (1977). *Geophys. J. R. Astron. Soc.* **49,** 29–48.

Bamford, D., Faber, S., Jacob, B., Kaminski, W., Nunn, K., Prodehl, C., Fuchs, K., King, R., and Willmore, P. (1976). *Geophys. J. R. Astron. Soc.* **44,** 145–160.

Bamford, D., Nunn, K., Prodehl, C., and Jacob, B. (1978). *Geophys. J. R. Astron. Soc.* **54,** 43–60.

Banda, E., Surinach, E., Aparicio, A., Sierra, J., and Ruiz De La Parte, E. (1981). *Geophys. J. R. Astron. Soc.* **67,** 779–789.

Barazangi, M., and Dorman, J. (1969). *Bull. Seism. Soc. Am.* **59,** 369–380.

Barazangi, M., and Isacks, B. (1971). *J. Geophys. Res.* **76,** 8493–8516.

Barberi, F., and Varet, J. (1975). *In* "Afar Depression of Ethiopia," (A. Pilger and A. Roesler, eds.), pp. 375–379. Schweizerbart, Stuttgart.

Barberi, F., Santacroce, R., and Varet, J. (1982). *In* "Continental and Oceanic Rifts," (G. Pálmason, ed.), Geodynamics Series 8, pp. 223–258. American Geophysical Union, Washington D.C.

Bartelsen, H., Lueschen, E., Krey, Th., Meissner, R., Schmoll, H., and Walther, Ch. (1982). *In* "The Urach Geothermal Project, Swabian Alb, Germany," (R. Haenel, ed.), pp. 247–262. Schweizerbart, Stuttgart.

Barton, P. J., and Matthews, D. H. (1984). *Ann. Geophys.* **2,** 663–668.

Barton, P. J., Matthews, D. H., Hall, J., and Warner, M. (1984). *Nature* **308,** 55–56.

Bath, M. (1974). "Spectral Analysis in Geophysics." Elsevier, Amsterdam.

Bell, J. S., and Gough, D. S. (1979). *Earth Planet. Sci. Lett.* **45,** 475–842.

Beloussov, V. V., Volovski, B. S., Volvovski, J. S., and Ryaboi, V. A. (1962). *Bull. Acad. Sci. USSR* Geophys. Ser. 8, pp. 662–669.

Ben-Avraham, Z., Nur, A., Jones, D., and Cox, A. (1981). *Science* **213**, 47–50.
Ben-Avraham, Z., Nur, A., and Jones, D. (1982). *J. Geophys. Res.* **87**, 3861–3867.
Berckhemer, H. (1968). *Schweizer Mineral. Petrogr. Mitt*, **48**, 235–246.
Berckhemer, H. (1969). *Tectonophysics* **8**, 97–105.
Berckhemer, H. (1970). *Z. Geophys.* **36**, 501–518.
Berckhemer, H., and Hsue, K. (eds.) (1982). "Alpine-Mediterranean Geodynamics," Geodynamics Series 7. American Geophysics Union, Washington, D.C. and Geological Society of America, Boulder, Colorado.
Berckhemer, H., Baier, B., Bartelsen, H., Behle, A., Burckhardt, H., Gebrande, H., Makris, J., Menzel, H., Miller, H., and Vees, R. (1975). *In* "Afar Depression of Ethiopia," (A. Pilger and A. Roesler, eds.), pp. 89–107. Schweizerbart, Stuttgart.
Berckhemer, H., Kampfmann, W., and Aulbach, E. (1982a). *In* "High Pressure Research in Geoscience," (W. Schreyer, ed.), pp. 113–132. Schweizerbart, Stuttgart.
Berckhemer, H., Kampfmann, W., Aulbach, E., and Schmeling, H. (1982b). *Phys. Earth Planet. Inter.* **29**, 30–41.
Berry, M. J. (1973). *Tectonophysics* **20**, 183–201.
Berry, M. J. and West, G. F. (1966). *Bull. Seism. Soc. Am.* **56**, 141–171.
Berry, M. J., and Fuchs, K. (1973). *Bull. Seism. Soc. Am.* **63**, 1393–1432.
Berry, M. J., and Mair, J. A. (1980). *In* "The Continental Crust and its Mineral Deposits," (D. W. Strangway, ed.), pp. 196–213. Special paper 20, Geological Association of Canada, Waterloo, Ontario.
Berthelsen, A. (1983). Internal Report, Central Geologic Institute, Kopenhagen University, Kopenhagen.
Bessanova, E. N., Fishman, V. M., Ryaboyi, V. Z., and Sitnicova, G. A. (1974). *Geophys. J. R. Astron. Soc.* **36**, 377–398.
Bickle, M. J. (1978). *Earth Planet. Sci. Lett.* **40**, 301–315.
Binns, R. A., Gunthorpe, R. J., and Groves, D. J. (1976). *In* "The Early History of the Earth," (B. F. Windley, ed.), pp. 303–313. Wiley, Chicester.
Birch, F. (1960). *J. Geophys. Res.* **65**, 1083–1102.
Birch, F. (1961). *J. Geophys. Res.* **66**, 2199–2224.
Bittner, R. (1983). "Auswertung seismischer Messungen auf dem Profil FENNOLORA 1979 in Mittelschweden," Diploma thesis, Institut für Geophysik, Universität Kiel.
Blanchet, R., and Montadert, L. (eds.) (1981). "Geology of Continental Margins," *Oceanolog. Acta* **4** Suppl. C3.
Blank, H. R., and Gettings, M. E. (1974). "Complete Bouguer Anomaly Map Yellowstone-Island Park Region." Open File Map 74-22, U.S. Geological Survey, Washington D.C.
Blundell, D. J. (1981). *In* "Petroleum Geology of the Continental Shelf of North-West Europe," pp. 58–64. Institute of Petroleum, London.
Blundell, D. J. (1983). *Phys. Earth Planet. Inter.* **31**, 377–381.
Böger, H. (1983). *Neues Jahrb. Geol. Palaeontol. Abh.* **165**, 185–227.
Bolt, B. A. (1978). "Earthquakes: A Primer." Freeman, San Francisco.
Boltwood, B. B. (1907). *Amer. J. Sci.* **23**, 77–88.
Bott, M. P. H. (1973). *In* "Implications of Continental Drift to the Earth Sciences," (D. H. Tarling and S. K. Runcorn, eds.). Academic, New York.
Bott, M. P. H. (1980). *In* "Dynamics of Plate Interiors," (A. Bally, P. Bender, T. McGretchin and R. Walcott, eds.), pp. 27–32. Geodynamics Series 1, American Geophysical Union, Washington, D.C. and Geological Society of America, Boulder, Colorado.
Bott, M. P. H. (1982). "The Interior of the Earth," 2nd ed. Arnold, London.
Bowen, N. L. (1928). "The Evolution of Igneous Rocks." Princeton Univ. Press, Princeton, New Jersey.

Boyd, F. R. (1973). *Geochim. Cosmochim. Acta* **27**, 2533–2546.

Brace, W. F. (1979). Paper presented at Conference on Magnitude of Deviatomic Stress in the Earth's Crust and Upper Mantle. U.S. Geol. Survey, Carmel, California.

Brace, W. F., and Orange, A. S. (1968). *J. Geophys. Res.* **73**, 1433–1445.

Brace, W. F., Pauling, B. W. J., and Scholz, C. (1966). *J. Geophys. Res.* **71**, 3939–3954.

Braile, L. W., and Smith, R. B. (1975). *Geophys. J. R. Astron. Soc.* **40**, 145–176.

Braile, L. W., Smith, R. B., Ansorge, J., Baker, M. R., Sparlin, M. A., Prodehl, C., Schlilly, M. M., Healy, J. H., Mueller, St., and Olson, K. H. (1982). *J. Geophys. Res.* **87**, 2597–2609.

Bredehoeft, J. D., Wolff, R. G., Kays, W. S., and Shuter, E. (1976). *Geol. Soc. Am. Bull.* **87**, 250–258.

Brewer, J. A. (1983). *First Break* **1**, 25–31.

Brewer, J. A., and Oliver, J. E. (1980). *Ann. Rev. Earth Planet. Sci.* **8**, 205–230.

Brewer, J. A., Brown, L. D., Steiner, D., Oliver, J. E., Kaufman, S., and Denison, R. (1981). *Geology* **9**, 569–575.

Brewer, J. A., Good, R., Oliver, J., Brown, L. D., and Kaufman, S. (1983a). *Geology* **11**, 109–114.

Brewer, J. A., Matthews, D. M., Warner, M. R., Hall, J., Smythe, D. K., and Wittington, R. J. (1983b). *Nature* **305**, 206–210.

Brotchie, J. F., and Sylvester, R. (1969). *J. Geophys. Res.* **86**, 3695–3707.

Brown, G. C., and Musset, A. E. (1981). "The Inaccessible Earth." Allen & Unwin, London.

Brown, L. D., Krumhansl, P. A., Chapin, A. R., C. E., Sanford, A. R., Cook, F. A., Kaufman, S., Oliver, J. E., and Schilt, F. S. (1979). *In* "Rio Grande Rift," (R. E. Riecker, ed.), pp. 169–184. Special Publication 23, American Geophysical Unions, Washington, D.C.

Brown, L. D., Oliver, J. E., Kaufman, S., Brewer, J. A., Cook, F. A., Schilt, F. S., Albough, D. S., and Long, G. H. (1981). *In* "Evolution of the Earth," (R. O'Connell and W. Fyfe, eds.), pp. 38–52. Geodynamics Series 5, American Geophysical Union, Washington, D.C. and Geological Society of America, Boulder, Colorado.

Brown, L. D., Jensen, L., Oliver, J., Kaufman, S., and Steiner, D. (1982). *Geology* **10**, 645–649.

Brown, L. D., Ando, C., Klemperer, S., Oliver, J., Kaufman, S., Czuchra, B., Walsh, T., and Isachsen, Y. W. (1983). *Geol. Soc. Am. Bull.* **94**, 1173–1184.

Bullen, K. E. (1975). "The Earth's Density." Chapman and Hall, London.

Buntebarth, G. (1980). "Geothermie." Springer-Verlag, Berlin and New York.

Buntebarth, G. (1982). *Earth Planet. Sci. Lett.* **57**, 358–366.

Buntebarth, G., and Rybach, L. (1981). *Tectonophysics* **75**, 41–46.

Burke, K., and Kidd, W. F. S. (1978). *Nature* **222**, 240–243.

Byerlee, J. D. (1967). *J. Geophys. Res.* **72**, 3639–3648.

Byerlee, J. D. (1978). *Pure Appl. Geophys.* **116**, 615–626.

Byerlee, J. D., and Bracc, W. F. (1938). *J. Geophys. Res.* **30**, 6031–6040.

Byerlee, J. D., and Brace, W. F. (1968). *J. Geophys. Res.* **73**, 6031–6040.

Cagniard, L. (1953). *Geophysics* **18**, 605–635.

Calganile, G., and Panza, G. F. (1979). *J. Geophys.* **45**, 319–332.

Calganile, G., Panza, G. F., and Knopoff, L. (1979). *Tectonophysics* **56**, 51–62.

Caloi, P. (1967). *In* "Advances in Geophysics," Vol. 12, (H. E. Landsberg and J. van Mieghem, eds.), pp. 80–211. Academic, New York.

Caloi, P., Marcelli, L., and Pannochia, G. (1949). *Ann. Geofis.* **2**, 347–358.

Caner, B. (1971). *J. Geophys. Res.* **76**, 7202–7216.

Cantwell, T., and Madden, J. R. (1960). *J. Geophys. Res.* **65**, 4202–4205.

Carr, M. H., Greely, R., Blasius, K., Guest, J., and Murray, J. (1977). *J. Geophys. Res.* **82**, 3985.

Carter, N. L., Anderson, D. A., Hansen, F. D., and Kranz, R. L. (1981). *In* "Mechanical Behavior of Crustal Rocks," (N. L. Carter, M. Friedman, J. M. Logan and D. W. Sterns, eds.), pp. 61–82. Geophysics Monograph Series 24, American Geophysical Union, Washington, D.C.

Cassel, B. R., Mykkeltveit, S., Kanestroem, R., and Husebye, E. S. (1983). *Geophys. J. R. Astron. Soc.* **72**, 733–753.

Casten, U., Hedebol, E., and Nielsen, P. (1975). *J. Geophys.* **41**, 357–366.

Cathless, L. M. (1975). "The Viscosity of the Earth's Mantle." Princeton Univ. Press, Princeton, New Jersey.

Cazenave, A., and Dominh, K. (1981). *Geophys. Res. Lett.* **8**, 1039–1042.

Cermák, V., and Haenel, R. (1982). "Geothermics and Geothermal Energy." Schweizerbart, Stuttgart.

Cermák, V., and Rybach, L. (eds.) (1979). "Terrestrial Heat Flow in Europe." Springer-Verlag, Berlin and New York.

Cervény, V., Molotkov, J., and Psencik, I. (1977). "Ray Methods in Seismology." Charles University, Prague.

Chapin, C. E. (1979). *In* "Rio Grande Rift: Tectonics and Magmatism," (R. E. Riecker, ed.), pp. 1–6. Special Publication 23, American Geophysical Union, Washington, D.C.

Chapman, D. S. (1985). *In* "Landolt Boernstein," (K. Fuchs and H Soffel, eds.), Group 5, Vol. 2, Subvolume b: "Geophysics of the Solid Earth, Moon and Planets," pp. 1–17. Springer-Verlag, Berlin and New York.

Chapman, D. S., Pollack, H. N., and Cermak, V. (1979). "Terrestrial Heat Flow in Europe," (V. Cermak, and L. Rybach, eds.), pp. 41–48. Springer-Verlag, Berlin and New York.

Chen, W. P., and Molnar, P. (1983). *J. Geophys. Res.* **88**, 4183–4214.

Chopra, P. N., and Paterson, M. S. (1981). *Tectonophysics* **78**, 453–473.

Christensen, N. I. (1965). *J. Geophys. Res.* **70**, 6147–6164.

Christensen, N. I. (1966). *J. Geophys. Res.* **71**, 3549–3556.

Christensen, N. I. (1974). *J. Geophys. Res.* **79**, 407–412.

Christensen, N. I. (1979). *J. Geophys. Res.* **84**, 6849–6857.

Christensen, N. I., and Fountain, D. M. (1975). *Geol. Soc. Am. Bull.* **86**, 227–236.

Christensen, N. I., and Ramananantoandro, R. (1971). *J. Geophys. Res.* **76**, 4003–4010.

Christensen, N. I., and Salisbury, M. H. (1975). *Rev. Geophys. Space Phys.* **13**, 57–86.

Claerbout, J. F. (1976). "Fundamentals of Geophysical Data Processing." McGraw-Hill, New York.

Clifford, T. N. (1974). Geol. Soc. Special Paper 156.

Cochran, J. R. (1981). *Oceanol. Acta* **4** Suppl. C3, 155–166.

Cohen, T., and Meyer, R. (1966). *In* "The Earth Beneath the Continents," (J. Steinhart and T. Smith, eds.), pp. 150–156. Geophysics Monograph Series 10, American Geophysical Union, Washington, D.C.

Connerney, J. E. R., Nekut, A., and Kuckes, A. F. (1977). *Geophys. Res. Lett.* **4**, 239–242.

Connerney, J. E. R., and Kuckes, A. F. (1980). *J. Geophys. Res.* **85**, 2615–2624.

Connerney, J. E. R., Nekut, A., and Kuckes, A. F. (1980). *J. Geophys. Res.* **85**, 2603–2614.

Cook, F. A., Brown, L. D., and Oliver, J. E. (1980). *Sci. Am.* **243**, 156–168.

Cook, F. A., Brown, L. D., Kaufman, S., Oliver, J. E., and Petersen, T. A. (1981). *Geol. Soc. Am. Bull.* **92**, 738–748.

Coulomb, C. A. (1773). *Acad. R. Sci. Mem. Math. Phys.* **7**, 343–382.

Cox, K. (1971). *In* "Understanding the Earth," 2nd ed. (I. G. Gass, P. J. Smith and R. C. L. Wilson, eds.), pp. 13–40. MIT Press, Cambridge, Massachusetts.

Crampin, St., Chesnokov, E. M., and Hipkin, R. A. (1984). *First Break* **2**, 9–18.

Crough, S. Th. (1983). *J. Geophys. Res.* **88**, 6449–6454.

Crowell, J. C. (1981). *Oceanol. Acta* **4** Suppl. C3, 137–142.

Dagnieres, M., Gallert, J., Banda, E., and Hirn, A. (1982). *Earth Planet. Sci. Lett.* **57**, 88–100.

Davydova, N. I. (1975). *In* "Seismic Properties of the Moho Discontinuity," (N. I. Davydova, ed.). Akademiya Nauk, SSSR.

De Beer, J. H., van Zigl, S. V., and Govgu, D. J. (1982). *Tectonophysics* **83**, 205–225.
Deichmann, N., and Ansorge, J. (1983). *J. Geophys.* **52**, 109–118.
Demenitskaya, R. M., and Belyaevsky, N. A. (1969). *In* "The Earth's Crust and Upper Mantle," (P. J. Hart, ed.), pp. 312–315. American Geophysical Union, Washington, D.C.
De Paolo, D. J. (1980a). *Science* **209**, 684–687.
De Paolo, d. J. (1980b). *Geochim. Cosmochim. Acta* **44**, 1185–1196.
De Paolo, D. J. (1981a). *In* "Evolution of the Earth," (R. J. O'Connell and W. S. Fyfe, eds.), pp. 59–68. Geodynamics Series 5, American Geophysical Union, Washington, D.C. and Geological Society of America, Boulder, Colorado.
De Paolo, D. J. (1981b). *EOS* **62**, 137–140.
De Paolo, D. J. (1983). *Geophys. Res. Lett.* **10**, 705–708.
Pe Paolo, D. J., and Johnson, R. W. (1979). Contr. Mineral. Petrol. **70**, 367–379.
Dix, C. H. (1965). *Geophysics* **30**, 1068–1084.
Dobrin, M. B. (1960). "Introduction to Geophysical Prospecting," 2nd ed. McGraw-Hill, New York.
Dohr, G. (1957a). *Erdoel Kohle* **10**, 278–281.
Dohr, G. (1957b). *Geol. Rundsch.* **46**, 17–26.
Dohr, G. (1981). "Applied Geophysics." Enke, Stuttgart.
Dohr, G., and Fuchs, K. (1967). *Geophysics* **32**, 951–967.
Dohr, G., and Meissner, R. (1975). *Geophysics* **40**, 25–39.
Dortman, N. B., and Magid, M. (1968). *Sovjetzsk. Geologya* **5**, 123–129.
Dott, R. H., and Batten, R. L. (1976). "Evolution of the Earth." McGraw-Hill, New York.
Dowling, F. L. (1970). *J. Geophys. Res.* **75**, 2683–2698.
Dupré, B., and Allègre, C. J. (1980). *Nature* **286**, 17–22.
Durham, W. B., and Goetze, C. (1977). *Tectonophysics* **40**, 15–18.
Dzievonski, A. M., Hales, A. L., and Lapwood, E. R. (1975). *Phys. Earth Planet. Inter.* **10**, 12–48.
East Pacific Rise Study Group (1981). *Science* **213**, 31–46.
Edel, J., Fuchs, K., Gelbke, C., and Prodehl, C. (1975). *Z. Geophys.* **41**, 333–356.
Eldholm, O., and Sundvor, E. (1979). *Tectonophysics* **59**, 233–238.
Eldholm, O., and Montadert, L. (1981). *Oceanol. Acta* **4** Suppl. C3, 7–10.
Engelhardt, L. (1978). *Erdoel Erdgas* **94**, 325–327.
Engelhardt, L. (1979). "Die Dämpfung seismischer Wellen und ihre Bestimmung aus Reflexionseismogrammen." Habil.-Schrift, Technische Universität, Braunschweig.
Etheridge, M. A., and Wilkie, J. C. (1979). *Tectonophysics* **58**, 159–178.
Faure, G. (1977). "Principles of Isotope Geology." Wiley, New York.
Faure, G., and Powell, J. L. (1972). "Strontium Isotope Geology." Springer-Verlag, Berlin and New York.
Feng, R., and McEvilly, T. V. (1983). *Bull. Seism. Soc. Am.* **73**, 1701–1720.
Ferrari, A. J., and Bills, B. G. (1979). *Rev. Geophys. Space Phys.* **17**, 1663–1677.
Finlayson, D. M. (1982). *J. Geophys. Res.* **87**, 10569–10578.
Finlayson, D. M., Collins, C. D., and Denham, D. (1980). *Phys. Earth Planet. Inter.* **21**, 321–342.
Fitch, T. J. (1979). *In* "The Earth: Its Origin, Structure and Evolution," (N. W. McElhiny, ed.), pp. 491–542. Academic, New York.
Flüh, E. R. (1982). "Geodynamische Entwicklung der nördlichen Anden." Doctorial thesis, Mathematics-Science Faculty, University of Kiel.
Flüh, E. R., and Milkereit, B. (1984). *In* "Interpretation of Seismic Wave Propagation in Laterally Heterogeneous Structures," (D. M. Finlayson and J. Ansorge, eds.), pp. 125–128. Report 258, Bureau of Mineral Resources, Geology, and Geophysics, Canberra.
Flüh, E. R., Milkereit, B., Meissner, R., and Meyer, R. P. (1981). *Zentralbl. Geol. Palaeontol.* **3/4**, 231–242.

Fountain, D. M. (1976). *Tectonophysics* **33**, 145–165.

Fountain, D. M., and Salisburg, M. H. (1981). *Earth Planet. Sci. Lett.* **56**, 261–277.

von Freese, R. R. B., Hinze, W. J., Sexton, J. L., and Braile, L. W. (1982). *Geophys. Res. Lett.* **9**, 293–295.

French, W. S. (1974). *Geophysics* **39**, 265–273.

French, W. S. (1975). *Geophysics* **40**, 961–980.

Frisch, W. (1978). *In* "Alps, Appenines and Hellenides" (H. Closs, D. Roeder, and K. Schmidt, eds.), pp. 167–172. Schweizerbart, Stuttgart.

Frisch, W. (1979). *Tectonophysics* **60**, 121–139.

Frisch, W. (1981). *Geol. Rundsch.* **70**, 402–411.

Fuchs, K., and Mueller, G. (1971). *Geophys. J. R. Astron. Soc.* **23**, 417–433.

Fuchs, K., and Vinnik, L. P. (1982). *In* "Continental and Oceanic Rifts," (G. Pálmason, ed.), pp. 81–98. Geodynamics Series 8, American Geophysical Union, Washington, D.C. and Geological Society of America, Boulder Colorado.

Fuller, M. (1983). *Rev. Geophys. Space Phys.* **21**, 593–598.

Fyfe, W. S. (1974). *Nature* **249**, 338–341.

Fyfe, W. S. (1978). *Chem. Geol.* **23**, 89–114.

Galfi, J., and Stegena, L. (1957). *Geol. Rundsch.* **146**, 26–29.

Galle, E. M., and Wilhoit, J. C. (1962). *J. Soc. Pet. Eng.* **225**, 145–160.

Garland, G. D. (1965). "The Earth's Shape and Gravity." Pergamon, Oxford.

Gass, I. G., Smith, P. J., and Wilson, R. C. L. (1971). "Understanding the Earth." Artemis Press, Sussex.

Gebrande, H. (1982). *In* "Landolt-Boernstein," (K. H. Hellwege, ed.), Group 5, Vol. 1, Subvolume b: "Physical Properties of Rocks," (G. Angenheister, ed.), pp. 1–96. Springer-Verlag, Berlin and New York.

Gee, D. G. (1978). *Tectonophysics* **47**, 393–419.

Gibbs, A. K., Payne, B., Setzer, T., Brown, L. D., Oliver, J., and Kaufman, S. (1984). *Geol. Soc. Am. Bull.* **95**, 280–294.

Giese, P. (1968). Geophysikalische Abhandlungen des Instituts für Geophysik und Meteorologie, Freie Universsität, Berlin, Abhandlung 1.

Giese, P. (1972). *Z. Geophys.* **39**, 395–409.

Giese, P. (1976a). *In* "Explosion Seismology in Central Europe," (P. Giese, C. Prodehl, and A. Stein, eds.), pp. 196–200. Springer-Verlag, Berlin and New York.

Giese, P. (1976b). *In* "Explosion Seismology in Central Europe," P. Giese, C. Prodehl, and A. Stein, eds.), pp. 201–214. Springer-Verlag, Berlin and New York.

Giese, P. (1983). *In* "Plateau Uplift," (K. Fuchs and H. Murawski, eds.), pp. 303–314. Springer-Verlag, Berlin and New York.

Giese, P., and Prodehl, C. (1976). *In* "Explosion Seismology in Central Europe," (P. Giese, C. Prodehl, and A. Stein, eds.), pp. 347–376. Springer-Verlag, Berlin and New York.

Giese, P., Prodehl, C., and Stein, A. (eds.) (1976). "Explosion Seismology in Central Europe." Springer-Verlag, Berlin and New York.

Giese, P., Reuter, K. H., Jacobshagen, V., and Nicolich, R. (1982). *In* "Alpine–Mediterranean Geodynamics" (H. Berckhemer and K. Hsue, eds.), pp. 39–74. Geodynamics Series 7, American Geophysical Union, Washington, D. C. and Geology Society of America, Boulder, Colorado.

Ginzburg, A., and Folkman, Y. (1980). *Earth Planet. Sci. Lett.* **51**, 180–190.

Ginzburg, A., Makris, J., Fuchs, K., Prodehl, C., Kaminski, W., and Amitai, V. (1979a). *J. Geophys. Res.* **84**, 1569–1582.

Ginzburg, A., Makris, J., Fuchs, K., Perathoner, B., and Prodehl, C. (1979b). *J. Geophys. Res.* **84**, 5605–5612.

Ginzburg, A., Makris, J., Fuchs, K., and Prodehl, C. (1981). *Tectonophysics* **80**, 109–119.

Girdler, R. W. (1964). *In* "Physics and Chemistry of the Earth," (S. K. Runcorn, ed.), Vol. 5, pp. 121–156. Pergamon, New York.

Gladwin, M. T., and Stacey, F. O. (1974). *Phys. Earth Planet. Inter.* **8**, 332–336.

Glocke, A., and Meissner, R. (1976). *In* "Explosion Seismology in Central Europe," (P. Giese, C. Prodehl, and A. Stein, eds.), pp. 252–256. Springer-Verlag, Berlin and New York.

Godin, Y. N., Volvowski, B. S., and Volvowski, I. S. (1960). *Dokl. Akad. Nauk. SSSR* **133** (in Russian).

Goodwin, A. M. (1981). *Science* **213**, 55–61.

Gough, D. I. (1983). *J. Geophys. Res.* **88**, 3367–3378.

Gough, D. I., and Bell, J. S. (1981). *Can. J. Earth Sci.* **18**, 638–645.

Gough, D. I., and Bell, J. S. (1982). *Can. J. Earth Sci.* **19**, 1358–1370.

Gough, D. I., Bingham, D. K., Ingham, M. R., and Alabi, M. O. (1982). *Can. J. Earth Sci.* **19**, 1680–1690.

Grant, F. S., and West, G. F. (1965). "Interpretation Theory in Applied Geophysics." McGraw-Hill, New York.

Greely, R. (1982). *J. Geophys. Res.* **87**, 2705–2712.

Griffith, A. A. (1921). *Philos. Trans. R. Soc. London A* **221**, 163–198.

Griffith, D. H. (1972). *Tectonophysics* **15**, 151–156.

Griggs, G. D., Turner, F. J., and Heard, H. C. (1960). *In* "Rock Deformation," pp. 39–104. Geological Society of America, Boulder, Colorado.

Griggs, D. T., Jackson, D. D., Knopp, L., and Shreve, R. L. (1975). *Science* **187**, 537–539.

Grow, J. A. (1973). *Geol. Soc. Am. Bull.* **84**, 2169–2192.

Grow, J. A. (1980). *In* "Geologic Studies of the COST B-3 Well," (P. A. Scholle, ed.), pp. 117–125. Circular 833, U.S. Geological Survey, Washington, D.C.

Grow, J. A., and Sheridan, R. E. (1981). *Oceanol. Acta* **4**, Suppl. C3, 11–20.

Grubbe, K. (1976). *In* "Explosion Seismology in Central Europe," (P. Giese, C. Prodehl, and A. Stein, eds.), pp. 268–282. Springer-Verlag, Berlin and New York.

Guest, J. E., and Greely, R. (1977). "Geology of the Moon." Wykeham, London.

Ginbuz, B. M. (1970). *Bull. Seism. Soc. Am.* **60**. 1921–1935.

Gutenberg, B. (1959). "Physics of the Earth's Interior." Academic, New York.

Gutenberg, B., and Richter, C. F. (1936). *Gerlands Beitr. Geophys.* **47**, 73–131.

Gutenberg, B., and Richter, C. F. (1954). "Seismicity of the Earth and Association Phenomena," 2nd ed. Princeton Univ. Press, Princeton, New Jersey.

Guterch, A. (1977). *Publ. Inst. Geophys. Pol. Acad. Sci. A* **4**(115), 347–357.

Guterch, A., Kovalski, T., Materzok, R., and Toporkiewicz, S. (1976). *Publ. Inst. Geophys. Pol. Acad. Sci. A* **2**(101), 347–357.

Guust, N., and Mueller, St. (1982). *Tectonophysics* **84**, 151–178.

Haak, V. (1977). *Acta Geodaetica Geophys. Montanist. Acad. Sci. Hung.* **12**, 7–10.

Haak, V. (1980). *Geophys. Surv.* **4**, 57–69.

Hadley, K. (1975). *EOS* **56**, 1060–1078.

Hadley, D., and Kanamori, H. (1977). *Bull. Seism. Soc. Am.* **88**, 1469–1478.

Haenel, R. (ed.) (1982). "The Urach Geothermal Project." Schweizerbart, Stuttgart.

Haggerty, S. E. (1978). *Geophys. Res. Lett.* **5**, 105–108.

Haimson, B. C., and Rummel, F. (1982). *J. Geophys. Res.* **87**, 6631–6649.

Hales, A. L., and Rynn, J. M. (1978). *Geophys. J. R. Astron. Soc.* **55**, 633–644.

Hall, D. H. and Brisbin, W. C. (1965). *Geophysics* **30**, 1053–1067.

Hall, D. H., and Hajnal, Z. (1973). *Bull. Seism. Soc. Am.* **63**, 885–910.

Hansen, F. D., and Carter, N. L. (1982). *EOS* **63**, 437.

Head, J. W., and Solomon, S. C. (1981). *Science* **213**, 62–75.

Healy, J. H., and Warren, D. H. (1969). *In* "The Earth's Crust and Upper Mantle," (P. Hart, ed.), pp. 208-220. Monograph 13, American Geophysical Union, Washington, D.C.

Heard, H. C., and Carter, N. L. (1968). *Am. J. Sci.* **266**, 1-42.

Heiskanen, W. A., and Vening-Meinesz, F. A. (1958). "The Earth and its Gravity Field." McGraw-Hill, New York.

Henshaw, P. C., and Merril, R. T. (1979). *Earth Planet. Sci. Lett.* **43**, 315-320.

Herrman, R. B. (1980). *Bull. Seism. Soc. Am.* **70**, 447-468.

Herzberg, C. T., Fyfe, W. S., and Carr, M. I. (1983). *Contrib. Mineral. Petrol.* **84**, 1-5.

Hilde, T. W. C. (1983). *Tectonophysics* **99**, 381-397.

Hinz, K., and Weber, J. (1976). *Compendium 1975/76, Erdoel Kohle, Erdgas, Petrochem.* 3-29.

Hinz, K., Dostmann, H., Hanisch, J., Kempter, E., and Wissmann, G. (1984). *In* "Statusbericht Rahmenprogram Rohstofforschung," pp. 413-430. Kernforschungsanlage Juelich G.M.B.H., Juelich.

Hirn, A., and Perrier, G. (1974). *In* "Approaches to Taphrogenesis," (H. Illies and K. Fuchs, eds.), pp. 329-340. Schweizerbart, Stuttgart.

Hirn, A., Steinmetz, L., and Fuchs, K. (1973). *Z. Geophys.* **39**, pp. 363-384.

Hirschleber, H. B., Lund, C. E., Meissner, R., Vogel, A., and Weinrebe, W. (1975). *J. Geophys.* **41**, 135-148.

Hood, L. L., and Cisowski, S. M. (1983). *Rev. Geophys. Space Phys.* **21**, 676-684.

Horwarth, F., and Berckhemer, H. (1982). *In* "Alpine Mediterranean Geodynamics," (H. Berckhemer and K. Hsue, eds.), 141-174. Geodynamics Series 7, American Geophysical Union, Washington, D.C. and Geological Society of America, Boulder, Colorado.

Horwarth, F., Doevényi, P., and Liebe, P. (1981). *Earth Evol. Sci.* **1**, 285-291.

Hsui, A. T., Marsh, B.D., and Toksöz, M. N. (1983). *Tectonophysics* **99**, 207-220.

Iher, H. M. (1979). *Tectonophysics* **56**, 165-197.

Iher, H. M., Oppenheimer, D. H., and Hitchcock, T. (1979). *Science* **204**, 495-497.

Illies, J. H. (1970). *In* "Graben Problems," (J. H. Illies and St. Mueller, eds.), pp. 1-26. Schweizerbart, Stuttgart.

Illies, J. H. (1974). *In* "Approaches to Taphrogenesis," (J. H. Illies and K. Fuchs, eds.), pp. 433-460. Schweizerbart, Stuttgart.

Illies, J. H. (1978). *In* "Tectonics and Geophysics of Continental Rifts," (J. B. Ramberg and E. R. Neumann, eds.), pp. 63-71. Reidel, Dordrecht.

Illies, J. H., and Baumann, H. (1982). *Z. Geomorphol. Suppl.* **42**, 135-165.

Illies, H., and Greiner, G. (1978). *Geol. Soc. Am. Bull.* **89**, 770-782.

International Society for Rock Mechanics, Commission on Standardization of Laboratory and Field Tests (1977). *Int. J. Rock Mech. Min. Sci. Geomech. Abstr.* **15**, 53-58.

Irvine, T. N. (1980). *In* "Physics of Magmatic Processes," (R. B. Hargraves, ed.), pp. 325-383. Princeton Univ. Press, Princeton, New Jersey.

Isacks, B., Oliver, J., and Sykes, L. R. (1968). *J. Geophys. Res.* **73**, 5855-5899.

Jaeger, J. C. (1969). "Elasticity, Fracture and Flow," 3rd ed. Methuen, London.

Jaeger, J. C., and Cook, N. G. W. (1976). "Fundamentals of Rock Mechanics," 2nd ed. Chapman and Hall, London.

Janle, P. (1981). *J. Geophys.* **49**, 57-65.

Janle, P., and Ropers, J. (1982). *Lunar Sci. Abstr., 13th*, 362-363. Lunar and Planetary Institute, Houston.

Jannsen, D., Voss, J., and Theilen, F. (1985). *Geophys. Prospect.* **33**, 479-497.

Jaoul, O., Tullis, J. A., and Kronenberg, A. K. (1984). *J. Geophys. Res.* **89**, 4271-4280.

Jarvis, J. G., and Campbell, I. H. (1983). *Geophys. Res. Lett.* **10**, 1133-1136.

Jarvis, J. G., and McKenzie, D. P. (1980). *Earth Planet. Sci. Lett.* **48**, 42-52.

Jeffreys, H. (1952). "The Earth," 3rd ed. Cambridge Univ. Press, London.

Jentsch, M. (1978/79). *Z. Geophys.* **45**, 335-372.

Joedicke, H. (1981). *In* "Vertikalbewegungen und ihre Ursachen am Beispiel des Rheinischen Schildes," pp. 193-197. DFG Protokoll, Neustadt/Weinstrasse.

Johnston, D. H., and Toksöz, M. N. (1981). *In* "Seismic Wave Attenuation," (M. N. Toksöz and D. H. Johnston, (eds.), pp. 6-12. SEG, Tulsa, Oklahoma.

Jones, T., and Nur, A. (1982). *SEG Tech. Prog.* 92-93. Dallas, Texas.

Jones, T., and Nur, A. (1983). *Geophys. Res. Lett.* **10**, 140-143.

Junger, A. (1951). *Geophysics* **16**, 499-505.

Kaila, K. L., Krishna, U. G., and Mall, D. M. (1981a). *Tectonophysics* **76**, 99-130.

Kaila, K. L., Murty, P. R. K., Rao, V. K., and Kharetchko, G. E. (1981b). *Tectonophysics* **73**, 365-384.

Kamptmann, W. (1984). "Laborexperimente zum elastischen und anelastischen Verhalten hochtemperierter magmatischer Gesteine im Frequenzbereich seismischer Wellen." Institut für Meteorologie und Geophysik der Universität Frankfurt, Federal Republic of Germany.

Kanamori, H., and Anderson, D. L. (1975). *Bull. Seism. Soc. Am.* **65**, 1073-1095.

Kármánn, Th. (1911). *Z. Ver. Dtsch. Ing.* **57**, 1749-1757.

Karig, D. E. (1971). *J. Geophys. Res.* **76**, 2541-2561.

Kasahara, K. (1981). "Earthquake Mechanics." Cambridge Univ. Press, London and New York.

Kay, R. W. (1980). *J. Geol.* **88**, 497-552.

Kay, S. M. (1983). *Can. Mineral* **21**, 665-681.

Kay, S. M., Kay, R. W., and Citron, G. P. (1982). *J. Geophys. Res.* **87**, 4051-4072.

Kay, S. M., Kay, R. W., Brueckner, H. K., and Rubenstone, J. (1983). *Contrib. Mineral. Petrol.* **82**, 99-116.

Keen, C. E., and Barett, D. L. (1971). *Can. J. Earth Sci.* **8**, 1056-1064.

Keen, C. E., Beaumont, C., and Boutilier, R. (1981). *Oceanolog. Acta* **4** Suppl. C3, 123-128.

Kern, H. (1978). *Tectonophysics* **44**, 185-203.

Kern, H. (1979). *Phys. Chem. Miner.* **4**, 161-171.

Kern, H. (1982a). *Phys. Earth Planet. Inter.* **29**, 12-23.

Kern, H. (1982b). *In* "High Pressure Researches in Geophysics," (W. Schreyer, ed.), pp. 15-45. Schweizerbart, Stuttgart.

Kern, H., and Fakhimi, M. (1975). *Tectonophysics* **28**, pp. 227-244.

Kern, H., and Richter, A. (1981). *J. Geophys.* **49**, pp. 47-56.

Khitarov, N. I., Lebedev, E. B., Dorfman, A. M., and Bagdasarov, N. S. (1983). *Geochimica* **9**, 1239-1246.

King, C. Y., Nason, R. D., and Tocker, D. (1973). *Philos. Trans. R. Soc. London A* **274** 355-360.

Kirby, S. H. (1983). *Rev. Geophys. Space Phys.* **21**, 1458-1487.

Kirsten, T. (1976). "Time and the Solar System." Max Planck Institute, Heidelberg.

Kiselev, A. I., Golovko, H. A., and Medvedev, M. E. (1978). *Tectonophysics* **45**, 49-60.

Kjartansson, E. J. (1979). *J. Geophys. Res.* **84**, 4737-4748.

Knopoff, L., and Chang, F. S. (1977). *J. Geophys.* **43**, 299-314.

Koch, P. R., Chustie, F. M., and Geoige, R. P. (1983). Unpublished report.

Kohlstedt, D. L. (1979). *Cornell Quarterly* **14**, 18-26.

Kohlstedt, D. L., and Goetze, C. (1974). *J. Geophys. Res.* **79**, 2045-2051.

Kohlstedt, D. L., and Weathers, M. S. (1980) *J. Geophys. Res.* **85**, 6269-6285.

Kohlstedt, D. L., Goetze, C., Durham, W. B., and van der Sande, J. (1976). *Science* **191**, 1045-1046.

Kohlstedt, D. L., Nichols, H. P. K., and Hornack, P. (1980). *J. Geophys. Res.* **85**, 3122-3130.

Kolmogorov, V. G., and Kolmogorova, P. P. (1978). *Tectonophysics* **45**, 101-105.

Kondorskaya, N., Slavina, L., Pivovarora, N., Baavadse, B., Alexidse, M., Gotsadse, S., Marusidse., G., Sicharulidse, D., Pavlenkova, N., Khromatskaya, E., and Krasnopertseva, G. (1981). *Pure Appl. Geophys.* **119**, 1157–1179.

Korhonen, H., and Parkka, M. T. (1981). *Pure Appl. Geophys.* **119**, 1093–1099.

Koshubin, S. J., Pawlenkova, N. J., and Yegorkin, A. V. (1984). *Geophys. J. R. Astron. Soc.* **76**, 221–226.

Kosminskaya, I. P. (1971). "Deep Seismic Sounding of the Earth's Crust and the Upper Mantle." Inst. Phys. of the Earth, Moscow.

Kosminskaya, I. P., and Pawlenkova, N. I. (1979). *Tectonophysics* **59**, 307–320.

Kosminskaya, I. P., Belyaerksy, N. A., and Volvorski, J. S. (1969). *In* "The Earth's Crust and Upper Mantle," (P. J. Hart, ed.), pp. 195–208. Geophysics Monograph 13, American Geophysical Union, Washington, D. C.

Kovach, R. L. (1978). *Rev. Geophys. Space Phys.* **16**, 1–18.

Krey, Th. (1976). *Geophys. Prospect.* **24**, 91–111.

Krey, Th. (1978). *Geophysics* **43**, 899–911.

Krey, Th. (1980). *Geophys. Prospect.* **28**, 359–371.

Kröner, A. (1977). *Tectonophysics* **40**, 101–135.

Kröner, A. (1979). *Geol. Rundsch.* **68**, 565–583.

Kröner, A. (ed.) (1981). "Precambrian Plate Tectonics." Elsevier, Amsterdam.

Kronenberg, A. K., and Tullis, J. A. (1984). *J. Geophys. Res.* **89**, 4281–4297.

Kuckes, A. F. (1973a). *Geophys. J. R. Astron. Soc.* **32**, 119–331.

Kuckes, A. F. (1973b). *Geophys. J. R. Astron. Soc.* **32**, 381–385.

Kushiro, I. (1980). *In* "Physics of Magmatic Processes," (R. B. Hargraves, ed.), pp. 93–120. Princeton Univ. Press, Princeton, New Jersey.

Lachenbruch, A. H. (1968). *J. Geophys. Res.* **73**, 6977–6989.

Langel, R. A. (1982). *Geophys. Res. Lett.* **9**, 239–242.

Langseth, M. G. (1965). *In* "Terrestrial Heat Flow," (W. H. K. Lee, ed.), pp. 58–73. American Geophysical Union, Washington, D.C.

Leeman, E. R. (1964). *J. S. Afr. Inst. Min. Metall.* **65**, 45–114 and 254–284.

Lehman, J. A., Smith, R. B., Schilly, M. M., and Braile, M. W. (1982). *J. Geophys. Res.* **87**, 2713–2730.

Le Pichon, X., Francheteau, J., and Bouin, J. (1973). "Plate Tectonics," Developments in Geotectonics 6. Elsevier, Amsterdam.

Levin, B. J. (1962). *In* "The Moon," (Z. Kopal and Z. K. Mikhailov, eds.), 157–167. Academic, New York.

Levis, J. S. (1974). *Sci. Am.* **230**, 50–65.

Li Y., Meissner, R., Xue En. (1983). *Phys. Earth Planet. Inter.* **33**, 213–218.

Liebscher, H. J. (1962). *Z. Geophys.* **28**, 162–184.

Liebscher, H. J. (1964). *Z. Geophys.* **30**, 51–96 and 115–126.

Liu, H. P., Anderson, D. L., and Kanamori, H. (1976). *Geophys. J. R. Astron. Soc.* **47**, 41–58.

Lorenz, V., and Nichols, J. A. (1984). *Tectonophysics* **107**, 25–26.

Lund, C. E. (1979). *Geol. Foeren. Stockholm Foerh.* **101**, 191–204.

Lyell, C. (1833). "Principles of Geology." Murray, London.

Maeusnest, O. (1982). *In* "The Urach Geothermal Project," (R. Haenel, ed.), pp. 157–160. Schweizerbart, Stuttgart.

Mahlburg-Kay, S., and Kay, R. W. (1985). *Geology* **13**, 461–464.

Makris, J. (1975). *In* "Afer Depression of Ethiopia," (A. Pilger and A. Roesler, eds.), pp. 379–390. Schweizerbart, Stuttgart.

Makris, J. (1982). "Seismic Investigation of the Northern Part of the Red Sea." Institut für Geophysik, Universität Hamburg, Hamburg, Federal Republic of Germany.

Makris, J., Mengel, H., Zimmermann, J., and Gouise, P. (1975). *In* "Afar Depression of Ethiopia," (A. Pilger and A. Roesler, eds.), pp. 135–144. Scheizerbart, Stuttgart.

Makris, J., Ben Abraham, Z., Behle, A., Ginzberg, A., Giese, P., Steinmetz, L., Whitmarch, R. B., Elefthesion, S. (1983). *Geophys. J. R. Astr. Soc.* **75**, 575–591.

Maluski, H., and Matte, Ph. (1984). *Tectonics* **3**, 1–18.

Manghnani, M. H., Ramananantoandro, R., and Clark, S. P. (1974). *J. Geophys. Res.* **79**, 5427–5446.

Marçais, J., Delany, F., Peivé, A. V., Nazivkin, D. V., Boydadov, A. A., and Khain, V. E. (1981). "Carte tectonique internationale de l'Europe et des regions avoisinantes," 2nd ed. L'Académie des Sciences de l'URSS et l'UNESCO, Moscow.

Massé, R. P., and Alexander, S. S. (1974). *Geophys. J. R. Astron. Soc.* **39**, 587–60.

Masurski, H., Eliason, E., Ford, P. G., McGill, G. E., Pettengill, G. H., Schaber, G. G., and Schubert, G. (1980). *J. Geophys. Res.* **85**, 8232–8260.

Mathur, S. P. (1974). *Tectonophysics* **24**, 151–182.

McCullogh, M. T., and Wasserburg, G. J. (1978). *Science* **200**, 1003–1011.

McElhinny, M. W. (1973). "Paleomagnetism and Plate Tectonics." Cambridge Univ. Press, London and New York.

McKenzie, D. (1978). *Earth Planet. Sci. Lett.* **40**, 25–37.

McLain, J. S. (1981). *Geophys. Res. Lett.* **8**, 1191–1194.

McMechan, G. A., and Ottolini, R. (1980). *Bull. Seism. Soc. Am.* **70**, 775–789.

McNutt, R. H., Crocket, J. H., Clark, A. H., Caelles, J. C., Famar, E., Haynes, S. J., and Zentilli, M. (1975). *Earth Planet. Sci. Lett.* **27**, 305–313.

Meissner, R. (1961). *Geophys. Prospect.* **9**, 533–543.

Meissner, R. (1966). *Geophys. Prospect.* **14**, 7–16.

Meissner, R. (1967). *Gerlands Beitr. Geophys.* **76**, 211–254 and 295–314.

Meissner, R. (1973). *Geophys. Surv.* **1**, 195–216.

Meissner, R. (1976). *In* "Explosion Seismology in Central Europe," (P. Giese, C. Prodehl, and A. Stein, eds.), pp. 380–384. Springer-Verlag, Berlin and New York.

Meissner, R. (1979). *Geojournal* **3.3**, 227–233.

Meissner, R. (1980). *In* "Earth Rheology, Isostasy and Eustasy," (N. A. Moerner, ed.), pp. 125–134. Wiley, New York.

Meissner, R. (1981a). *Naturwissenschaften* **68**, 437–442.

Meissner, R. (1981b). *In* "Multidisciplinary Approach to Earthquake Prediction," (A. M. Isikara and A. Vogel, eds.), pp. 563–575. Vieweg, Braunschweig, Wiesbaden.

Meissner, R. (1983). *Ann. Geophys.* **1**, 121–127.

Meissner, R., and Fakhimi, M. (1977). *Geophys. J. R. Astron. Soc.* **49**, 133–143.

Meissner, R., and Flüh, E. R. (1979). *J. Geophys.* **45**, pp. 349–352.

Meissner, R., and Janle, P. (1984). *In* "Landolt Boernstein," (K. Fuchs and H. Soffel, eds.), Group 5, Vol. 2, Subvolume a: "Geophysics of the Solid Earth, Moon, and Planets," pp. 379–417. Springer-Verlag, Berlin and New York.

Meissner, R., and Lange, M. (1977). *Proc. Lunar Sci. Conf. 8th* pp. 551–562.

Meissner, R., and Lueschen, E. (1983). *First Break* **1**, pp. 1–6.

Meissner, R., and Stegena, L. (1977). "Praxis der seismischen Feldmessungen und Auswertung." Gebrüder Bornträger, Berlin and Stuttgart.

Meissner, R., and Strehlau, J. (1982). *Tectonics* **1**, pp. 73–89.

Meissner, R., and Theilen, F. (1983). Special Paper 3, World Petr. Congr. London 1983, pp. 363–379. Wiley, Chichester.

Meissner, R., and Vetter, U. (1974). *In* "Approaches to Taphrogenesis," (H. Illies and K. Fuchs, eds.), pp. 236–243. Schweizerbart, Stuttgart.

Meissner, R., and Vetter, U. (1976a). *Tectonophysics* **35**, 135–148.

Meissner, R., and Vetter, U. (1976b). *In* "Explosion Seismology in Central Europe," (P. Giese, C. Prodehl, and A. Stein, eds.), pp. 396-400. Springer-Verlag, Berlin and New York.

Meissner, R., and Vetter, U. (1979). *J. Geophys.* **45**, 147-158.

Meissner, R., and Wever, Th. (1985). *In* "Reflection Seismology: A Global Perspective," (M. Barazangi and L. D. Brown, eds.), pp. 75-93. Geodynamics Series 13. American Geophysical Union, Washington, D.C. and Geological Society of America, Boulder, Colorado.

Meissner, R., Berckhemer, H., Wilde, R., and Poursadec, M. (1970). *In* "Graben Problems," (H. Illies and St. Mueller, eds.), pp. 184-189. Schweizerbart, Stuttgart.

Meissner, R., Bartelsen, H., Glocke, A., and Kaminski, W. (1976a). *In* "Explosion Seismology in Central Europe," (P. Giese, C. Prodehl, and A. Stein, eds.), pp. 245-251. Springer-Verlag, Berlin and New York.

Meissner, R., Berckhemer, H., and Glocke, A. (1976b). *In* "Explosion Seismology in Central Europe," (P. Giese, C. Prodehl, and A. Stein, eds.), pp. 303-312. Springer-Verlag, Berlin and New York.

Meissner, R., Flueh, E. R., Stibane, F., and Berg, E. (1976c). *Tectonophysics* **35**, 115-136.

Meissner, R., Bartelsen, H., and Murawski, H. (1980). *Tectonophysics* **64**, 59-84.

Meissner, R., Bartelsen, H., Krey, Th., and Schmoll, J. (1982). *In* "Geothermal Energy," (V. Cermák, and R. Haenel, eds.), pp. 285-292. Schweizerbart, Stuttgart.

Meissner, R., Lueschen, E., and Flueh, E. R. (1983a). *Phys. Earth Planet. Inter.* **31**, 363-376.

Meissner, R., Springer, M., Murawski, H., Bartelsen, H., Flüh, E. R., and Duerschner, H. (1983b). *In* "Plateau Uplift," (K. Fuchs and H. Murawski, eds.), pp. 276-287. Springer-Verlag, Berlin and New York.

Meissner, R., Springer, M., and Flüh, E. R. (1984). *In* "Variscan Tectonics of the North Atlantic Region," (D. H. W. Hutton and D. J. Anderson, eds.), pp. 23-32, Special Publication 14, Geological Society of London. Blackwell, Oxford, London, Edinburgh.

Mendes Victor, L. A., Hirn, A., and Veinante, J. L. (1980). *Ann. Geophys.* **36**, 469-476.

Mercier, J. C. (1980). *J. Geophys. Res.* **85**, 6293-6303.

Mercier, J. C., and Carter, N. L. (1975). *J. Geophys. Res.* **80**, 3349-3362.

Mereu, R. F., and Hunter, J. A. (1969). *Bull. Seism. Soc. Am.* **59**, 147-155.

Messil, R. H. (1967). U.S. Bureau of Mines Report of Investment RI7015.

Meyer, J. (1980). *In* "Elektromagnetische Tiefenforschung, Kolloquium in Berlin," (V. Haak and J. Homilius, eds.), pp. 215-220. Geophys. Inst., Freie Universität, Berlin.

Milkereit, B. (1984). "Bearbeitung seismischer Weitwinkeldaten im Interceptzeit-Strahlparameter-Raum." Doctor thesis, Math.-Sci. Faculty, Universität Kiel.

Miller, H., Mueller, St., and Perrier, G. (1982). *In* "Alpine-Mediterranean Geodynamics," (H. Berckhemer and K. Hsue, eds.), pp. 175-203. Geodynamics Series 7, American Geophysical Union, Washington, D.C. and Geological Society of America, Boulder, Colorado.

Mintrop, L. (1910). *Int. Kongr. Bergbau Huettenwesen. Mech. Prakt. Geol.* pp. 98-112. Selbstverlag.

Mishra, D. C. (1982). *Earth Planet. Sci. Lett.* **57**, 415-420.

Moerner, A. (1980). *In* "Earth Rheology, Isostasy and Eustasy," (A. Moerner, ed.), pp. 535-554. Wiley, New York.

Mohorovičić, A. (1909). *Jb. Met. Obs. Zagreb*, **9**, 1-63.

Mohr, O. (1900). *Z. Verein. Dtsch. Ing.* **44**, 1524-1530 and 1572-1577.

Mohr, P. (1982). *In* "Continental and Oceanic Rifts" (G. Pálmason, ed.), pp. 293-309. Geodynamic Series 8, American Geophysical Union, Washington, D.C. and Geological Society of America, Boulder, Colorado.

Moik, J. K. (1980). Report SP-431. NASA, Washington, D.C.

Molnar, P., and Tapponier, P. (1975). *Science* **189**, 419-426.

Molnar, P., and Tapponier, P. (1978). *J. Geophys. Res.* **83**, 5361-5375.

Montadert, L., Roberts, D. G., De Charpal, O., Guennoe, P., and Sibuet, J. C. (1979a). *In* "Initial Reports of the DSDP, 48," pp. 1025-1060. U.S. Govt. Printing Office, Washington, D.C.

Montadert, L., De Charpal, O., Roberts, D. G., Guennoc, P., and Sibuet, J. C. (1979b). *In* "Deep Drilling Results in the Atlantic Ocean," M. Talwani, W. Hay, and W. B. Ryan, eds.), pp. 437-452. Maurice Ewing Series 3, American Geophysical Union, Washington, D.C.

Mooney, H. M., and Bleifuss, R. (1953). *Geophysics* **18**, 383-393.

Mooney, W. D. (1981). *EOS* **62**, 19.

Mooney, W. D., and Prodehl, C. (1978). *Z. Geophys.* **44**, 573-601.

Moorbath, S. (1976). *In* "The Early History of the Earth," (B. F. Windley, ed.), pp. 351-360. Wiley, London.

Morris, G. B., Raitt, R. W., and Shor, G. G. (1969). *J. Geophys. Res.* **74**, 4300-4316.

Mostaanpour, M. M. (1984). "Einheitliche Auswertung krustenseismischer Daten in Westeuropa. Darstellung von Krustenparametern und Laufzeitanomalien," Berliner geowiss. Abh. B (Geophysik). Reimer, Berlin.

Muckelmann, R. (1982). *Tech. Prog. Abstr.*, pp. 141-143. SEG, Dallas, Texas.

Mueller, St. (1977). *In* "The Earth's Crust," (J. G. Heacock, ed.), pp. 289-317. Geophysical Monograph Series 20, American Geophysical Union, Washington, D.C.

Mueller, St. (1978). *In* "Tectonics and Geophysics of Continental Rifts," (I. B. Ramberg, and E. R. Neumann, eds.), pp. 11-28. Reidel, Dordrecht.

Mueller, St. (1982). "Mountain Building Processes," (U. J. Hsue, ed.), pp. 181-199. Academic, New York.

Mueller, St., Ansorge, J., Egloff, R., and Kissling, E. (1980). *Eclogae Geol. Helv.* **73**, 463-483.

Murawski, H. (1984). *Mitt. Geol. Palaeontol. Inst. Univ. Hamburg Festband,* **56**, 185-204.

Murawski, H., Albers, H. J., Bedo, P., Berners, H. P., Duerr, St., Huckfriede, R., Kauffmann, G., Kowalzyk, G., Meiburg, P., Muller, A., Ritzkowski, S., Schwab, K., Semmel, A., Stapf, K., Walter, R., Winter, K. P., and Zantel, H. (1983). *In* "Plateau Uplift," (H. Illies, K. Fuchs, K. von Gehlen, H. Maelzer, H. Murawski, and A. Semmel, eds.), pp. 9-38. Springer-Verlag, Berlin and New York.

Murrell, S. A. F. (1964). *Br. J. Appl. Phys.* **15**, 1195-1210, 1211-1223.

Musgrave, A. W. (1961). *Geophysics* **26**, 738-753.

Mutter, J. C., Talwani, M., and Stoffa, P. L. (1982). *Geology* **10**, 353-357.

Mutter, J. C., Talwani, M., and Stoffa, P. L. (1984). *J. Geophys. Res.* **89**, 483-502.

Nafe, J. E., and Drake, D. C. (1959). *J. Geophys. Res.* **64**, 1545-1555.

Nakamura, Y., Latham, G. U., Dorman, H. J., and Duennebier, F. K. (1976). *Proc. Lunar Sci. Conf., 7th* pp. 3113-3121.

Nekut, A., Connerney, J. E. P., and Kuckes (1977). *Geophys. Res. Lett.* **4**, 239-242.

Nelson, K. D., Lillie, J. R., de Voogd, B. M., Brewer, J. A., Oliver, E., Kaufman, S., Brown, L., and Viele, G. W. (1982). *Tectonics* **1**, 413 430.

Nerzerow, I. L., Seminova, A. N., and Simbirewa, J. G. (1969). "Physical Basis of Foreshocks," (in Russian). Nauka Publ., Moscow.

Neugebauer, H. J., and Spohn, T. (1978). *Tectonophysics* **50**, 275-305.

Neukum, G., and Hiller, K. (1981). *J. Geophys. Res.* **86**, 3097-3121.

Nur, A. (1983). *Rev. Geophys. Space Phys.* **21**, 1779-1785.

Nur, A., and Ben-Avraham, Z. (1978). *J. Phys. Earth* **26**, 21-29.

Nur, A., and Ben-Avraham, Z. (1982). *J. Geophys. Res.* **87**, 3644-3661.

Nuttli, O. W. (1973). *J. Geophys. Res.* **78**, 876-885.

Ocola, L. C., and Meyer, R. P. (1973). *Geol. Soc. Am. Bull.* **84**, 3387-3404.

O'Connell, R. J., and Fyfe, W. S. (eds.) (1981). "Evolution of the Earth." Geodynamics Series 5, American Geophysical Union, Washington, D.C. and Geological Society of America, Boulder, Colorado.

Oliver, J. (1980). *Am. Sci.* **68**, 676–683.

Oliver, J., Dobrin, M., Kaufman, S., Meyer, R., and Phinney, R. (1976). *Geol. Soc. Am. Bull.* **87**, 1537–1546.

Olsen, K. H., Keller, G. R., and Stewart, J. N. (1978). "Crustal Structure Near the Rio Grande Rift from Seismic Refraction Measurements." Informal Report, Scientific Laboratory, Los Alamos.

Olsen, K. H., Keller, G. R., and Steward, J. N. (1979). *In* "Rio Grande Rift," (R. E. Riecker, ed.), pp. 127–144. Special Publication 23, American Geophysical Union, Washington, D.C.

Ord, A. (1981). "Determination of Flow Stress from Microstructures of Mylonitic Rock." Ph.D. Thesis, University of California, Los Angeles, California.

Pálmason, G. (ed.) (1982). "Continental and Oceanic Rifts." Geodynamics Series 8, American Geophysical Union, Washington, D.C. and Geological Society of America, Boulder, Colorado.

Panza, G. F. (1980). *In* "The Solution of the Inverse Problem in Geophysical Interpretation," (R. Cassinis, ed.), pp. 39–78. Plenum, New York.

Panza, G. F., Calganile, G., Scandone, P., and Mueller, St. (1984). *In* "Spektrum der Wissenschaft: Ozeane und Kontinente," (P. Giese, ed.), pp. 132–143. Spektrum-Verlag, Heidelberg.

Parrish, D. K., Krivz, A. L., and Carter, N. L. (1976). *Tectonophysics* **42**, 75–110.

Parson, B., and Sclater, J. G. (1977). *J. Geophys. Res.* **82**, 803–827.

Pawlenkova, N. J. (1973). "Wave Fields and Models of the Earth's Crust," (in Russian). Naukova Dumka, Kiev.

Pawlenkova, N. J. (1979). *Tectonophysics* **59**, 381–390.

Pawlenkova, N. J. (1980). *Proc. Assembly ESC, 17th*, Budapest, pp. 539–543.

Pawlenkova, N. J. (1984). *Gerlands Beitr. Geophys.* **93**, 125–132.

Payo, G. (1972). *Geophys. J. R. Astron. Soc.* **30**, 85–99.

Pennington, W. D., Mooney, W. D., Hissenhoven, R. V., Meyer, H. J., and Ramirez, J. E. (1979). *Geophys. Res. Lett.* **6**, 65–68.

Phillips, R. J., and Lambeck, K. (1980). *Rev. Geophys. Space Phys.* **18**, 27–76.

Phinney, R. A., and Odom, R. I. (1983). *Rev. Geophys. Space Phys.* **21**, 1318–1332.

Pilger, A., and Rösler, A. (eds.) (1975). "Afar Depression of Ethiopia." Schweizerbart, Stuttgart.

Piper, J. D. A. (1978). *In* "Tidal Friction and the Earth's Rotation," (P. Brosche and J. Suendermann, eds.), pp. 197–234. Springer-Verlag, Berlin and New York.

Pitman III, W. C. (1983). *Rev. Geophys. Space Phys.* **21**, 1520–1527.

Porsgay, K., Albu, I., Petrovics, I., and Ráner, G. (1980). "Character of the Earth's Crust and Upper Mantle on the Basis of Seismic Reflection Measurements in Hungary." Presented on the Workshop meeting of the IASPEI Commission on Controlled Source Seismology in Park City, Utah.

Press, F., and Ewing, M. (1952). *Bull. Seism. Soc. Am.* **42**, 212–228.

Press, F., and Siever, R. (1982). "Earth." Freeman, San Francisco.

Priestley, K., and Orcutt, J. (1982). *J. Geophys. Res.* **87**, 2634–2642.

Prodehl, C. (1979). "A Re-Interpretation of Seismic Refraction Measurements Made from 1961 to 1963 and a Comparison with the Crustal Structure of Central Europe." Paper 1034, U.S. Govt. Printing Office, Washington, D.C.

Prodehl, C. (1981). *Tectonophysics* **80**, 255–269.

Prodehl, C., and Pakiser, L. C. (1980). *Geol. Soc. Am. Bull.* **91**, 147–155.

Puzyrev, N. N., Mandelbaum, M., Krylov, S., Mishenkin, B., Krupskaya, G., and Petrik, G. (1973). *Tectonophysics* **20**, 85–95.

Puzyrev, N. N., Mandelbaum, M., Krylov, S., Mishenkin, B., and Petrik, G. (1978). *Tectonophysics* **45**, 15–22.

Raheim, A., and Green, D. H. (1974). *Contrib. Petrol.* **48**, 179–203.

Raikes, S. A. (1980). *J. Geophys.* **48**, 80–83.

Raikes, S. A., and White, R. E. (1984). *Geophys. Prospect.* **32**, 892–919.

Ramberg, I. B., Spjeldnas, N. (1978). *In* "Tectonics and Geophysics of Continental Rifts," (I. B. Ramberg and E. R. Neumann, eds.), pp. 167–194. Reidel, Dordrecht.

Rayleigh, Lord (1887). *Proc. London Math. Soc.* **17**, 4–11.

Research Group for Explosion Seismology (1966). *In* "The Earth Beneath the Continents," pp. 334–348. Geophysics Monograph 10, American Geophysical Union, Washington, D.C.

Reymer, A., and Schubert, G. (1984). *Tectonics* **3**, 63–78.

Riecker, R. E. (ed.) (1979). "Rio Grande Rift: Tectonics and Magmatism." Special Publication 23, American Geophysical Union, Washington, D.C.

Rikitake, T. (1966). "Electromagnetism and the Earth's Interior." Elsevier, Amsterdam.

Rikitake, T. (1976). "Earthquake Prediction." Development in Solid Earth Geophysics Series 9, Elsevier, Amsterdam.

Rinehart, J. S. (1980). "Geysers and Geothermal Energy." Springer-Verlag, Berlin and New York.

Rinehart, J. E., Sanford, A. R., and Ward, R. M. (1979). *In* "Rio Grande Rift: Tectonics and Magmatism," (R. E. Riecker, ed.), pp. 237–252. Special Publication 23, American Geophysical Union, Washington, D.C.

Ringwood, A. E. (1969). *In* "The Earth's Crust and Upper Mantle," (P. J. Hart, ed.), pp. 1–17. Geophysics Monograph Series 13, American Geophysical Union, Washington, D.C.

Ringwood, A. E. (1974). *J. Geol. Soc. London* **130**, 183–204.

Ringwood, A. E. (1975). "Composition and Petrology of the Earth's Mantle." McGraw-Hill, New York.

Rokityanski, I. I., Amirov, V. K., Kulin, S. N., Loovinov, I. M., and Shuman, V. N. (1976). *In* "The Electric Conductivity Anomaly in the Carpathians in Geoelectric and Geothermal Studies," (A. Adám, ed.), pp. 604–613. Akadémiai Kiadó, Budapest.

Roller, J., and Jackson, W. (1966). *In* "The Earth Beneath the Continents," (J. S. Steinhart and T. J. Smith, eds.), pp. 270–275. Geophysics Monograph Series 10, American Geophysical Union, Washington, D.C.

Ronov, A. B., and Yaroshevsky, A. A. (1969). *In* "The Earth's Crust and Upper Mantle," (P. J. Hart, ed.), pp. 37–57. Geophysics Monograph Series 13, American Geophysical Union, Washington, D.C.

Rothé, J. P. (1947). *C.R. Acad. Sci.* **224**, 1295–1297.

Rothé, J. P., and Peterschmitt, E. (1950). *Ann. Inst. Phys. Globe Strasbourg* **5**, 3–28.

Roy, R. F., Blackwell, D. D., and Decker, E. R. (1972). *In* "The Nature of the Solid Earth," (E. C. Robertson, ed.), pp. 506–543. McGraw-Hill, New York.

Royden, L., Sclater, J. G., von Herzen, R. P. (1980). *Am. Assoc. Pet. Geol. Bull.* **64**, 173–187.

Ruegg, J. C. (1975a). *Ann. Géophys.* **31**, 329–360.

Ruegg, J. C. (1975b). *In* "Afar Depression in Ethiopia," (A. Pilger and A. Roesler, eds.), 120–134. Schweizerbart, Stuttgart.

Rummel, F. (1973). "SFB 77, Jahresbericht 1972," pp. 74–94. Deutsche Forschungsgemeinschaft, Bonn-Bad Godesberg.

Rummel, F. (1974). *In* "Advances in Rock Mechanics," *Proc. Int. Congr. Rock Mech.*, Denver, 517–523.

Rummel, F. (1978). BMFT-Research Contri. ET-3023 A-1, Rep. 3.

Rummel, F., and Alheid, H. J. (1979). *Proc. Int. Res. Conf. Intra-Cont. Earthquakes* pp. 33–65. *Inst. Earthquake Eng. Seism.*, Scopje.

Rummel, F., and Fairhorst, C. (1970). *Rock Mech.* **2**, 189–204.

Rummel, F., and Frohn, C. (1982). *In* "High Pressure Researches in Geophysics," (W. Schreyer, ed.), pp. 103–111. Schweizerbart, Stuttgart.

Rummel, F., Baumgaertner, J., and Alheid, H. J. (1982). *Proc. Workshop Hydraul. Fract. Stress Meas. 17* (M. D. Zoback and B. C. Haimson, eds.), pp. 256–278. University Press, Menlo Park, California.

Rybach, L. (1973). *Beitr. Geol. Schweig Geotech. Ser.* **51**.

Rybach, L. (1976). *Pure Appl. Geophys.* **114**, 309–318.

Rybach, L., and Buntebarth, G. (1982). *Earth Planet. Sci. Lett.* **57**, 367–376.

Sanford, A. R., Aleptckin, O. S., and Toppozada, T. R. (1973). *Bull. Seism. Soc. Am.* **63**, 2021–2034.

Sanford, A. R., Mott, R. P., Shuleski, P. J., Rinehart, E. J., Caravella, E. J., Ward, R. M., and Wallace, T. C. (1977). *In* "The Earth's Crust," (J. G. Heacock, ed.), pp. 385–403. Geophysics Monograph Series 20, American Geophysical Union, Washington, D.C.

Sapin, M., and Hirn, A. (1974). *Ann. Geophys.* **30**, 181–202.

Sapin, M., and Prodehl, C. (1973). *Ann. Geophys.* **29**, 127–145.

Sass, J. H. (1971). *In* "Understanding the Earth," (J. G. Gass, P. J. Smith, and R. C. L. Wilson, eds.), 2nd ed (1972), pp. 81–87. MIT Press, Cambridge, Massachusetts.

Sattlegger, J. (1964). *Geophys. Prospect.* **12**, 115–134.

Schilt, S., Oliver, J., Brown, L., Kaufman, S., Albough, D., Brewer, J., Cook, F., Jensen, L., Krumhansl, P., Long, G., and Steiner, D. (1979). *Rev. Geophys. Space Phys.* **17**, 354–368.

Schmucker, U. (1970). *Bull. Scripps Inst. Oceanog.* **13**.

Schmucker, U. (1973). *Phys. Earth Planet. Inter.* **7**, 365–378.

Schmucker, U. (1974). Protokoll Koll. "Erdmagnetische Tiefensondierungen," pp. 313–343. Deutsche Forschungsgemeinschaft, Bonn-Bad Godesberg.

Schmucker, U. (1980). *In* "Elektromagnetische Tiefenforschung, Kolloquium Berlin 1980," (U. Haak and J. Homilius, eds.), pp. 317–320. Institut für geophysikalische Wissenschaften der Freien Universität, Berlin.

Schneider, G. (1975). "Erdbeben." Enke, Stuttgart.

Schneider, G. (1984). *In* "Landolt Boernstein," (K. Fuchs and H. Soffel, eds.), Group 5, Vol. 2, Subvolume a: "Geophysics of the Solid Earth, Moon and Planets," pp. 47–82. Springer-Verlag, Berlin and New York.

Scholz, C. H., Barazangi, U., and Skar, M. L. (1971). *Geol. Soc. Am. Bull.* **82**, 2979–2990.

Schulze, G. A. (1949). *Erdöl Tektonik*, **2**, 282–285. Amt für Bodenforschung, Hannover-Celle.

Schulze, G. A. (1974). *In* "Zur Geschichte der Geophysik," (H. Birett, K. Helbig, W. Kertz, and U. Schmucker, eds.), pp. 89–98. Springer-Verlag, Berlin and New York.

Sclater, J. G., and Christie, P. A. (1980). *J. Geophys. Res.* **85**, 3711–3739.

Seeber, L., Armbruster, J. G., and Quittmeyer, R. C. (1981). *In* "Zagros, Hindukusch Himalaya. Geodynamic Evolution," (H. K. Gupta and F. M. Delang, eds.), pp. 215–242. Geodynamics Series 3, American Geophysical Union, Washington, D.C. and Geological Society of America, Boulder, Colorado.

Semenov, A. M. (1969). *Izv. Acad. Sci., USSR, Phys. Solid Earth* (*Engl. Transl.*) **3**, 245–248.

Sengoer, A. M. C., and Kidd, W. S. F. (1979). *Tectonophysics* **55**, 361–376.

Sharpe, H. N., and Peltier, W. R. (1979). *Geophys. J. R. Astron. Soc.* **59**, 171–203.

Shankland, T. J., and Chung, D. H. (1974). *Phys. Earth. Planet. Inter.* **8**, 121–129.

Shelton, G., and Tullis, T. A. (1981). *EOS* **62**, 396.

Sheridan, R. E. (1974). *In* "The Geology of Continental Margins," (C. A. Burke and L. L. Drake, eds.), pp. 391–408. Springer-Verlag, Berlin and New York.

Sheridan, R. E., Grow, J. A., Behrendt, J. C., and Bayer, K. C. (1979) *Tectonophysics* **59**, 1–26.

Shive, P. N. (1983). *Rev. Geophys. Space Phys.* **21**, 665–671.

Sibson, R. H. (1974). *Nature* **249**, 542–545.

Sibson, R. H. (1982). *Bull. Seism. Soc. Am.* **72**, 151–163.

Sibson, R. H. (1984). *J. Geophys. Res.* **89**, 5791–5800.

Singh, S., and Herrmann, R. B. (1983). *J. Geophys. Res.* **88**, 527-538.

Sipkin, S., and Jordan, T. (1979). *Bull. Seism. Soc. Am.* **69**, 1055-1079.

Sleep, N. H. (1971). *Geophys. J. R. Astron. Soc.* **24**, 325-350.

Smith, A. G., Hurley, A. M., and Briden, J. C. (1982). "Paläokontinontale Weltkasten des Phanesozoikums." Enke-Verlag, Stuttgart.

Smith, R. B. (1978). *Mem. Geol. Soc. Am.* **152**, 111-141.

Smith, R. B. (1982). *J. Geophys. Res* **87**, 2581ff.

Smith, R. B., Schilly, M. M., Braile, L. W., Ansorge, J., Lehman, J. L., Baker, M. R., Prodehl, C., Healy, J. H., Mueller, St., and Greensfelder, R. W. (1982). *J. Geophys. Res.* **87**, 2583-2596.

Smith, R. J. (1973). "Topics in Geophysics." MIT Press, Cambridge, Massachusetts.

Smith, T., Steinhart, J., Aldrich, L. (1966). *In* "The Earth Beneath the Continents," (J. S. Steinhart and T. J. Smith, eds.), pp. 181-197. Geophysics Monograph Series 10, American Geophysical Union, Washington, D.C.

Smithson, S. B., and Brown, S. K. (1977). *Earth Planet. Sci. Lett.* **35**, 134-144.

Smithson, S. B., Brewer, J., Kaufman, S., Oliver, J., and Hurich, C. (1979). *J. Geophys. Res.* **84**, 5955-5972.

Smythe, O. K., Dobinson, A., McQuillin, R., Brewer, J. A., Matthews, D. H., Blundell, D. J., and Kelk, B. (1982). *Nature* **299**, 338-340.

Sollogub, V. B. (1969). *In* "The Earth's Crust and Upper Mantle," (P. J. Hart, ed.), pp. 189-194. Geophysics Monograph Series 13, American Geophysical Union, Washington, D.C.

Sollogub, V. B., and Chekunov, A. V. (1983). *First Break* **1**, 9-17.

Sollogub, V. B., Prosen, D. Dachev, C., Petkov, I., Velchev, T., Mihailov, S., Mituch, E., Porsgoy, K., Militzer, H., Knothe, G., Uchman, I., Constantinescu, P., Cornea, I., Subbotiu, S. I., Chekunov, A. V., Garkalenko, I. A., Khain, V. E., Slavin, V. I., Beranek, B., Weiss, J., Hrdlička, A., Dudek, A., Zounkova, M., Suk, M., Feifar, M., Milovanović, B., and Roksandić, M. (1973). *Tectonophysics* **20**, 1-33.

Sollogub, V. B., Pavlenkova, N. J., and Chekunov, A. V. (1968). *In Proc. Assem. Ens. Seismol. Comm. 8th Budapest*, pp. 252-260.

Sollogub, V. B., Grin, N. E., and Gontovaya, L. I. (1975). *Boll. Geofis. Teor. Appl.* **17**, 79-84.

Solomon, S. C., Stephens, S. K., and Head, J. W. (1982). *J. Geophys. Res.* **87**, 7763-7771.

Sparlin, M. A., Braile, L. W., and Smith, R. B. (1982). *J. Geophys. Res.* **87**, 2619-2633.

Spencer, J. W. (1981). *J. Geophys. Res.* **86**, 1803-1812.

Spetzler, H., Mizutani, H., and Rummel, F. (1982). *In* "High Pressure Research in Geophysics," (W. Schreyer, ed.), pp. 85-93. Schweizerbart, Stuttgart.

Spohn, T., and Neugebauer, H. J. (1978). *Tectonophysics* **50**, 387-412.

Springer, J. E., and Thorpe, R. K. (1982). "Borehole Elongation versus *in situ* Stress Orientation," *Inst. Conf. In Situ Testing Rock Soil Masses*, pap. UCRL-87018, Santa Barbara, California.

Stacey, F. D. (1977a). "Physics of the Earth," 2nd ed. Wiley, New York.

Stacey, F. D. (1977b). *Phys. Earth Planet. Inter.* **15**, 341-348.

Steckler, M. S., and Watts, A. B. (1978). *Earth Planet. Sci. Lett.* **41**, 1-13.

Stegena, L. (1976). *Acta Geodaet. Geophys. Montanist. Acad. Sci. Hung.* **11**, 377-397.

Stegena, L. (1982a). *Z. Geol. Wiss.* **10.3**, 349-356.

Stegena, L. (1982b). *Tectonophysics* **83**, 91-99.

Stegena, L. (1984). JASPEI Assembly Hyderabad/India, Abstract Volume.

Stegena, L., Geozy, B., and Horvath, F. (1975). *Tectonophysics* **26**, 71-90.

Steinbrugge, K. V., and Zacher, E. G. (1960). *Bull. Seism. Soc. Am.* **50**, 389-396.

Steinhart, J., and Meyer, R. (1961). "Explosion Studies of Continental Structure." Publication 622, Carnegie Institute, Washington, D.C.

Stesky, R. M., Brace, W. F., Riley, D. K., and Robin, P. Y. F. (1974). *Tectonophysics* **23**, 177-203.

Stesky, R. M., and Brace, W. M. (1973). *J. Geophys. Res.* **78**, 7614–7621.

Stille, H. (1944). "Geotektonische Gliederung der Erdgeschichte." Abhandlungen der Preussischen Akademie dei Wissenschaften, Mathematisch–Naturwiss enschaftliche Klasse 3, Berlin.

Strangway, D. W. (ed.) (1980). "The Continental Crust and its Mineral Deposits." Special Paper 20, Geological Association of Canada, Waterloo, Ontario.

Strehlau, J., and Meissner, R. (1986). *J. Geophys. Res.* In press.

Sundvoll, B. (1978). *In* "Tectonics and Geophysics of Continental Rifts," (I. B. Ramberg and E. R. Neumann, eds.), pp. 289–296. Reidel, Dordrecht.

Suttil, R. J. (1980). *Earth Planet. Sci. Lett.* **49**, 132–140.

Talwani, M., and Eldholm, O. (1972). *Geol. Soc. Am. Bull.* **83**, 3575–3606.

Talwani, M., and Langseth, M. (1981). *Science* **213**, 22.

Talwani, M., Worzel, J. L., and Landisman, M. (1959). *J. Geophys. Res.* **64**, 40–59.

Talwani, M., Hay, W., and Ryan, W. B. F. (eds.) (1979). "Deep Drilling Results in the Atlantic Ocean: Continental Margins and Paleoenvironment." Maurice Ewing Series 3, American Geophysical Union, Washington, D.C.

Talwani, M., Mutter, J., and Eldholm, O. (1980). *Oceanol. Acta* **4** Suppl. C3, 23–30.

Taner, M. T., and Koehler, F. (1969). *Geophysics* **34**, 859–881.

Tapponier, P., and Molnar, P. (1977). *J. Geophys. Res.* **82**, 2905–2930.

Tarkov, A. P., and Basula, J. V. (1983). *Phys. Earth Planet. Inter.* **31**, 281–292.

Tarling, D. H. (1978). "Evolution of the Earth's Crust." Academic, New York.

Tarling, D. H. (1983). "Principles and Applications of Palaeomagnetism." Chapman and Hall, London.

Taylor, S. R., and McLennan, S. M. (1981). *Philos. Trans. R. Soc. London A* **301**, 381–399.

Telford, W. M., Geldart, L. P., Sheriff, R. E., and Keys, D. A. (1976). "Applied Geophysics." Cambridge Univ. Press, London and New York.

Theilen, F. (1982). *Tech. Prog. Abstr.* pp. 145–146. Society of Exploration Geophysicists, Dallas, Texas.

Theilen, F., and Meissner, R. (1979). *Tectonophysics* **61**, pp. 227–242.

Thellier, E., and Thellier, O. (1959). *Ann. Geophys.* **15**, 285–297.

Thiessen, J. (1979). "Seismische und gravimetrische Untersuchungen der Krustenstrucktur am marokkanischen Kontinentalrand," Diploma thesis. Institut für Geophysik, Universität Hamburg.

Tittmann, B. R. (1977). *Philos. Trans. R. Soc. London A* **285**, 475–479.

Tittmann, B. R., Clark, V. A., Richardson, J., and Spencer, T. W. (1980). *J. Geophys. Res.* **85**, 5199–5208.

Toksöz, M. N., and Hsui, A. T. (1978a). *Tectonophysics* **50**, pp. 177–196.

Toksöz, M. N., and Hsui, A. T. (1978b). *Icarus* **34**, 537–547.

Toksöz, M. N., and Johnston, D. H. (eds.) (1981). "Seismic Wave Attenuation." Society of Exploration Geophysicists, Tulsa, Oklahoma.

Toksöz, M. N., Press, F., Dainty, A. M., and Anderson, K. R. (1974a). *Moon* **9**, 31.

Toksöz, M. N., Dainty, A. M., Solomon, S. C., and Anderson, K. R. (1974b). *Rev. Geophys. Space Phys.* **12**, 539–564.

Toksöz, M. N., Johnston, D. H., and Menear, J. (1975). "Lunar Science VI," pp. 815–817. The Lunar Science Institute, Houston.

Tozer, D. C. (1977). *Sci. Prog. Oxford* **64**, 1–27.

Trappe, H. (1983). "Auswertung von Weitwinkelreflexionen des Projekts Urach," Diploma thesis. Institut für Geophysik der Universität Kiel.

Turcotte, D. L., and Ahorn, J. L. (1977). *J. Geophys. Res.* **82**, 3762–3766.

Turner, F. J. (1968). "Metamorphic Petology." McGraw-Hill, New York.

Ungemach, P. (1982). *In* "Geothermics and Geothermal Energy" (V. Cermak and R. Haenel, eds.), pp. 219–240. Schweizerbart, Stuttgart.

U.S. National Report, Contributions in Geomagnetism and Palaeomagnetism 1979–1982, pp. 593–693. American Geophysical Union, Washington, D.C.

Vacquier, V., Steenland, N. C., Henderson, R. G., and Zietz, I. (1951). "Interpretation of Aeromagnetic Maps." Memoir 47, Geological Society of America, Boulder, Colorado.

Van Zijl, J. S. V. (1969). *Geophysics* **34**, 450–462.

Van Zijl, J. S. V. (1977). *In* "The Earth's Crust," (J. G. Heacock, ed.), pp. 470–500. Geophysical Monograph Series 20, American Geophysical Union, Washington, D.C.

Van Zijl, J. S. V., and Joubert, S. J. (1975). *Geophysics* **40**, 657–663.

Veizer, J., and Jansen, S. L. (1979). *J. Geol.* **87**, 341–370.

Vendenskaya, A. V. (1956). *Izv. Akad. Nauk. USSR, Ser. Geofiz.* pp. 227–284.

Verhoogen, J., Turner, F. J., Weiss, L. E., and Wahrhaftig, C. (1970). "The Earth." Holt, Rinehart and Winston, New York.

Vetter, U., and Meissner, R. (1976). *In* "Explosion Seismology in Central Europe," (P. Giese, C. Prodehl, and A. Stein, eds.), pp. 189–193. Springer-Verlag, Berlin and New York.

Vetter, U., and Meissner, R. (1977). *Tectonophysics* **42**, 37–54.

Vetter, U., and Meissner, R. (1979). *Tectonophysics* **59**, 367–380.

Vetter, U., and Minster, J. B. (1981). *Bull. Seism. Soc. Am.* **71**, 1511–1530.

Vine, F. J., and Matthews, D. H. (1963). *Nature* **199**, 947–949.

Volvovsky, B. S., Volvovsky, I. S., and Kombarov, N. Sh. (1983). *Phys. Earth Planet. Inter.* **31**, 307–312.

Voss, J. (1978). *Lunar Sci. Conf. Abstr., 8th* pp. 958–960.

Voss, J., Weinrebe, W., Schildknecht, F., and Meissner, R. (1976). *Proc. Lunar Sci. Conf. 7th* pp. 3133–3142.

Walcott, R. I. (1973). *Ann. Rev. Earth Planet. Sci.* **1**, 15–37.

Wänke, H. (1984). Workshop of the Mainz Consortium "Accretion and Differentiation of the Planet Earth", Abstract Volume. Deutsche Forschungsgemeinschaft, Bonn-Bad Godesberg.

Wasilewski, P., and Fountain, D. M. (1982). *Geophys. Res. Lett.* **9**, 333–336.

Wasilewski, P., and Mayhew, M. A. (1982). *Geophys. Res. Lett.* **9**, 329–332.

Waters, K. H. (1978). "Reflection Seismology." Wiley, New York.

Watts, A. B. (1982). *Nature* **297**, 469–474.

Watts, A. B., and Ryan, W. R. F. (1976). *Tectonophysics* **36**, 25–44.

Watts, A. B., Bodine, J. H., Ribe, N. M. (1980). *Nature* **283**, 532–537.

Weaver, B. L., and Tarney, J. (1984). *Nature* **310**, 575–577.

Weber, K. (1981). *Geol. Mijnbouw* **60**, 149–159.

Weertman, J. (1970). *Rev. Geophys. Space Phys.* **8**, 145–168.

Weertman, J., Weertman, J. R. (1975). *Annu. Rev. Earth Planet. Sci.* **3**, pp. 293–315.

Weigel, W., Wissmann, G., and Goldflam, P. (1982). *In* "Geology of the Northwest African Continental Margin," (V. von Rad, K. Hinz, M. Sarnthein, and E. Seibold, eds.), pp. 132–159. Springer-Verlag, Berlin and New York.

Weinrebe, W. (1981). *Pure Appl. Geophys.* **119**, 1107–1115.

Werner, D., Kahle, H. G., Ansorge, J., and Mueller, St. (1982). *In* "Continental and Oceanic Rifts," (G. Pálmason, ed.), pp. 283–292. Geodynamics Series 8, American Geophysical Union, Washington, D. C. and Geological Society of America, Boulder, Colorado.

Wesson, R. L., Burford, R. Q., and Elsworth, W. L. (1973). *In* "Proceedings of the Conference on Tectonic Problems of the San Andreas Fault System" (R. Kovach and A. Nur, eds.), pp. 303–321. Geol. Sci. 13, Stanford University, Stanford, California.

West, G. F. (1980). *In* "The Continental Crust and its Mineral Deposits," (D. W. Strangway, ed.), pp. 117–148. Geol. Assoc. Canada Special Paper No. 20, Waterloo, Ontario.

Wever, Th. (1984). "Statistische Auswertung von Steilwinkelreflexionen aus Gebieten der kontinentalen Kruste und ihre Anwendung auf die Analyse der Krustenstrukturen," Diploma thesis. Institut für Geophysik, Universität Kiel.

Wiese, H. (1965). "Geomagnetische Tiefentellurik." **Abhandlung 36**, Geomagnetisches Institut, Potsdam.

Will, M. (1976). *In* "Explosion Seismology in Central Europe," (P. Giese, C. Prodehl, and A. Stein, eds.), pp. 168–177. Springer-Verlag, Berlin and New York.

Wilson, J. T. (1966). *Nature* **211**, 676–681.

Wimmenauer, W. (1982). *In* "The Urach Geothermal Project," (R. Haenel, ed.), pp. 161–163. Schweizerbart, Stuttgart.

Windley, B. F. (1977). "The Evolving Continents." Wiley, New York.

Windley, B. F. (1983). *J. Geol. Soc. Am.* **140**, 849–865.

Winkler, K. W., and Nur, A. (1982). *Geophysics* **47**, 1–15.

Wohlenberg, J. (1982). *In* "Landolt-Boernstein," Group 5, Subvolume a (G. Angenheister, ed.), pp. 66–119. Springer-Verlag, Berlin and New York.

Woollard, G. P. (1969). *In* "The Earth's Crust and Upper Mantle," (P. J. Hart, ed.), pp. 320–340. American Geophysical Union, Washington, D.C.

Worzyk, P. (1978). "Protokolle Elektromagnetische Tiefenforschung, Neustadt/Weinstrasse, DFG-Kolloquium." Deutsche Forschungsgemeinschaft, Bonn-Bad Godesberg.

Wyllie, P. J. (1971). "The Dynamic Earth." 416 p. Wiley, New York.

Yamashina, K. (1978). *Tectonophysics* **51**, 139–154.

Yoshi, T. (1979). *Tectonophysics* **55**, 349–360.

Zamarayev, S. M., and Ruzhich, V. V. (1978). *Tectonophysics* **45**, 41–47.

Zener, C. (1948). "Elasticity and Inelasticity of Metals." University Press, Chicago.

Ziegler, P. A. (1978). *Geol. Mijnbouw* **57**, 589–626.

Ziegler, P. A. (1982). "Geological Atlas of Western and Central Europe." Elsevier, Amsterdam.

Ziegler, P. A. (1984). *Geol. Mijnbouw* **65**, 93–108.

Ziegler, P. A., and Louverens, C. J. (1979). *In* "The Quaterny History of the North Sea," (E. Oele, R. T. E. Schittenhelm, and A. G. Wiggers, eds.), pp. 7–22. Acta Univ. Ups. Symp., Univ. Ups. Annum Quingentesimum Celebrantis 2.

Zoback, M., and Roller, J. C. (1979). *Science* **206**, 445–447.

Zoback, M. D., and Zoback, M. L. (1981). *Science* **213**, 96–104.

Index

A

Aachen thrust fault, 297–298
Absorption band model, attenuation, 153
Accreted terranes, 353–355
Accretional peak, 8
Accretional tail, 372
Acoustic emissions, 166
Activation energies, 194
Adirondacks, 90, 285
Aeromagnetic surveys, 118
Afar triangle, advanced rifts, 332–334
African Atlantic margin, 343–345
Age dating, *see also* Radioactive decay
  absolute, 195–212
  closed systems, 201
  deviations from average initial isotopic ratios, 209–212
  growth of isotopic ratios of Pb, Sr, and Nd, 208
  heat production as function of time, 243–244
  history, 195–196
  laboratory measurement principles, 201–203
  meaning of time, 201
  methods, 203–209
  Nd, isotopic evolution, 209
  nonradiogenic daughters, 199
  Pb–Pb method, 204
  physical background, 196–201
  radiogenic daughters, 198–200
  Rb–Sr method, 200, 205–207
  segregation of crustal complexes, 209–212
  Sm–Nd method, 207–209
  U–Pb method, 204, 207
Aleutian island arc, 314–315
Alkaline basalts, 224
Alpha-particles, 196
Amphibolites, 232–233, 362
Anadarko basin, 270
Anatexis, 228
Andes
  as continental arc, 353
  crustal cross-section, 311–312
  high electric conductivity zones, 312
  middle sector cross-section, 311–312
  seismic and gravity information, 30–31
  Sr–Sr values versus age, 312–314
  very low velocity crust, 311
Anisotropy
  granulite, 151
  minerals, 216–217
  P-waves, 148–151
Anorthosite, formation, 227
Apennines, dispersion data, 41
Appalachians
  crustal cross-section, 286

DSS profiles, 275
kinematic model of main collisional events, 300
line drawings, 284, 286
reflection density, 284–285
situation map, 283
zone of detachment, 283–285
Arabian shield
crustal cross-section, 254
situation map, 254
V–z profiles, 251–252, 254
Archean, 373–376
activity periods, 373
high-grade gneisses, 374
Asperities, 163–164
Asthenosphere, definition, 6–7
Attenuation, 151–162
defined, 152–153
determination from seismic signals, 156–158
dissipation function, 153
frequency dependence, 192–193
as function of temperature and frequency, dunitic sample, 159, 161
measurement technique, 155–156
modulus effect, 153
P-wave and S-wave terms, 154
relationship between Q_P and Q_S measurements, 154–155
relationship of laboratory and field study results, 159, 161–162
relationship with temperature, 194
relaxation peak, 152
results, laboratory measurements, 158–161
rise time, 150–157
specific attenuation, 151
spectral ratio, 150–157
values, cross-section through subduction zone, 161
wavelet modeling, 157–158
Aulacogen, 257, 259
Auxiliary plane, 51, 54

B

BA, *see* Bouguer anomalies
Back-arc basin, 352
Baikal rift system, 322–324
Baltic shield, 251–252, 274

Baltimore Canyon, 341–342
Bar wave velocity, 142
Basalts
arc, 381
magmas, 224, 227, 351
tholeiitic, 224, 289
Basification, 334
Basin and Range Province
Cenzoic extension, 329, 331
highly conducting layer, 90
line drawings, 330
reflection profiles, 329–330
as rift, 329
situation map, 330
zones of detachment, 329
Basin and Range topography, 329
Batholith, Proterozoic, 383
Bay of Biscay margin, 339
Beta-particles, 196
Bipolarity of surfaces, 16–18
Birch's law, 28
Blake plateau, 342
Blue Norma, 337
Blueshists, 375
Body waves, 35, 38
Borehole breakouts, 100–101
Bouguer anomaly, 25, 27–28
Andes, 311
correlation with young mountain ranges, 27
large positive, 329
relief map, United States, 34
Yellowstone, 360
Bouguer correction, 27
Britain
LISPB profiles, 273, 275
MOIST profiles, 287
WINCH profiles, 286–287
Brittle–ductile transition, 162, 171
Byerlee's law, 109, 167–168

C

Cagniard's method, 85
Caledonides
crustal cross-section, 274–275
histograms, 291
LISPB profiles, 273, 275
locations and references, standardized $V(z)$ functions, 273

MOIST line, 286–287
reflection density, 290–291
situation map, 287
standardized velocity–depth functions, 276
WINCH lines, 286–289, 364
Canadian shield, velocity–depth functions, 252, 255–256
Cataclasite, 365
Cenozoic mountain belts, 301–316
    continental magmatic and island arcs, 310–316
    continent–continent collisions zones, 302–310
Center of mass–center of figure offset, 15–16
Central Graben, 340
Chondritic uniform reservoir, 208
CM–CF offset, see Center of mass–center of figure offset
Compressional stress, horizontal principal, 100
Compressional waves, see P waves
Continental arc, 353, 381
Continental crust, 4–5
    accretional and recycling process schematic, 392
    average element composition, 217, 220
    basaltic magmas, 227
    composition, 213–241
    depth in suture zones, 390
    evolution, 368–393
        active margins, 377–378
        Archean, 373–376
        banded iron formation, 376–377
        continent growth, 374
        convection, 370
        critical Rayleigh number, 369–370
        crustal growth, 370–371, 376
        crustal patches, 373
        crustal volume, 371
        formative phase, 372–373
        magma density, 371
        melting process, 371
        models, 377–378
        oceanic plates, 377
        Phanerozoic, 378–379
        physical and petrological boundary conditions, 369–372
        Proterozoic, 376–378
        radioactive heat production, 369
    tectonic pattern change, 377–378
    thermotectonic events, 375
    thickening of crustal parts, 375
    igneous rocks, see Igneous rocks
    margin subsidence, 237
    metamorphic facies, 230–233
    reaction series and magmatic activity, 225–228
    rheology, see Rheology, continental crust
    rock-forming minerals, 215–220
    state of stress, 35
    strongly differentiated, 392
    structural–compositional models, 214
    thickness versus age, 393
Continental magmatic arcs, 310–312, see also Andes
Continental structure, cross-section, 6
Continent–continent collision, 355, 385
    Himalaya-type, 386
    suture zones, 390
    zones, 302–310
        Alpine DSS profile network, 302–303
        Alpine geophysical cross-section, 302, 304–305
        compilation of standardized velocity–depth functions, 310
        Himalaya–Tibet area, 307–308
        locations and references, standardized $V(z)$ function, 309
        Pannonian basin, crustal–lithospheric cross-section, 305–306
Cooling half-space model, 348
Cordillera-type mountains, 343
Creep, 169–181, see also Dislocations; Rheology, continental crust
    boundary between dislocation glide and diffusional creep, 180
    Coble, 179
    constant, 180, 185
    deformation mechanism map, 179
    diffusional, 179
    effective viscosity, 178
    experimental relationships, 177–181
    experimental techniques, 171–173
    fault zones, 190
    general equation, 178
    hydrolytic weakening, 173
    in situ conditions, 178–179
    long-term average Newtonian creep, 182
    low-T-plasticity, 178

Nabarro – Herring, 179
power-law, 179 – 180, 184
rate of Maxwell body, 182 – 183
recrystallization, 177
relations between viscosity and attenuation, 192 – 195
rupture cycles, 182 – 184
small grain sizes, 180
strain hardening, 171
strain – time relationship, constant stress, 173
stress – strain relationships and extrapolation, 170
triaxial testing machine, 172
Crust
  brittle part, 50
  continental, 4 – 5
  definitions, 1 – 4
  deformation of rising plume, 349 – 350
  formation, 10
  general crustal structure, 4
  heating by intrusions, effects, 356 – 357
  hot, 355
  oceanic, 4 – 5
  postrift stage, 347
  as product of planetary differentiation process, *see* Planetary differentiation process
  seismic view, 83 – 84
  sinking, 349 – 350
  spreading, 349 – 350
  stretching, 347 – 348
  structures, 122
  two-layered, travel time curves, 36 – 37
Crustal growth
  ancient mafic lower crust, 387 – 388
  arc basalts, 381
  concept, 379
  continental arcs, 381
  crustal shortening, 385
  density of magma types, 382
  differentiation, 387 – 388
  horizontal accretion, 385 – 386
  hot spots, 384
  island arcs, 381
  magmatic arc formation, 382
  mechanism, 379 – 393
  recycling, 386, 389 – 393
  rising magmas, 381 – 382
  schematic, 380

  transition of island arc to crustal origin, 383
  vertical accretion, 381 – 384
Crustal root, 21 – 22
Crustal stresses, from fault plane solutions, 51 – 54
Crystallization, fractional, 224
Crystal settling, 224
Curie depth, 115, 119 – 120, 122
Curie point, position, 139

D

Dead Sea rift, 321 – 322
Deep seismic sounding, 59 – 60
  continental magmatic and island arcs, 310 – 312
  Dnjepr – Donetz depression, 257 – 259
  platforms, 257 – 259
  reduced record section, 60, 62 – 63
  refraction/wide-angle interpretation, 65 – 76
    apparent velocity versus depth, 68
    asymptotic velocity, 69
    critical angle, 67
    cross section along profile, 74 – 75
    dipping interface geometry, 73
    negative counting method, 69, 71
    p–τ method, 71 – 72
    stages, 66
    travel time and amplitude behavior, 69 – 70
    travel time diagrams, 76
    V–z model, 71 – 72
    Wiechert – Herglotz integral, 69
  shields, 246 – 257
    Arabian shield, 251 – 252, 254
    Baltic shield age provinces, 251 – 252
    Canadian shield, V–z profiles, 252, 255 – 256
    Eurasian shields, 247
    Fennoscandian shield, V–z profiles, 251, 253
    Indian shield, field record and V–z profiles, 249 – 250
    low-velocity zones, 248, 255
    P-wave velocity and petrological meaning, 247 – 248
    Ukrainian shield cross-section, 250
    Voronesh shield, V–z profiles, 250 – 251
Dehydration reactions, 362

Density, 25–26, 28–29
Detrital magnetization, 116–117
Diagenesis, 234
Differentiation, 387–388
Dilatation phenomena, initiation, 168
Direct-current-conduction method, electric conductivity, 85
Dislocations
  Burgers vector, 176
  decorated, 175
  edge, 176
  glide, 176, 180
  motion along slip plane, 174
  plastic deformation, 174
  quartz grains, 177
  screw, 176
  types, 174–177
Dissipation function, 153
Dix formula, 80
Dnjepr–Donetz depression, 257–259
Dolomite, as chemical sediments, 235
Double couple concept, 52
Ductile formation, 162
Dynamic frictional stress, 55

E

Earth
  crustal cross-sections after seismic assessment, 10
  differentiation, 9
  magnetic dipole field, 113
Earthquakes, see also Seismicity
  body waves, 36–38
  concentration, mechanisms, 43
  depth distributions, 48–51
  frequency of occurrence and magnitude, 56–57
  maximum horizontal stress, 45
  occurrence, 45
  stress behavior at time of rupture, 55
  stress drop–depth distribution, 50
  surface waves, 38–41
East African rift system, 320
East Coast Magnetic Anomaly, 341
Eastern Europe, 31–32
Eclogites, 233
Electric conductivity, 84–90
  apparent resistivity, 86
  complex inductive scale length, 86
  conductivity–depth profiles, 91
  field methods, 85–87
  geomagnetic induction arrows, 86–87
  magnetic variation components, 86–87
  magnetite stringers, 87
  Wiese relationship, 87
Electric resistivity
  as function of temperature, 159, 161
  water-saturated igneous rocks, 166
Elevation, relief map, United States, 34
Ensialic orogenies, 375
Europe
  crustal depth, 261–262, 265
  Eastern, 31–32
  simplified tectonic map, 264
  Western, 31–32

F

FAA, see Free air anomalies
Faille du Midi, 297–298
Fault
  creep, 190
  listric, 107–108
  surface area, relationship to seismic moment, 56
  thrust, 107–108
  types, 44
Fault plane, 53
  discrimination from auxiliary plane, 54
  projection, 53
  Schmidt net, 52–53
  two-dimensional geometry, 102
  Wuff net, 52–53
Fault zone, 105–108, 362–367, see also specific faults
  Caledonides, 364
  low-velocity, 362
  mylonitization, 365
  path for magma ascent, 192
  reverse faulting, 106–107
  role, 190–192
  schematics, 365–366
  shallow dipping, 363
  steeply dipping, 363
  stresses, 107, 191
  strike–slip, 106–107, 367
  subhorizontal, 363
  traces, 107–108
  viscosity behavior, 192–193

width, 365–367
Wind River thrust system, 363
Feldspar
  plagioclase, 216
  proportion in crust, 3
  solid solutions, 216–217
  structure, 215
  variation from felsic to mafic, 221
Felsic rocks, 221
Ferromagnetic susceptibility, 115
Flat-jack method, *in situ* stress measurements, 96–97
Flexural rigidity, 23
Fracture, 102–105
  angle of, 102–103
  experiments, 163–167
    acoustic emission, 166
    granitic sample, constant strain rate, 165
    microcracks, 165
    serpentinite, 166
    stick–slip behavior, 164
Free air anomaly, 25, 27–28, 34
Friction experiments, 167–169

G

Gabbro, three-dimensional picture, 161
Gamma-particles, 196
Garnet, Fe–Mg distribution ratio, 133
Geomagnetism, 112–123, *see also* Paleomagnetism
  basement assessment, 119–120, 122
  continental position history, 120–121
  crustal structures, 122
  Curie depth assessment, 119–120, 122
  magnetic anomaly maps, 122–123
  magnetic dipole field, 113
  magnetic measurements, 117–119
  relationship between latitude and magnetic inclination, 114
Geosyncline, 238–239
Geothermal areas, 356–361
Geothermics, 124–140
  average crustal geotherms, 139
  core formation, 124
  Curie point position, 139
  formulas for crustal temperature assessment, 126–129
  geochemical methods, 132–133
  geophysical methods, 134–137

heat flow, 124, 127, 129–130, 137–140
heat of tectonic processes, 125–126
heat production, 128, 130–132
Laplace's differential equation, 127
metamorphic facies, 232
Poisson's differential equation, 127
P-wave velocity, 134–135, 137
radiogenic sources, 125
relationship between surface heat flow and age of last tectonic event, 125–126
  heat production, 129
sensitivity of P-wave velocities to temperature, 136
temperature–depth function calculation, 128
thermal conductivity measurement, 130
tidal friction, 125
upper crust, 133
Geothermometers, 132–133
Germany, *see also* Western Germany
  focal mechanism of earthquakes, 46
  high-conductivity layers, 88
  occurrence of earthquakes and maximum horizontal stress, 45
Gneiss, 134–135, 374
Gouges, behavior in friction experiments, 167
Graben, *see also* specific grabens
  defined, 316
  development, 347
  plate recycling, 389
Granite
  P-wave velocity, 134–135, 144–145
  Sr isochrones, 207
Granulate, P-wave and S-wave velocity as function of temperature, 146
Granulite, 151, 232–233
Gravity
  behavior across mountain ranges, 23
  edge effect, 22–23
  maps, 33–35
  measurements, history, 21
  profile, across active margin, 30
  regional, 27
Great Glen fault, 364
Green Mountains, crustal cross-section, 285–286
Greenstone belts, 373
Gulf of Aden, 336–337
Gulf of Biscay, 331–333
Gulf of California, 336

Gulf of Mexico, 342–343
Gulf of Suez rift, 321, 322

H

Heat conduction, partial differential equation, 126
Heat flow, 124, 127, 129–130, 137–140
Heat production
  abundance of U, Th, and K, 130–131
  correlation with
    density and cation packing density, 131–132
    P-wave velocity, 132
  decreasing, 128
  measurement, 130–132
  relationship to surface heat flow, 129
Himalaya–Tibet area, 307–308
Histograms, 268, 285, 291, 298
Hot crust, 355
Hungary, conductivity data, 88–90
Hydraulic fracturing methods, 98
Hydrolytic weakening, 173
Hypsographic curves, 16–18

I

Iceberg, gravity behavior, 22
Igneous rocks, 220–228
  alkali–silica diagram with MgO as parameter, 221
  anatexis, 228
  andesitic magmas, 228
  anorthosite formation, 227
  differentiation products, 227
  extrusives, 221
  felsic, 221
  formation, 223–225
  fractional crystallization, 224
  intrusives, 221
  mafic, 221
  mineralogical composition, 222–223
  reaction series and magmatic activity, 225
Indian–Eurasian collision, preceding collisions, 385
Indian shield, field record and $V–z$ profiles, 249–250
Induction methods, electric conductivity, 85, 87
Iron, banded formation, 376–377

Island arcs, 312–316
  compilation of standardized velocity depth functions, 316
  high temperatures, 314
  location and references, standardized $V(z)$ functions, 316
  structure, 314, 351–352
  vertical crustal accretion, 381
Isostatic models, 21–25
  gravity effect of topography, 22–23
  summary, 24
Isotopes, decay scheme, 197
I-values, 205

J

Japanese island arc, 351–352

K

Komatiites, 370–371
  basalts, 224

L

Laboratory experiments
  age dating, 201–203
  attenuation experiments, *see* Attenuation
  creep experiments, *see* Creep
  fracture experiments, 163–167
  friction experiments, 167–169
  limitations, 141
Laplace's differential equation, 127
L_g waves, 41
Limagnegraben, 320
Limestone, as chemical sediments, 235
Listric faults, 107
Lithosphere, definition, 6–7
Low-velocity body, 293, 296, 357–358
Low-velocity layers, 135
Low-velocity zones, 361–362
  Yellowstone, 359–360
Lunar crust, 12–14, *see also* Moon

M

Mafic rocks, 221
Magmas
  activity, reaction series and, 225–228
  andesitic, 228
  basaltic, 16, 224, 227, 351

crustal contamination, 390
density, 384
primitive density, 371
rising, 381–382
Magmatic arc, 382–383
Magnetic anomaly maps, 122–123
Magnetic basement, assessment, 119–120, 122
Magnetic dipole field, schematic, 113
Magnetite, stringers, 88
Magnetization, 115–116
Magnetometers, 117–118
Magnetotelluric method, electric conductivity, 85–86
Margins
    active, 351–355, 377–378
    development from rifts, 346–351
    mature, 350
    passive continental, 335–346
    subsidence, 236–237, 344, 347–348
    young to middle-aged, 349–350
Mars, 10, 14, 18
Mass number, 196
Mass spectrometer, 202–203
Maxwell body approach, 181–184
    models, 182
McKenzie stretching model, 346
Mercury, 14, 17–18
Metamorphic facies, 230–233
Metamorphism, 229–230
Metasediments, 234
Micas, structure, 215–216
Michigan basin, 267, 269
Microcracks, 165
Mid-oceanic ridge basalts, 211
Minerals, rock-forming, 215–220
    anisotrophy, 216–217
    classes of rocks formed, 215
    physical properties, 216–217
Minnesota, 266–268
Mississippi embayment, intracontinental earthquakes, 47
Mohr's circle, Byerlee's relationship and, 168
Mohr's diagram, 92, 94, 103
Mohr's envelopes, 103–104
Moon
    cooling and thickening of surface plates, 373
    crustal cross-sections after gravity assessment, 10
    differentiation, 9
    hypsographic curves, 17–18
    lowland and highland areas, 13
    lunar crust, 12–14
Mountain ranges, 21–23
Moveout time, 77–78, 80
Muskox intrusion, 224–225
Mylonites, 365
Mylonitization, 190–191, 365

N

Nd–Sm method, deviation from standard curves, 210
Near-vertical reflection, 61, 64–65, *see also* Paleozoic areas
    correlation with velocity–depth function, 61
    interpretation, 76–83
        Dix formula, 80
        migration process, 81–82
        moveout times, 77–78, 80
        muting, 80
        reflection density histograms, 83
        stacking velocities, 77–79
        trial stack, 78–79
        unmigrated record, 81–82
    passive continental margins, 335
    shields and platforms, 266–272
        Anadarka basin, 270
        Michigan basin line drawings, 267, 269
        Minnesota COCORP profiles, 266–268
        U.S. COCORP reflection lines, 266–267
        Wind River thrust fault, 269
    streamers, 335
Negative counting method, 69, 71
New Jersey shelf, 341–343
Nodal plane, 51
North Australian craton, V–z profiles, 252, 255
North Sea, 339–341
Norwegian continental margin, 336–338
Norwegian Sea, 336–337

O

Ocean floor, 113, 211
Oceanic crust, 4–5
Oceanic leaking cycle, 391
Oceanic plate, breaking, 389

Oceanic Platelets, accretion, 385
Oceanic structure, cross-section, 6
Olivine, 179, 216, 224
Orogenic belts, *see* Cenozoic mountain belts
Oslo graben, 317–318
Overcoring-diagram method, *in situ* measurements of strain, 97
Overcoring-strain gauge method, *in situ* stress measurements, 97
Overthrusting, 362

P

p-τ method, 71–72
Paleomagnetism, 112
    physical basis, 115–117
    platelet movement, 114
Paleozoic areas, *see also* Caledonides; Western Germany
    early
        near-vertical reflection, 281, 283–290; *see also* Appalachians
        refraction measurements, 272–276
    graben, situation map, 318
    late, near-vertical reflection, 290, 292–301
    middle to late, refraction measurements, 276–281
Pannonian basin, crustal-lithospheric cross-section, 305–306
Pb–Pb method, 204–205
Pendulum measurements, 20–21
Phanerozoic, 378–379
Planets
    cross-section, 9
    hypsographic curves, 16–18
    physical and geological data, 11
Platelets, movement and attachment to continents, 114
Plate tectonic cycle, 239–241, *see also* Wilson cycle
Platform
    deep seismic soundings, 257–259
    defined, 237
    near-vertical reflection, 266
    sediments, 257
    standardized velocity–depth functions, 260–261
Plume models, 348–349
Poisson's differential equation, 127
Planetary differentiation process, 8–19

bipolarity of surfaces, 16–18
CM–CF offset, 15
crust formation, 10
crust on Mercury, Mars, and Venus, 14–15
lunar crust, 12–14
planetary outlook, 18
Precambrian shield areas and platforms, 243–246
    crustal thickness, 259, 261–266
    deep seismic sounding, *see* Deep seismic sounding
    locations and references, standardized $V(z)$ functions, 262
    magnetic processes, 245
    near-vertical reflection, *see* Near-vertical reflection
    old cratons, 243–244
    United States, crustal depth below sea level, 259, 263
Proterozoic, 376–378
Proximity effect, 22
P-waves, 35, *see also* Attenuation; Seismic P- and S-wave experiments
    anistropy, 148–151
P-wave velocity
    advanced rift structures, 329
    Afar triangle, 334
    calculation from travel time, 143
    change as a function of temperature difference, 137
    Baikal rift, 323–324
    control for lowermost crust, 149
    correlation with heat production, 132
    as function of pressure and temperature, 144–145
    as function of temperature, 144, 146–147
    as function of temperature and frequency, dunitic sample, 159, 161
    granitic sample at constant strain rate, 165
    petrological meaning, 247–248
    premelting behavior, 146
    Red Sea, 334
    reversals, 134–135
    sensitivity to temperature, 136
    Yellowstone, 359–360
Pyroxene, 133, 215

Q

Quartz grains, dislocation structure, 177

Quartzite, P-wave and S-wave velocity as function of temperature, 146

R

Radioactive decay, 196–198
Radioactive heat production, 369
Rare earth elements, mass spectrometry, 203
Rayleigh number, 369–370
Rayleigh waves, dispersion curves, 39
Rb–Sr method, 200, 205–207
Reaction series, magmatic activity and, 225–228
Recycling, 386, 389–393
    amount, globally recycled material, 391
    locations of, 389
    ultramafic residue, 389
Red Sea
    advanced rifts, 332–334
    Moho, 334
    pattern of opening, 336
    rift and graben structures, 320–321
    velocity-depth functions from rifts and grabens, 322
Reflections
    density, histograms, 83
    near-vertical, 61, 64–65
    seismic prospecting, 58
Reflectivity method, 72
Refraction shooting, wide-angle events, 60
Regional, 27
Rheology
    continental crust, 181–195
        crust–mantle boundary, 188
        curves of dry and wet solidus, 186
        dry lower crust, STRESSMAX, 188
        fault zone role, 190–192
        temperature curves, 186
        temperature–depth function, 184, 186
        viscosity as function of depth, 189
    Maxwell body approach, 181–184
    stresses as function of temperature, 184–188
    viscosity as function of temperature, 188–190
Rhinegraben, 298, 318–320
Rhonegraben, 320
Rifts, *see also* specific rifts
    advanced, 329–335
        summary, 334–335

characteristic features, 327–328
    common features, 317
    development, 316–317
    development into margins, 346–351
    failed, 345, 349
    locations and references, standardized $V(z)$ functions, 327
    prerift crustal thinning, 347
    Red Sea, 320–321
    structures
        advanced, 329–335
        compilation of standardized velocity–depth functions, 328
    valley, defined, 316
Rift-subduction cycle, 391
Rio Grande rift, 324–327
Rock
    brittle behavior, 162
    dehydration effects on P-wave and S-wave velocities, 141, 146–147
    detrital permanent magnetization, 116–117
    electrical conductivity, 84
    failure, rate of external stress application, 164
    ferromagnetic susceptibility, 115
    induced magnetization, 115
    magnetism, physical basis, 115–117
    natural remanent magnetization, 115
    sedimentary rocks, 158, 234
    thermoremanent magnetization, 115–116
Royden intrusion model, 346–347
Rupture cycle, 182–184

S

Saar–Nahe trough, 298
San Andreas fault zone, strike–slip fault cross-section, 367
Sandstones, 235
Schistosity, 230
Schmidt net, 52–53
Schmucker's curves, 85
Sediment
    chemical, 235–236
    clastic, 234–235
    controlled by biological processes, 235
    cycle, 391–392
    geosynclines, 238–239
    island arc formation, 314

platforms, 237, 257
prerift, 345
role in plate tectonic cycle, 239–241
structure of margins, 335
subsidence of margins, 236–237
trapping, 389–390
Sedimentary deposits, 233–234
Sedimentary environments, 236
Sedimentary rocks, 158, 234
Sedimentation, 238, 344
Seismicity, 41–58
    correlation between zones and long-wave-
        length gravity anomalies, 47
    defined, 41, 43
    worldwide distributions, 42–43
Seismic moment, 54, 56–57
Seismic refraction, across active margin, 30,
    352
Seismic velocity, correlation with densities,
    28–29
Seismological methods, 35–41
Seismological parameters, 54–58
Seismology
    controlled-source, 58–60
    crustal study development, 64
Serpentinite, 165–166, 171
Shear modulus, as function of temperature,
    159, 161
Shear strength, calculation, 108–112
    maximum stresses, 111–112
    overview, 110
Shear stress, 91
    angle of most probable fracture, 109
    Byerlee's law, 109
    fracture, 102
    as function of θ, 93
    greatest, 93–94
    principal, 94
Shear waves, see S waves
Shields, standardized velocity–depth func-
    tions, 260–261
    histograms, 268
Silicates, 215, 218–219, see also Feldspar;
    Quartz
Skin effect, 85
Slip plane, 174
Sm–Nd method, 207–209
Socorro rift, 324
South American margin, 353–354
Sr–Sr method, versus age, Andes, 312–314

Seismic P- and S-wave experiments, 142–
    151
    bar wave velocity, 142
    cross-section, high-p, high-T apparatus,
        144
    data from measurements, 143–147
    measurements at constant temperature and
        varying confirming pressure, 143
    measurement technique, 142–143
    P-wave anisotropy, 148–151
Stacking velocity, 77–79
Stick–slip behavior, 164
Strain, 165, 171
Stress, see also Fracture; Shear stress
    apparent, 55
    behavior at time of rupture, 55
    components, 92
    definition, 91–96
    differential, versus strain, 165
    drop, 55
    as function of temperature, 184–188
    in situ measurement, 96–102
        borehole breakouts, 100–101
        flat-jack method, 96–97
        hydraulic fracturing method, 98
        injection pressure versus time, 99–100
        overcoring-diameter method, 97
        overcoring-strain gauge method, 97
    internal friction, 104
    maximum, as function of depth, 184, 187,
        189
    Mohr's diagrams, 92, 94
    normal, 91, 93
    orientation, 93
    principal, 92
    principal deviatory, direction and magni-
        tude, 95
    saturation, as function of creep rate, 183
    shape, 105
    shear, see Shear stress
    temporal behavior, differential viscosities,
        183
    vertical, 100
STRESSMAX, 188
Stretching factor, 348
Strike–slip fault zone, cross section, 48
Strontium, 206–207
Surface heat flow, 127, 129
Surface-wave magnitude, 54, 57
Surface waves, 35, 38–41

S-waves, 35, *see also* Attenuation; Seismic P-
and S-wave experiments
S-wave velocity
calculation from travel time, 143
as function of temperature, 144, 146
as function of temperature and frequency,
dunitic sample, 159, 161

T

Taphrogenesis, 316
Tectonic stress, *see* Stress
Terranes, accreted, 353–355
Tesseyre–Thornquist lineament, 317
Texas, Hardeman County, 270–271
Thermal conductivity, measurement, 130
Thermal contraction models, 348
Tholeiitic basalt, 224, 389
Triaxial testing machine, 172

U

Ukrainian shield, cross-section, 250
United States
COCORP reflection lines and age provinces, 266–267
crustal depth below sea level, 259, 263
postrifting subsidence, 341
western, heat flow maps, 139
U–Pb method, 204, 207
decay schemes, 198
Upper mantle, 5–6, 40
Urach I
composition of single-record reflections,
294
geothermal anomaly, 356–357
line drawing, 295
low-velocity body, 293, 296, 358
occurrence of reflections, 293, 298
polarity of reflecting boundaries, 293, 297

V

Variscan orogeny, 276, 300
Variscides
compilation of standardized velocity-depth
functions, 282

composite line drawing, 297, 299
histograms, 298
locations and references, standardized $V(z)$
functions, 281
situation map, 292
Venus, 14–15, 18
Viscosity
effective, versus quality factor, 195
fault zone, 192–193
as function of depth, 189
as function of temperature, 188–190
minimum in lower crust, 188
relations with
attenuation, 192–195
temperature, 194
viscosity, 192–195
subcrustal lithosphere, 194
values, cross-section through subduction
zone, 161
Volcanism, mare, 12
Volcano, development, 349–350
Vøring plateau, 338–339
Voronesh shield, $V–z$ profiles, 250–251
$V–z$ model, 71–72

W

Western Europe, 31–32
Western Germany, *see also* Variscides
average crustal velocities, 279
composite line drawing, 297, 299
crustal depth, 278
DSS lines, 277–278
travel time diagrams, 76
Wet bulk density, 25
Wichita Mountains, 270
Wiechert–Herglotz integral, 69
Wiese relationship, 87
Wiese vector method, electric conductivity,
85–86
Wilson cycle, 239–241
Wind River thrust system, 269, 363
Wulff net, 52–53

Y

Yellowstone Snake River Plain, 358–361